COAL & EMPIRE

COAL &
EMPIRE

*The Birth of Energy Security
in Industrial America*

PETER A. SHULMAN

Johns Hopkins University Press
Baltimore

© 2015 Johns Hopkins University Press
All rights reserved. Published 2015
Printed in the United States of America on acid-free paper

Johns Hopkins Paperback edition, 2019
2 4 6 8 9 7 5 3 1

Johns Hopkins University Press
2715 North Charles Street
Baltimore, Maryland 21218-4363
www.press.jhu.edu

The Library of Congress has cataloged the hardcover edition of this book as follows:

Shulman, Peter A., 1979–
Coal and empire : the birth of energy security in industrial America /
Peter A. Shulman.
pages cm
Includes bibliographical references and index.
ISBN 978-1-4214-1706-6 (hardcover : alk. paper)—ISBN 978-1-4214-1707-3
(electronic)—ISBN 1-4214-1706-5 (hardcover : alk. paper)—ISBN 1-4214-1707-3
(electronic) 1. Coal trade—Political aspects—United States—History—19th
century. 2. Energy policy—United States—History—19th century.
3. Industrialization—Political aspects—United States—History—19th century.
4. United States—Foreign relations—1865– I. Title.
HD9546.S58 2015
338.2'7240973—dc23   2014035090

A catalog record for this book is available from the British Library.

ISBN-13: 978-1-4214-3636-4
ISBN-10: 1-4214-3636-1

Special discounts are available for bulk purchases of this book.
For more information, please contact Special Sales at 410-516-6936 or
specialsales@press.jhu.edu.

Johns Hopkins University Press uses environmentally friendly book materials,
including recycled text paper that is composed of at least 30 percent post-consumer
waste, whenever possible.

To Marian Kuperberg, Betty Shulman, and Eli Shulman,
always the best reasons to visit the archives,

and Irving Kuperberg, who would have loved this stuff

CONTENTS

This project began in graduate school, and I retain a debt to my classmates and colleagues who not only shaped its early stages but my scholarly development more generally. You all actually made graduate school fun. Thank you to Bill Turkel, Shane Hamilton, Jenny Smith, Candis Callison, Etienne Benson, Nick Buchanan, Rebecca Woods, Orkideh Behrouzan, Anita Chan, Kieran Downes, Brendan Foley, Xaq Frolich, Nate Greenslit, Chihyung Jeon, Shekhar Krishnan, Richa Kumar, Wen-Hua Kuo, Dave Lucsko, Lisa Messeri, Eden Miller-Medina, Natasha Myers, Jamie Pietruska, Rachel Prentice, Sophia Roosth, Michael Rossi, Aslihan Sanal, Ryan Shapiro, Kaushik Sunder Rajan, Livia Wick, Tim Wolters, Sara Wylie, and Cheng-Pang Yeang. My classmates, Meg Hiesinger, Esra Ozkan, Anne Pollock, and Anya Zilberstein inspired me from the time we first met on registration day. Sandy Brown was an occasional office mate and constant colleague and friend. I wish we still lived on the same side of the globe.

The HASTS faculty showed me what it meant to be a scholar and guided me when I began. Special thanks to Chris Capozzola, John Durant, Mike Fischer, Deborah Fitzgerald, Hugh Gusterson, Evelynn Hammonds, Meg Jacobs, David Kaiser, Evelyn Fox Keller, Ken Keniston, Leo Marx, Bruce Mazlish, Anne McCants, David Mindell, Peter Perdue, Harriet Ritvo, Susan Silbey, Roe Smith, and Roz Williams. Dave, Meg, Chris, and Harriet, the original members of my committee, continued to support both me and this project long after their formal responsibility to do so had ended, and for that I am forever grateful.

Some of us stuck around academia in part owing to a general lack of concern about money, but it turns out it still matters. Along the way, I've been fortunate to receive a series of grants and fellowships that have supported me and this research. Funding from MIT's HASTS Program and the Kelly/Douglass Humanities fund at MIT allowed me to present early stages of this work at conferences and receive valuable feedback. MIT's Center for International Studies supported me for two summers, funding a research visit to Anchorage, Alaska, and its coal-bearing environs. The Dibner Institute for the History of Science and Technology, now sadly shuttered, provided me a home at MIT for two years. An Albert J. Beveridge Grant from the American Historical Association supported research in Washington, DC, while a Stuart L. Bernath Dissertation

Grant from the Society for the History of American Foreign Relations funded research on Appalachian coal mining at Virginia Tech, as well as a trip to Pocahontas, Virginia, and a tour of the town's historic coal mine. The Naval War College's Edward Miller Fellowship in Naval History allowed me months of research in Newport, Rhode Island, where Professor John Hattendorf and Dr. Evelyn Cherpak provided a welcome intellectual home. The Naval Historical Center's Rear Admiral John D. Hays Pre-Doctoral Fellowship in U.S. Naval History and the Hugh Hampton Young Memorial Fund Fellowship from MIT helped me finish writing the dissertation and move on to reconceptualizing this project as a book. Once I was in the book stage, support from a summer stipend from the National Endowment for the Humanities, a Legislative Archives Fellowship from the National Archives' Center for Legislative Archives (generously provided by the Foundation for the National Archives), and funding from Case Western Reserve University allowed me to travel to multiple archives and complete my writing.

On a Mellon Fellowship at Johns Hopkins, I had the privilege of learning from an extraordinary group of scholars at an early stage of reworking this project. Thank you to Robert Kargon, Sharon Kingsland, Bill Leslie, Larry Principe, Dan O'Connor, and especially Gabrielle Spiegel, who brought a range of young humanists together in a warm and stimulating environment.

A fellowship at the National Archives' Center for Legislative Archives in 2011 helped redirect this project, a result as much of the people who worked there as the documents they helped me find. A special thank you to Richard Hunt, Bill Davis, Rod Ross, and especially Richard McCulley. The colloquium Richard put together to workshop an early version of chapter 1 was the most challenging intellectual hour since my generals exam in grad school. For making it so, thank you to Matt Wasniewski, Betty Koed, Neelesh Nerurkar, Tim Francis, Jeff Stine, Kristin Ahlberg, David Nickles, Bart Hacker, David Painter, Margaret Vining, John McNeill, Michael Schiffer, and Michelle Krowl. Thanks also to David Ferriero, chief custodian of the official record of American history, for not only supporting my work and that of other scholars but working tirelessly to make our past come alive for the general public.

No historian works without the often hidden labors of librarians and archivists. I've had the good fortune to work with extraordinarily helpful and knowledgeable professionals, many of whom went out of their way to track down errant documents. Special thanks to the staff at the National Archives and the Library of Congress for always making me feel at home. Thanks too, to Nan Card, for making a summer of weekly treks to the Rutherford B. Hayes Library a pleasant experience.

In addition to sharing portions of this work at conferences, I have also benefited from presenting parts of it at a number of institutions, including a Sawyer

Seminar on Energy and Society at Boston University's Pardee Center for the Study of the Longer-Range Future, Caltech (thank you to my friend Erik Snowberg for making this happen), an inaugural Cultures of Energy program at Rice University, the University of Guelph's Hammond Lecture Series on Energy and the Environment, a Harvard Conference on Intellectual Foundations of Global Commerce and Communications, and the Wish Symposium here at Case Western. The feedback I received at these programs has been invaluable.

I'm fortunate to know a number of terrific scholars who share my interests in the history of energy, communications, technology, and policy, and with whom I have had the good fortune of discussing or sharing portions of this work. Thank you especially to Richard John, Christopher Jones, Richard Hirsh, Ann Greene, Heidi Tworek, Simone Müller-Pohl, Naomi Oreskes, Robert Lifset, Louis Galambos, Theresa Ventura, Robert Vitalis, Joseph Pratt, Ann Johnson, and John McNeill.

At Case Western, I have the great fortune of working in academe's most supportive and collegial history department. Thank you to Molly Berger, Dan Cohen, Ananya Dasgupta, John Flores, Jay Geller, John Grabowski, David Hammack, Marixa Lasso, Miriam Levin, Beth Todd, and Rhonda Williams for your insights and support over the years. Ken Ledford, you're a fantastic mentor. Alan Rocke is the model of a scholar, teacher, and citizen of the university. Jonathan Sadowsky makes it easy to be friends with the boss. I've learned more about authentic teaching from Renee Sentilles than anyone—I look forward to sharing more classrooms together in the future. Ted Steinberg unfailingly reminds me to focus on the main issues. Gillian Weiss helps me remember that history began before 1800. As dean, Cyrus Taylor unfailingly supports our intellectual endeavors. Francesca Brittan has heard more about the travails of writing and given more (great) advice than nearly anyone. Thanks especially to John Broich, Wendy Fu, Noelle Giuffrida, Andrea Rager, and Kevin Dicus for making our corner of Mather House a pleasure to work in. Thanks, too, to research assistants Barrett Sharpnack and Stephanie Liscio, and to Bess Weiss, Emily Sparks, and the much-missed Marissa Ross and Kalli Vimr for keeping everything running.

Portions of chapter 6 appeared in an earlier form in Peter A. Shulman, " 'Science Can Never Demobilize': The United States Navy and Petroleum Geology, 1898–1924," *History and Technology* 19, no. 4 (2003): 365–85.

At Johns Hopkins University Press, Bob Brugger has supported this project even when I was in a muddle in the midst of revisions. Thanks to Bob, Catherine Goldstead, Kathryn Marguy, Hilary Jacqmin, and Juliana McCarthy for making the publication process as constructive and painless as possible. Also, many thanks to MJ Devaney, whose copyediting skills have immeasurably improved this manuscript.

My family around Washington, DC, provided constant support with meals, beds, and transportation during my frequent archival visits. Thanks especially to the Perlmutters, Lisa, Jeff, Michael and Jennifer, and the Marks family, Ellen, Kenny, Julie, Lindsay, and Greg. My grandparents have never stopped being my first cheerleaders. Honey and Pop, thank you for Tilghman and teaching me about politics, family, and the Chesapeake Bay. Mama, your mandel bread will never be surpassed—you are missed every day. My parents, Judi and Mark Shulman, always put education first and instilled in me a love of learning that I can only repay by trying to pass it along to the next generation. My brother, Matthew, may never quite understand the appeal of academic life, but he has unflinchingly supported my intellectual pursuits. Thank you also to my Shy/Bilick family, who have known about this book for almost as long as they've known me. Thank you, Toni Shy and Tzvi Bilick, Shalom Shy and Mariela Dalov, Rami and Helenka Shy, Yael Shy, Daviel Shy, and Breetel and Isaac Graves.

To Malachai and Josiah, you have been unfailingly patient, curious, and supportive but most of all eager to finally hold this book in your hands. I hope someday you will think it worth the wait (and Malachai—thank you especially for solving my organizational problem at the end). For many years this project has been something of a family pet—it required both daily maintenance and structured our vacations. I think it's time to get that dog.

To Trysa, you're my first and best critic. You're also an inspiration. From the moment you first asked why I ever wanted to study oil, we've been in this together. I love you, always.

COAL & EMPIRE

IN WHICH THE PRESIDENT SEEKS AN AUDIENCE WITH THE KING

War demonstrates that adequacy of oil supply is an important element in military strategy and may mean defeat or victory in battles and campaigns.

Herbert Feis, *Petroleum and American Foreign Policy*

When they came to write their accounts of Franklin Roosevelt's historic meeting with Saudi King Abdul Aziz bin Abdul Rahman al-Faisal al-Saud on February 14, 1945, American witnesses emphasized the extraordinary. It was a trip to "exotic parts of the world with exotic people," noted Roosevelt's Secret Service detail Michael Reilly.[1] William D. Leahy, fleet admiral and Roosevelt's chief of staff, cataloged the king's retinue—"the royal fortuneteller, the royal food taster, the chief server of the ceremonial coffee" among over forty others—and could only say it was "like something transported by magic from the middle ages."[2] Captain John S. Keating, who escorted the king from Arabia to his rendezvous with Roosevelt in the Suez Canal's Great Bitter Lake, observed that in this corner of the world so unfamiliar to Americans, his crew was forced to rely on antique British Admiralty maps from 1834 and that his ship, the USS *Murphy*, was the first American naval vessel ever to anchor in the port of Jidda.[3]

Describing their experiences, Americans returned again and again to tropes of the mysterious and marvelous. Everyone mentioned the sheep, enclosed in an improvised pen of ropes wound between depth charges at the *Murphy*'s stern. After the animals were slaughtered, their carcasses were hung from the ship's flagstaff. On deck, observers beheld the spectacle of the king's giant tent—the "big top"—its center supported by the massive barrel of a five-inch gun, tilted vertically.[4] After the *Murphy* rendezvoused with the president's ship, the USS *Quincy*, Roosevelt secreted himself away on deck to privately observe the arriving entourage. "This is fascinating," he exclaimed repeatedly to no one in particular. "Absolutely fascinating."[5]

Yet for a meeting that prompted so many witnesses to record such evocative details, we know with certainty very little about what the two heads of state actually met to discuss. Roosevelt spoke no Arabic and ibn Saud no English, so the dialogue was conducted through translator William Eddy, an ex-marine colonel, academic, spy, and the American minister to Saudi Arabia. In the official account of the conversation, Eddy and his Saudi counterpart recorded conversations about the Jewish settlement of Palestine, French involvement in Syria and Lebanon, and Roosevelt's interest in promoting agricultural irrigation in the Arabian desert. Left out was any mention of the reason most subsequent commentators assumed the two figures must have met: oil.[6]

Whether discussed that day or not, the subject of oil certainly loomed large in the background. Two American firms, the Standard Oil Company of California and the Texas Company, shared an exclusive oil concession from the king. In 1938, their geologists had discovered petroleum deposits, along with evidence of unprecedented underground reserves. The companies' joint venture, known after 1944 as the Arabian American Oil Company, or ARAMCO, had been lobbying the Roosevelt administration since 1941 to provide direct aid to the fragile Saudi government and had already advanced the crown some $6.8 million against future royalty payments themselves. For ARAMCO, shoring up American-Saudi relations was an investment in its own future, since oil production in the country demanded a stable government. For their part, the Saudis sought to ensure the finances needed to continue ruling their only recently united kingdom. State revenues from annual hajj pilgrimages had shrunk since the global depression of the 1930s, and the war had caused nascent oil production there to dwindle.[7]

As for the Roosevelt administration, the State, War, and Navy departments each perceived multiple American interests. Geopolitically, a stable Saudi government would provide a buffer between the United States and the Soviet Union in an already volatile region of the world. Militarily, the country could offer logistical advantages for the placement of airfields and transit rights for Allied aircraft bound for the Pacific. Materially, the massive oil concession—to American firms—might serve as a vital source of world supplies, thus taking pressure off production in the Western hemisphere. Taken together, argued Secretary of State Edward Stettinius to the president in December 1944, the government had an imperative to aid the Saudis. "If such help is not provided by this Government," Stettinius warned, "undoubtedly it will be supplied by some other nation which might thus acquire a dominant position in that country inimical to the welfare of Saudi Arabia and to the national interest of the United States."[8]

In the years since, this meeting has become a kind of set piece in writings about the emerging postwar geopolitics of energy. In these accounts, the relationship forged between Roosevelt and ibn Saud functions as a marker in U.S.-Mid-East relations, providing an indication of the rapid expansion of U.S. involvement in a region hitherto peripheral to American foreign policy. The exoticism of the encounter underscored the apparent novelty of American actions. Driven by new interests in energy supplies and global security that the war had given rise to, the two leaders met as equals but also as representatives of potential producers and potential consumers of oil. While historians have noted that the meeting was neither the true beginning of American engagement in the region nor the first time Americans sought to develop foreign oil fields, they have nevertheless described it as signaling a new phase in American foreign relations. New interests in fossils fuels and their potential global strategic significance appeared to result in new policies. "The meeting between Roosevelt and Abdel Aziz had lasting symbolic and practical importance," writes political scientist Rachel Bronson, noting that the meeting proved far more politically consequential than U.S. diplomatic recognition of Saudi Arabia back in 1933. According to Mid-East scholar Aaron David Miller, "Roosevelt's conversations with the Saudi king seemed to reflect the United States' growing awareness of its interests in Saudi Arabia—if not the entire Middle East."[9]

There was indeed much that was new in how the Americans who followed Roosevelt approached the geopolitics of energy. But there was also much that was not so new at all. Almost exactly a century before the two leaders met, in March 1845, an American navy lieutenant named William C. Chaplin rowed up the Brunei River for an audience with Sultan Omar Ali. Chaplin served aboard an American naval frigate, the USS *Constitution*, that had been dispatched to Brunei to secure newly discovered coal deposits there for the use of future lines of American steamships. At the time, these coal deposits were the subject of a brief, but intense, contest between the British and Americans to determine which nation would gain a commercial and strategic advantage in the vast potential markets of southeast Asia. Unlike Roosevelt's meeting with ibn Saud a century later, Chaplin's negotiation with Omar Ali did not result in a new and lasting geopolitical arrangement around energy. Still, the meeting reflected but one way Americans were learning to think about fossil fuels in terms of security and the national interest. This book is about the path between these two meetings.

Readers in the early twenty-first century hardly need reminders of the link between energy, national interests, and global security. Since the 1970s, phrases like "energy independence," "the oil weapon," and "energy security" have become entrenched in our popular political lexicon. The decade brought oil embargoes, gas lines, and price hikes. For the United States, it saw declining domestic oil production and increasing reliance on foreign crude. Since this time, the security of the world oil supply—and the security of the nation that depended so much upon it—has become a familiar element of American diplomatic and defense policy. Learning to think in these terms, however, did not begin in the 1970s.[10]

Edward Stettinius wrote in 1944 of "the national interest" of the United States in forging an Arabian alliance and Arabian oil. "The national interest" was an old phrase, but one into which politicians, policy makers, and political scientists breathed new life in the 1930s and '40s.[11] The national interest was employed "as a kind of iron necessity which binds governments and governed alike," according to Charles Beard in 1933.[12] Unlike the various and selfish "interests" targeted by Progressives at the dawn of the century, the "national interest" unified the country. Justifying government action in the name of the national interest transformed choices into imperatives. The concept collapsed diverse options into a single, self-evident, collective obligation. It lent to the contingencies of public policy an air of necessity and inevitability. For Beard, the national interest had emerged as the modern world's "pivot of diplomacy," the ultimate justification for international relations. Why did states act in the modern world? Beard answered simply that they acted to pursue and protect their interests, be they economic, ideological, or in defense of their security.[13]

One of those interests was energy. Historians have long noted that since the early twentieth century, Americans have connected both their physical and economic security to the availability of energy sources, oil in particular. They have pointed to the U.S. Navy's conversion from coal to fuel oil as precipitating the security link, reflected in the establishment of naval petroleum reserves like Wyoming's Teapot Dome and the passage of conservation measures through the 1920s and beyond to ensure future military and naval supplies. Historians have described the manifest importance of oil during World War I, when airplanes, tanks, trucks, ships, and submarines all depended on the fuel, with U.S. and U.S.-controlled Mexican production overwhelmingly supplying Allied demand. And they have explored the diplomatic maneuvers that the United States (along with the United Kingdom and France) deployed in the 1920s and 1930s to help its nationals secure oil supplies in places like Mesopotamia, Persia, Mexico, Venezuela, the Dutch East Indies, and elsewhere. Collectively, historians have explained the centrality of security concerns within a nascent federal oil policy and diplomacy as a natural consequence of increased consumption, particularly in the navy and the broader industrial economy.[14]

This interpretation makes a great deal of sense, as the rise in oil use in the early twentieth century was indeed dramatic. Between 1912 and 1919, naval fuel oil use alone grew from 360,000 barrels to 5.8 million a year, while military fuel consumption in tanks and airplanes likewise took off.[15] At the same time, domestic gasoline consumption exploded, as automobiles, trucks, and tractors became characteristic features of the American landscape. By 1920, yearly domestic oil consumption exceeded 530 million barrels, more than double the consumption for 1914. The machine age also employed petroleum products in a range of specialized uses. Asphalt linked city and countryside. Naphtha dry-cleaned the suits of the growing number of Americans engaged in office work. Lubricating oils quite literally kept American industry running.[16] After World

War I, the availability of oil had become so intertwined with the functioning of the economy and the machinery of defense that the maintenance of oil supplies had become of paramount importance. As Calvin Coolidge's Commission on Oil Reserves put it in 1924, the nation required a "definite policy of conservation in aid of national security."[17]

World War II only cemented oil as the preeminent modern war materiel. For centuries, food had most constrained an army's movements. Now it was motive fuel. During the war, Americans still shipped vast quantities of food abroad, but they sent nearly sixteen times that volume in oil, comprising in total more than 60 percent of all military shipments overseas. At the peak of military fuel consumption during the war, American forces devoured nearly 1.4 million barrels of oil a day, over 30 percent of all domestic production. As early as 1948, military planners anticipated that a future war would demand twice as much fuel. The subject of oil was, according to Secretary of Defense James Forrestal, "one of profound importance to the security of our country and to its economic existence."[18] This vision continued to inform American foreign and economic policy for the remainder of the century. From protecting the suppliers of Mid-East oil to writing a tax code that subsidized domestic production, American policy collectively sought to provide for the extraordinary growth of energy consumption for both military and economic purposes.[19]

Yet contrary to most historical accounts, Americans did not discover the connection between energy, national interests, and security in the twentieth century but during the nineteenth, when the widespread adoption of fossil fuels first presented opportunities and challenges for foreign relations, economic expansion, national defense, and naval strategy. Then, it was not oil but coal that posed new problems. Steamships powered by coal offered new advantages of speed, power, and the prospect of travel independent of winds and waves. Yet unlike sail, coal tethered vessels to shore like never before. Fleets of steamships, both naval and commercial, depended on coaling stations, supply routes, and even the peculiar chemical properties of fuel mined from certain fields. Until the middle of the nineteenth century, coal had always been just another commodity, rarely a subject of special government concern. The adoption of steam power for naval vessels and government-subsidized mail steamers turned that commodity into a substance whose availability could direct the course of war or shape commercial opportunity. Even with these stakes, however, the precise federal responsibility for ensuring fuel supplies—even for the navy—remained unclear for a nation organized around what Brian Balogh has called "a government out of sight." Nineteenth-century Americans, according to Balogh, preferred governance by authorities other than the federal one, be they state or local governments, "mixed" public-private enterprises, or simply the less visible but still essential institutions of the federal judiciary, customs system, or exploring expeditions. As America slowly industrialized, it was not obvious how the nation would integrate the infrastructure of fossil energy into its existing approaches to diplomacy, trade policy, and war planning. The importance of security was plain enough, as was coal itself, but how exactly Americans would find security in, or for, new sources of energy was far from obvious.[20]

Even as they came to think of sources of power, particularly fossil fuels, as being of special national significance, Americans had to address a range of fundamental questions at the intersection of naval strategy, foreign policy, and technological change. Would the adoption of new technology enhance or constrain Americans' opportunities in the world? Would the infrastructure necessary to support steam power be built via international cooperation or unilateral action? Would technology drive choices in foreign affairs, or would the desires of American policy makers, merchants, and naval officers catalyze the development of new technologies? Would particular choices of motive power predetermine particular geographies of expansion? Would policy for shipping, postal communication, and naval defense be organized by markets, politics, or technocratic experts? Would American fuel needs be met by domestic supplies or foreign dependence? Would, and should, guaranteeing the global infrastructures of energy for trade and defense become a function of national authority? And if so, how?[21]

These questions, of course, were not raised in a vacuum. As Americans puzzled over fossil fuels and the new technology of steam power, they were also forging a new role for the United States in the world. Beginning in the 1840s, steam power gave policy makers new tools to pursue old goals, all the while introducing new questions and problems into American foreign relations. These questions included how to use steamships to construct international communications networks, how to support the challenge of fueling a globe-spanning navy and merchant marine, and how to ensure steady access to the coal, and later oil, that became the lifeblood of industrial America. By the end of the century, Americans laid claim to a global empire, raising still more problems of security, strategy, and what came to be known as logistics. Over time, policy makers found that the adoption of oceangoing steamships led to questions about larger networks of fuel that stretched back to Appalachian mines and forward to island coaling stations. All these ostensibly technological questions were also hotly political ones, as the technological aspects of naval policy and foreign relations were shaped by domestic commercial, sectional, and partisan interests. When seen from the perspective of coal, the great process of industrialization and the emergence of the United States as a global power unfolded at the same time as intertwined processes.[22]

Historians have, of course, discussed topics like coaling stations in the late nineteenth century. In fact, the subject is pervasive in the foreign relations literature for the simple reason that most of the overseas locations Americans sought to annex after the Civil War, whether unsuccessfully like Haiti or successfully like Hawaii, were desired and justified at least in part for their value as strategic coaling depots for the navy. Yet too often, historians have framed discussions of island coaling stations, and of the constraints of fossil energy more generally, around the narrow question of whether they did or did not contribute to the emergence of an overseas American empire. On the one hand are scholars like the strategic theorist Bernard Brodie, who pointed to the steamship's dependence on distant coaling stations and declared in 1941 that "to recount the effects of the steamship from this point of view would be to retell much of the diplomatic history of the world since the middle of the nineteenth century." On the other hand are those like James A. Field Jr., who emphasized the limited success Americans had until 1898 in acquiring offshore naval facilities, the failures they faced in developing bases that already existed, and the often rudimentary nature of overseas coaling stations even after the turn of the twentieth century.[23]

But the questions raised by new sources of energy and new technologies of steam power were not merely, or even primarily, questions of empire. They were more fundamentally about how Americans conceived of their place in the world and acted on those conceptions. As these technologies developed, they became new tools for addressing old problems as well as new ones. With the new capabilities of steam engines—speed, power, access to previously unavailable routes—Americans returned again and again to anxieties about global commercial opportunities, sectional cohesion, territorial vulnerabilities to invasion or isolation, and only later to the opportunities and burdens of managing a controversial empire. For each of these anxieties, steam power and the coal to fuel it were from an early date seen as somehow tied to American security, and how Americans made sense of energy and security must be seen in its own terms. Americans of the nineteenth century would likely have found perplexing both Brodie's view that steamships remade global diplomacy and Field's assertions that the pursuit of coaling stations was not especially significant in mobilizing support for territorial acquisitions in the 1890s. When nineteenth-century Americans thought about steamships they were not typically thinking about empires and imperial administration, but mail routes, black colonization, or new networks to boost commerce. Simultaneously, even in instances when arguments about the security importance of coal proved unsuccessful in securing new territories, they nevertheless turned out to have lasting significance, helping structure how Americans made policy around energy and security through the latter part the twentieth century. Understanding how is one of the objectives of this book.

In the chapters that follow, I focus on three groups who helped shape the American relationship between energy and security. The first two are familiar subjects of diplomatic and naval history: naval administrators and officers, who played central roles in articulating the significance of coal for the navy and its importance for American security more generally, and politicians and policy makers, who debated different visions of America and fought to turn those visions into national policy. These figures worked inside presidential administrations and also, significantly, in the key congressional committees that oversaw subjects like naval affairs, national security, postal communication, and natural resources. Drawing on personal correspondence as well as vast congressional committee holdings at the U.S. National Archives, this book emphasizes the central role played by Congress in shaping the American approach to energy. The third group, scientists and engineers, are less often found in discussions of American security, at least in the nineteenth and early twentieth centuries. But boiler designers, geologists, and mining engineers were crucial figures, who both responded to the imperatives of national security and who themselves helped shape the perception of those imperatives.

Together, these Americans had to learn to see energy as integral to national security and a vital responsibility of the federal government. They had to learn to think about energy as a subject for foreign relations, naval strategy, and national security. They had to discover the national interest. It was a gradual process that unfolded over a century, and it forced Americans to make choices about their place in the world. They had to wrestle with the perceived demands imposed by the new technology of steam power and either struggle to meet those demands or pursue still newer technologies with fewer, or at least different, constraints. Without question, this process has a long history.

This book advances three arguments. First, Americans did not begin thinking about energy in terms of security around oil in the early twentieth century but rather around coal in the nineteenth. During this century, Americans argued about the role of coal and the meaning of security itself. They debated how sources of power and ideas about security appropriately reflected the role of the United States in the world. Furthermore, these American debates around coal before World War I shaped how Americans came to think about oil, which increasingly appeared as a significant strategic fuel beginning in the 1910s. As late as 1901, French Chadwick, then president of the Naval War College, could declare that oil "is so closely allied to coal that it may be regarded in the same category and the two thus dealt with as one."[24] As one, because in 1901, few people concerned with national security imagined consuming oil as a fuel in ways very different from how they already consumed coal. From the perspective of national security, powerful continuities connected coal to oil, even though the two fuels were typically produced by entirely different industries through different labor arrangements and under different economic circumstances.

The second argument focuses specifically on the relationship between energy and empire. Contrary to many accounts, the supposed security need for distant American coaling stations in the late nineteenth century did not play a significant role in catalyzing the emergence of an American island empire around 1898. Instead, the reverse happened: the establishment of that empire created entirely unprecedented demands for coal and coaling stations because Americans suddenly needed ways to protect their new, distant colonies from external threat and overcome internal resistance to American rule. Prior to 1898, Americans had typically taken different approaches to solving their coal problems—from entering into diplomatic agreements to inventing new technologies for conserving coal to developing new navigation techniques for making routes more direct. Perhaps most significantly, before 1898, few Americans anticipated elaborate naval operations far from American shores and still less the need for bureaucratic methods for directing them—the very circumstances created during the post-1898 Philippine war and, to an even greater degree, World War I. There had certainly been debates over coaling stations between 1865 and 1898, but they had principally been about economic speculation, not security. After 1898, coal, coaling stations, and national security took on new meanings.

The final argument of this book is that technological change was not an independent variable in American foreign relations but an integral element of it. Industrialization and the emergence of the United States as a global power were intertwined historical processes. Specifically, the arguments and actions of American politicians, naval officers, and business leaders drew consciously on the innovations of fossil-fueled steam power in articulating roles for the United States in the world. Americans debated their role in the world with knowledge of what steam power made possible or on the basis of what they wished and believed it made possible. The design of engines, selection of fuels, organization of lines of supply—in short, the entire infrastructure of the global fossil fuel network—all helped structure the broader vision of the United States in the world. At the same time, mechanical and bureaucratic changes reflected pressures from outside American borders, as Americans pursued new approaches to energy in response to perceived threats to their security.

These three arguments are woven together throughout the book, each chapter addressing a different theme as the book advances chronologically, with some overlap, between the 1840s and 1940s. The first three chapters trace the emergence of steam power and fossil fuels as subjects for federal policy makers and international relations in the two decades before the Civil War. During this period, questions of fuel, state power, and the global geography of energy remained contested, but the outlines of nearly every policy choice available in later decades may already be found.

Chapter 1, "Empire and the Politics of Information," explores the origins of international steam politics by considering debates that took place in nineteenth-century America over the changing political economy of global information. During this period, new transoceanic steam vessels became as valuable for quickly conveying news and correspondence as for their potential advantages in war. In the 1840s, the United States inaugurated a policy of providing federal subsidies for mail-carrying private steamships, ships of a size then unthinkable for private capital to finance alone. For some promoters, steam power provided an engine for transcending sectional antagonism by opening markets in the exotic lands of South America, Africa, and the Far East—an engine to conquer what mail steamer advocate William Henry Seward called "the ultimate empire of the ocean." As these congressionally funded lines began plying the oceans, however, their operators—both private contractors and U.S. Navy officers—began encountering obstacles related to fueling. These problems ranged from lack of adequate supply to low quality to high cost. Because the vessels depended on public funding for ostensibly national purposes, the fueling problems became subjects of public debate.[25]

The next two chapters explore how Americans began addressing these fueling problems in commercial steamers as well as the new navy steamers built to protect that commerce. Chapter 2, "Engineering Economy," moves from legislators and steam contractors to the engineers, mechanics, and scientists who helped build new steam vessels and develop the fueling infrastructures that allowed them to run. The United States was known in the mid-nineteenth century to be rich with coal, but no two coals were identical and Americans eagerly sought to identify which varieties and which coal-bearing regions best suited American steam engines. American chemists and geologists conducted a range of analyses on North American varieties, while engineers proposed innovations to Congress for use aboard public vessels. This chapter argues that the principal way Americans thought about solving the new challenges of global steam power was not by hunting for an empire of foreign coaling stations but instead by pursuing what nineteenth-century Americans called "economy." Economy was an expansive concept, embracing not only the design and practical operation of steam engines but the chemistry of particular kinds of coal and the construction of networks that connected production with consumption. Through this research, scientists and engineers directly responded to debates over national policy.

Chapter 3, "The Economy of Time and Space," continues the investigation of how the concept of economy guided American thought on building global steam networks by moving from laboratories to the geographic imagination. Historians have long noted how nineteenth-century Americans spoke of the "annihilation" of time and space to describe their changing perceptions of travel, distance, and communication. But in the 1840s and 1850s, they spoke more commonly about how steam power could *economize* time and space, not destroy it altogether. Steam made certain distant places seem closer and newly important, even before Americans could use the new technology to get there practicably. The result was a series of government expeditions both to help locate possible foreign supplies of fuel and to establish depots for refueling, all to support American commerce overseas. These expeditions to places like Borneo, Formosa, and Japan, undertaken alongside the technical investigations detailed in Chapter 2, brought Americans into a new world of diplomatic engagements and raised questions about the wisdom, purpose, and propriety of securing territory overseas.

The last three chapters of the book explore the evolution of the fuel problem and how coal, and later oil, both shaped American thinking about security and foreign relations and was, in turn, shaped by them. Chapter 4, "The Slavery Solution," uses the lens of coal to internationalize the American Civil War. Both the North and the South struggled to obtain coal supplies for their naval and industrial operations. The constraints of steam power proved particularly troubling in the Caribbean, both materially and diplomatically. Lincoln and his cabinet faced this challenge just as they began pursuing another policy Lincoln had long supported: the colonization of free blacks outside the United States. Lincoln oversaw the only moment in American history when colonization became an expressed policy of the federal government. This chapter focuses on how ideas about coal and labor influenced his favored colonization project, a colony in Chiriquí in what is now the border region between Panama and Costa Rica. The colony, a product of a decade of prior colonization attempts, was intended to employ free American blacks to mine coal for naval use. Though ultimately a failure, the attempt reveals the complex ways Americans thought about the new challenges of industrial energy. This chapter details not only why the project failed but also the ways the Union ultimately succeeded in fueling its war with domestic supplies.

Chapter 5, "The Debate over Coaling Stations," investigates the period between 1865 and 1898 and asks what Americans meant when they argued (or denied) that steam power required foreign coaling stations. Arguments over these stations were a near constant in post–Civil War diplomacy, as Americans sought to justify the acquisition of islands from the Danish West Indies to Santo Domingo to Hawaii for their essential value as refueling depots. Read backward from the years after 1898 and the emergence of a global American island empire and the militarization of American foreign policy, these efforts have appeared as ominous precursors. But the reasons through which Americans justified the need for coaling stations in the 1860s were different from the reasons they offered in the 1890s. In the earlier period, appeals to commercial expansion dominated, and many sites sought for these stations were constrained by developments in international law that meant they could only be useful in peacetime. Later, Americans more frequently expressed a sense of territorial insecurity and predicated the need for coaling stations on the need for national defense. Yet crucially, as in the antebellum period, most Americans who pondered the question of how to fuel American security thought more about technological ways to solve problems than ones involving the pursuit of foreign territory. By 1898, Americans seized the opportunity to build an overseas empire— precisely at the moment naval strategists had concluded that developments in engineering had made the pursuit of most island territory unnecessary and unwise.

Chapter 6, "Inventing Logistics," moves from diplomacy to the revolutionary changes in naval organization that helped modernize the Navy Department and ultimately made possible its complex global operations in the twentieth century. Since the 1970s, historians have underscored the way the history of warfare has often turned less on much-studied strategies or battle tactics than on the unglamorous calculations of logistics. But while supplying armies and navies have indeed always been a part of warfare, our modern notion of logistics has not. In the United States, there was simply no organized study or practice of logistics until the turn of the twentieth century, when even the word "logistics" itself was unfamiliar. Gradually, as the pressures of defending distant colonies and fueling and outfitting a two-ocean navy increased, American officers and academics developed an original science of logistics. Logistics thinking inverted the causality of empire building. Before 1898, the loudest expansionists declared that the need for coaling stations demanded annexing foreign territory; afterward, logistics planners explained that the existence of foreign territory demanded the establishment of coaling stations. In this sense, logistics thinking was largely shaped by concern over fuels, and as that concern grew, it helped strategic planners think anew about the resources essential for modern warfare. No longer content to rely on markets, they began pursuing greater control over the potential sources of their supplies. This chapter explores the development of this science of logistics, moving between the institutions whose work helped create it: the classrooms of the Naval War College in Newport, Rhode Island, where lectures, exam problems, and war game simulations taught midcareer officers how to plan for war; the navy's Bureau of Supplies and Accounts, where the field unfolded in a practical way; and the naval policy-making General Board in Washington, DC, where the theory and practice were synthesized into policy. The chapter then traces how these developments in logistics guided the navy's pursuit of a coal mining operation in Alaska's Matanuska Valley and helped shape the scandal of Teapot Dome.

A brief conclusion carries the narrative from the mid-1920s through the outbreak of World War II and provides a reflection on the themes brought up in the book along with an argument about their relevance today.

In sum, this book reveals how energy did not suddenly become critical to American security in 1945; rather the central role it assumed after World War II was the culmination of a history stretching back over a century. It was an element of the national interest that had to be learned, but this education hardly took place in direct or obvious ways. Recovering the history of coal and empire forces us to recognize that the complex political problems we today identify with oil—from fractious geopolitics to unstable global markets—are not simply characteristics of an oil economy but social and political challenges that would confront us and our integrated world regardless of our sources of power. But if history can help us understand our problems, we can hope that it can help us envision solutions to them as well.

CHAPTER ONE

EMPIRE AND THE POLITICS
OF INFORMATION

Thou shalt make mighty engines swim the sea,
    Like its own monsters—boats that for a guinea
Will take a man to Havre–and shalt be
    The moving soul of many a spinning-jenny,
And ply thy shuttles, till a bard can wear
As good a suit of broadcloth as the mayor.

William Cullen Bryant, "A Meditation on Rhode Island Coal"

"The mail is in; here is the 'Straits' Times!' " came the call in Canton. The year was 1853. Pressed on "a half sheet of foolscap" in Singapore, the *Straits' Times* was a digest of newspapers freshly delivered from Europe and the United States. These western papers reached southeast Asia by steamships and across the overland route connecting the Mediterranean and Red seas. The ships, which delivered mail as well as news, connected Canton and Boston in a scant sixty-five days. "Such speed is almost incredible even now," exclaimed an American expatriate visiting the bustling Chinese port. Only twenty years earlier, the fastest American and British clippers might have taken twice as long, and the future augured even greater speeds. "Boston and Canton will be still more closely approximated in point of time," he continued, "when a railroad connects the Atlantic and Pacific coasts of the United States and a system of steam navigation is established across the Pacific, between California and China."[1] The pursuit of this system—a global network of postal communication powered by steam and at least partially under American control—frustrated and tantalized Americans throughout the middle of the nineteenth century.

The United States was a commercial nation in 1853, and commerce depended on information. Business letters connected capital to its investments, traveling merchants to their houses, and traders to the producers and consumers of their goods. Diplomatic orders linked remote ministers to their home departments. Newspapers alerted farmers to commodity prices in international markets. Personal correspondence bound emigrants to their distant families. International postal exchange formed the tenuous thread that kept modern institutions operating, and as the example of the *Straits' Times* suggests, nowhere in the mid-nineteenth century were Americans more dependent on this exchange than in China. There, over six thousand miles and an ocean away from home, American trading houses jostled to exchange opium, ginseng, lead, and cotton goods for teas and silk.[2] "Early information as to the changing condition of the markets in Europe and America is very important to merchants in China," related the expatriate. Control over this information meant "to a considerable extent, the advantages of a limited monopoly of the trade." For traders, it could mean the difference between profit and bankruptcy.[3]

The expatriate's account of how steamships had transformed the global flow of information was a common narrative during this period. So too was a corollary he did not need to mention: that the burden of building this network of steam communication devolved in large measure on the state. Midcentury, no amount of private capital alone could be mustered to build and maintain the largest steamships and, just as importantly, no private company could be counted on to meet public needs. As one memorialist to Congress succinctly explained in 1858, "one of the responsibilities lying most heavily upon the Governments of commercial countries is that of maintaining the written intercourse of their people, not only among themselves, but with friendly nations with whom they have the relations of commerce."[4] In the United States, governmental responsibility for domestic correspondence had been widely acknowledged since the passage of the Post Office Act in 1792. That governments also bore a responsibility to ensure *international* communication was a radically new idea.

New, and pursued by the United States between 1845 and 1860 through a policy of federal subsidies for mail-carrying steamships. Historians have long portrayed this experiment as little more than a footnote to the industrialization of transportation, an irrational pursuit of national pride in the face of less sanguine economic realities, and an example of a failed government intervention in the private economy.[5] Contemporaries saw things differently. "American Atlantic steam navigation was wisely and even necessarily undertaken," exclaimed William Henry Seward as he successfully argued for increasing American subsidies to the Collins Line of mail steamers, "to maintain our present commercial independence, and the contest for the ultimate empire of the ocean."[6] Americans like Seward worried about ceding carriage of international mail to the British because of a widely shared conviction about the domestic and international importance of information. Domestically, Americans saw global flows of information as crucial tools of interstate and intersectional commercial rivalries. Simultaneously, improved steam communication could benefit the entire country and forge a nation from a mere union of states. Internationally, Americans were the victims of Britain's persistent failure to circulate American diplomatic correspondence, business letters, and newspapers in a timely fashion. They watched anxiously as Europeans built steam communication links to Latin

America, encouraging the expansion of trade. Meanwhile, trade stagnated between Latin America and the United States, where such lines were lacking. Americans in the mid-nineteenth century expressed a sophisticated understanding of the role of global information, one that suggests our contemporary belief in the novelty of living in the "information age" and "network society" is not so new after all.[7]

Recent investigations of the domestic U.S. political economy of information have revealed a complex and evolving system integral to both public and private life. Between 1840 and 1860, argues David Henkin, Americans developed a new "culture of the post," as changing expectations, practices, and ideas about mail communication altered their "perceptions of time, space and community."[8] What's more, as Richard John has demonstrated, the American postal system hardly formed an invisible infrastructure nestled in the background of American society. Instead, it was for many Americans not only their principal mode of interacting with an agency of the national government but itself a perennially debated political subject.[9] Other studies have examined how different varieties of information bolstered social power and have explored how new technologies helped nationalize the news consumed in a continent-sized nation.[10] Few of these studies, however, have investigated how the same Americans seeking cheaper postage between New York and Chicago simultaneously demanded it between New York and Liverpool, or how the economic and cultural changes brought about by a new infrastructure of domestic postal communications were also shaped by the commercial, diplomatic, and personal exchanges newly possible around the globe.

The project to build a global network of American mail communication was based on steam. These mail steamers were not isolated artifacts but machines embedded in complex networks of private capital, international markets, government regulation, naval strategy, diplomatic relations, and, as this chapter shows, material constraints. In the 1840s and 1850s, many Americans had ideas about how to construct these mail steam networks, but communicating information required an infrastructure of transportation, and steamers needed coal. As Americans pursued international steam communication, they were forced to confront new challenges, ones that brought the strategic significance of fossil fuels to the attention of the government in Washington for the first time. Pursuing better communication, Americans discovered the international challenges of securing coal.

The Information Economy

Competition to control commercial information in the early nineteenth century began in the Atlantic. In the two decades after the end of the Napoleonic Wars, American packet boats came to dominate communications between the Old

World and the New. Up to this point, postal correspondence had been intermittent, dependent on the irregular schedules of commercial vessels, but on January 1, 1818, New York merchants launched the first regular shipping line to bring mail, passengers, and cargo across the ocean. This line, known as "the Black Ball" for the distinctive discs adorning the ships' fore topsails, sailed between New York and Liverpool on a monthly schedule. Other lines followed, and by 1840 there were forty-eight packets crossing the ocean. American sailing vessels, according to maritime historian Robert Albion, "were *the* ships of the North Atlantic."[11]

As the packet lines came to dominate international postal exchange, American and British shipping firms and engineers began experimenting with using steam for ocean travel. In 1819 the American steamship *Savannah* crossed from the eponymous city in Georgia to Liverpool in twenty-seven days, though most of the journey was, in fact, powered by sail. It would be another fourteen years before the Canadian *Royal William* would complete the Atlantic voyage entirely by steam. In 1838, the British *Sirius* began carrying mail across the ocean, followed only days later by Isambard Brunel's *Great Western*. That year, the British government began accepting proposals for regular steam mail routes, and the following year, it entered into a contract with Samuel Cunard of Halifax, Nova Scotia, for seven years worth of service between Liverpool, Halifax, and Boston. Despite these technological and organizational innovations, for a while, American packet boats continued to grow in number and speed, competing with steam for carrying the mail. For correspondents, more frequent sailings began making Atlantic communication both faster and more predictable than ever before. Over the coming years, the British government subsidized additional mail steam services along other routes of commercial interest, from the Mediterranean to India, China, and the western coast of South America. As the cost of sending mail decreased and its delivery became more regular, British steamers began capturing the American mail that had once traveled in American packet ships. As the British post office subsidized each additional line, constructing the connective tissue that linked the British empire together, it increased its control over international communications.[12]

Even after the establishment of the packet lines, however, global information did not flow freely, especially for Americans. Some American firms, like those conducting business in South America, were forced to rely on informal networks of mail conveyance. This obstacle to communication was painfully true for the vast American whaling industry, as ship owners were typically unable to contact their captains—or capital—for years at a time. "Many instances have occurred," observed one memorialist to Congress supporting an American steam line connecting the Pacific coasts of Panama and Valparaiso, "in which bad management, or want of necessary information has occasioned serious losses, or in which unfaithful masters or agents have wasted or embarrassed large amounts of property, and which might have been prevented, by seasonable communication of intelligence and of instructions."[13] More poignant was the plight of the whalers themselves. "One commander of a whale ship from Nantucket recently informed me" wrote Albert Gallatin Jewett, an American chargé in Lima, "that during his last voyage, of more than three years, his wife and family wrote him one hundred and forty letters, of which he received only fifteen: that, on his return, as he cast anchor on his own shore, expecting to meet them all, he was informed that his wife had been dead for nearly a year."[14]

The fragility of American postal correspondence by sea continued well after steamships began replacing the old packet lines. National lines did not serve all correspondents equally, and as the steam networks of other nations expanded, Americans found that the absence of reliable communications from their country hampered their opportunities for trade. In the 1850s, American merchants seeking trade with Brazil and the West Indies found it increasingly difficult to compete with European rivals, for British and French steam lines were able to deliver the mail of European traders between twelve to eighteen days sooner than U.S. ships could deliver American intelligence. Since there was no direct, regular steam line to commercial cities in South America, mail bags and passengers traveling from the United States first had to voyage east to England, Portugal, or West Africa and then board a European line bound for South America, a circuitous route covering two or three times the distance of a direct line. A direct line, however, did not yet exist.[15]

Closer to home, American merchants faced other obstacles. In the early 1850s, New York traders could not simply dispatch letters to their agents in the West Indies. Even though an American line, the United States Mail Steamship Company, by then connected New York with Cuba, and British lines linked Havana with ports in Barbuda, Trinidad, Guiana, and the great entrepôts of St. Thomas and Curaçao, the British lines saw to the interests of British merchants by refusing to integrate their service with the American one. New York merchants complained of having to send their mail to the American consul in Havana, who would then personally forward the correspondence to various destinations aboard British lines. The only alternative was relying on American sailing vessels that carried valuable cargos of sugar and molasses to the United States, but these ships sailed irregularly, often without touching any Caribbean port for months after leaving with a cargo. Like thousands of his fellow merchants, Theodore G. Schomburg believed the solution rested with steam. Considering the hindrances, he wrote, "the immence [sic] value of this new means of rapid and regular inter-communication will be more fully appreciated." For merchants in Philadelphia and New York, the growth of European commerce with South America was an affront, for it violated their belief that the United

States, not distant Europe, constituted the "natural market" for the tropical produce of the western hemisphere.[16] Moreover, fretted one American promoter, with its growing network of subsidized steamers, "England is made the great centre from which all political information radiates, and it comes, not unnaturally, tinged with the coloring acquired in its transmission," further disposing potential trading partners to view the United States with unease. In this view, steam lines constituted technological subversions of economic law and American rights.[17]

The growing British control of Atlantic postal communication posed special problems for American diplomacy. In addition to formal dispatches, the State Department and its ministries abroad regularly exchanged newspapers, congressional publications, and other printed material—sources of information as essential to government representatives as they were to American merchants. The British custom house and post office, however, frequently confiscated American mailbags and applied arbitrary and discriminatory charges on official correspondence. When mailbags contained an obvious mix of costly (and light) dispatches and cheaper (and heavier) newspapers, the British custom house weighed the lot and levied the higher charges on everything. The only alternatives were to ensure that American messengers—usually ordinary passengers—personally carried the sealed public mail on every transatlantic voyage or to separate newspapers from dispatches and send the latter by the regular, open mail. Opting for the open mail approach required posting the dispatches a second time upon reaching Britain—as if they had in fact originated there—thus incurring additional charges and delays. "If we choose to make use of their mail conveyances," grumbled the American minister to the Court of St. James, Edward Everett, "we must pay anything they think proper to charge us; but I think they have no right to force our despatches into their mail." Even using messengers presented challenges, however, for these private citizens were often unreliable couriers. In his dispatches to Washington, Everett acknowledged that most American messengers served their country's interests well, but he complained of others who abused their status as conveyors of public mail by smuggling dutiable merchandise or who were more preoccupied with their own affairs than with the task they'd been assigned. As a result, official mailbags could languish in Liverpool for days before they finally reached the ministry. And if that weren't enough, Americans still contended with overzealous British customs agents who confiscated mailbags under some pretense or another.[18]

These minor indignities could conceivably have been borne, but British postal policy had more insidious effects on American communication. While British newspapers circulated within the United Kingdom without charge and in the United States at rates equal to those charged American newspapers, American papers in Britain and those forwarded through Britain to Europe faced prohibitive letter rates of five or six shillings apiece. These charges choked the global flow of information and the extent of American influence on the continent. Charles A. Wickliffe, serving as postmaster general, first proposed, unsuccessfully, a set of "mutual arrangements" in 1844 to remedy this situation with Britain, but to no avail. It would be another three more years before the obstacles to transatlantic correspondence were overcome.[19]

Throughout the decade, Americans remained desperate for international news but struggled to receive it. When Herman Melville's older brother Gansevoort served as the secretary to the American legation in London in 1846, he was assigned responsibility for preparing the dispatch bags for America. For one voyage in January, he packed just twelve letters but forty-three newspapers, from the *London Illustrated News* to *Punch*, each addressed to recipients as varied as his mother, brother Herman, senators Lewis Cass and Reverdy Johnson, and President Polk. Two months later, Melville sent forty newspapers and not a single letter. Access to news and newspapers was the essential prize of transatlantic steam communication.[20]

The American vulnerability to Britain's steam-powered monopoly on international postal traffic became most clear when the U.S. Congress attempted to establish government subsidized steam lines of its own. President John Tyler had urged the establishment of such lines in his 1844 annual message, and under an act passed in March of 1845, Congress directed the postmaster general to contract with private investors to create two steam mail lines, one of them for transporting printed matter to and from Europe.[21] In 1847, the United States entered into a postal contract with Edward Mills, who began a promising mail steamer service between New York and Bremen through the English port of Southampton. Mills's Ocean Steam Navigation Company, or Bremen Line, would join the five ships of the British Cunard Line (Liverpool to Halifax and Boston) and the one French Line vessel (Havre or Cherbourg to New York) already in service on the Atlantic.[22]

Americans turned to Bremen because the Prussian government promised to subsidize the line and because the economics of shipping and taxation there was attractive. In the 1840s, the states of the German Zollverein, or customs union, taxed imports at levels higher than nearby France, driving European imports and exports to the French port of Havre. Yet American trade prospects elsewhere in Europe appeared promising. Goods from Switzerland, Italy, and parts of Germany alone comprised almost a third of the total package traffic between Havre and New York in 1846. The American minister in Berlin, Andrew Donelson, noted that if the new line from Bremen to New York could recapture those packages alone, they would completely fill thirteen of the anticipated twenty yearly voyages of the new Bremen Line of mail steamers. This prospect of greater trade led the king of Prussia to commit some $100,000 toward

Bremen Line ships, and the company anticipated additional aid from allied German states. Still, the port of Bremerhaven remained difficult to reach by sea in winter, and Bremen's commercial infrastructure was insufficient to handle traffic of too large a scale. These facts together served as warnings of the financial risks of the venture, at least in the short term. As it turned out, however, the line failed for other reasons.[23]

The Bremen Line's first ship, the *Washington*, limped into Southampton on June 15, 1847, traversing the Atlantic in just under fourteen days. Church bells pealed to announce her arrival, while residents flocked to the harbor to greet the new vessel. What they found was a ship askew. Only one of the *Washington*'s two great paddlewheels touched the water; its coal supplies had been displaced during an Atlantic storm, giving the vessel a pronounced tilt that alarmed its 112 passengers. It was an inauspicious beginning. Over the following year, the *Washington* and her sister ship, the *Hermann*, would each go on to suffer a succession of mechanical failures and service delays, while managerial conflict would dog their owner, the Ocean Steam Navigation Company.[24] Yet while the Bremen Line ultimately played a minor role in developing transatlantic communication, its inauguration ignited a new diplomatic conflict between Britain and the United States.

When the Bremen Line was launched, the supporters of Cunard sought to ensure it would pose no serious competition to their profitable monopoly. Midway through the *Washington*'s maiden voyage, when only its passengers and crew knew of its mechanical difficulties, George Bancroft, then American minister in London, received notice that the British post office had instructed its agents to charge the letters on the *Washington* the full ocean rate—the cost of carriage by a Cunard steamer—even though the British post office played no role and incurred no expense in ferrying this mail across the sea. When pressed by Bancroft, Lord Clanricarde, the British postmaster general, initially denied that the new charges were intended to protect Cunard and British revenues but soon admitted what Bancroft considered obvious. Bancroft understood the stakes immediately: not only the survival of the American mail steamers but the likelihood of escalating retaliations by the United States and Britain that threatened to suffocate transatlantic communication altogether. Indeed, by September, the American government had resolved to begin levying an additional—and oppressive—shilling on any British letter passing through America on its way to Canada and anticipated doing so as long as the British post office refused to withdraw its charges on transatlantic American mail. Hoping domestic pressure would force the British government to change its policy, Bancroft anticipated that "British merchants will not approve obstacles in the way of free correspondence any more than we do."[25]

British merchants did not approve of obstacles to correspondence, but resolving the postal dispute consumed far more time and effort than Bancroft had imagined. Beyond charging the additional shilling transit rate on mail to Canada, the American postmaster general, Cave Johnson, was constrained by law from imposing other retaliatory measures on British mail to the United States. For his part, Bancroft believed that following Britain and instituting double postage on mail carried by Cunard steamers was too mild a response. Instead, he demanded that Congress allow the postmaster general to turn away any foreign mail steamer, a power Bancroft suggested Johnson use to enjoin half the Cunard fleet from touching American ports until the British post office conceded.[26] Under no circumstances, however, should Congress refuse to act somehow. Submission to British rates, Bancroft insisted, would only humiliate the United States, and he reported that British public opinion would surely side with America against the post office.[27]

Even into October 1847, Bancroft anticipated a swift resolution of the postal conflict.[28] His optimism proved unfounded, however, and the subject drew the American government deeper into the new problems created by mail steamers. The two countries not only disagreed on what letter rates would apply and how postal revenues would be divided between them but also on the crucial topics of both newspaper rates and transit rates—the costs for conveying American letters through Britain to continental Europe. Bancroft considered these terms imperative not merely for the Bremen Line or any future line "but still more because it would be an immense advantage to our merchants and to all Americans who travel on the Continent of Europe."[29]

The Polk administration acted as forcefully as it could without further congressional empowerment, though it possessed few tools besides strong rhetoric designed to pressure London. In his annual message, the president warned that without the elimination of double postage on American mail, "it will become necessary to confer additional powers on the Postmaster-General in order to enable him to meet the emergency and to put our own steamers on an equal footing with British steamers engaged in transporting the mails between the two countries, and I recommend that such powers be conferred."[30] In his own annual report, Cave Johnson referred to "the obnoxious order of the British post office," suggesting that the levy amounted to theft. "The British government, by their order of June last," he wrote, "appropriates the American steamship Washington to their own use, so far as postage is to be derived from it, as fully as if it were her own, established and maintained at her own expense; and this for the avowed purpose of protecting the British mail steamers against those of the United States."[31] Still, as January 1848 came to a close, Bancroft still lacked any authority to further threaten the British post office. "Has congress acted?" he wondered to Cave Johnson, "Will it act?"[32]

The problem, Johnson replied, was not congressional opposition to retaliation but unfortunate timing. The post office had already drafted a bill setting postage rates on a nation's foreign mail as equal to the rate that nation imposed on American letters in American steamers. The department had submitted the bill to Congress, and the post office committees in both houses were considering it. But because the country was in a presidential election year, during which fierce partisan passions had already been aroused, swift action on even uncontroversial measures was not possible. "The approaching Presidential election swallows up every thing else," Johnson confessed in February. A month later, little had changed. "The presidency is now the engrossing subject," he wrote Bancroft. "Even peace or war is but secondary, not much is likely to be done which does not in some way connect itself with it." As legislators had no idea how to use foreign postal rate schedules for partisan advantage, Bancroft was left without additional leverage.[33]

In the United States, frustrated merchants tried to bypass the public postal systems. As the transatlantic communications network began to break down in 1847, for instance, a private mail carrier emerged in Montreal that covertly shuttled Canadian correspondence directly to Cunard vessels. Since unauthorized competition with the post office was illegal, Cave Johnson ordered the arrest and prosecution of both the operators and patrons of the line. Authorities arrested at least one Canadian express messenger on April 30, 1848, as he was traveling to New York City with a box of letters marked "Admiralty." Johnson planned to send the box to Lord Clanricarde with no charge, cannily "expressing of course the opinion that his officers in Canada or Great Britain can not be in any wise connected with the petty violation of our laws."[34]

The stalemate over a postal treaty stretched into 1848. Throughout, Lord Clanricarde persisted in charging letters in American steamers double ocean postage. Finally, Congress broke through its paralysis, fearing that further inaction would destroy its nascent mail steamer project. On April 12, the House passed a measure permitting the postmaster general to impose retaliatory rates on mail carried by foreign vessels when the foreign country applied such rates to mail in American vessels destined for that country. After a delay (possibly motivated by personal tensions between the postmaster general and the chairman of the Senate Committee on the Post Office and Post Roads), the Senate followed on May 29 and amid the distractions of the election year summer, President Polk signed the measure on June 27.[35]

In response, British merchants quickly put pressure on the British post office, and the House of Lords took up the matter.[36] The merchants and their representatives in Parliament took a dim view of their government's refusal to concede to American demands for equal treatment of British and American mail at reduced costs.[37] In August, members of the House of Commons from the commercial cities of Liverpool and Manchester pressed the government to explain why no progress had been made since February; Lord Palmerston, the prime minister, remained evasive.[38] When confronted with the frustrations of the commercial class by George Bancroft, Palmerston lashed out at the United States. "The origin of all the difficulty is with you," he insisted. "Your government set up Steam Boats without being requested to do so by England—We never asked you to do it." Minutes later, when Bancroft warned him how British obstinacy might be received in America, the prime minister exploded. Recounting the recent history of transatlantic diplomacy, Palmerston claimed that on every issue—the northeast boundary, the disposition of Oregon, British actions in Ireland—Britain felt besieged and threatened by the United States. Bancroft replied simply that if Britain maintained its aggressive postal policy, the United States would take the extreme step of seeking its manufactures from somewhere else in Europe.[39]

Ultimately, after months of stalling, the combination of pressure from the United States and merchants on both sides of the Atlantic forced the British government to agree to a postal treaty. The process had consumed over a year and a half, and even when finally agreed to, Lord Clanricarde confessed that he remained against an agreement. The treaty, with complex provisions fixing inland and ocean rates as well as establishing how postal revenues were to be divided between the two countries, effectively reduced the cost of posting a letter overseas from nearly anywhere in the United States to anywhere in Britain to twenty-four cents, a rate equal to what had previously been the ocean rate alone and lower than the postage for the much shorter distance between London and Paris. The agreement slashed newspaper rates, paving the way for vastly greater flows of commercial and political information, as both countries agreed to forego any charges for ocean transit and opted merely for a token two cents (or no more than a British penny) for each paper entering or leaving its borders. The treaty eliminated ocean charges for periodicals and pamphlets and removed obstacles to the transit of letters and newspapers through Britain to the continent, as this mail would instead be charged reduced ocean rates along with the minimal postage to Europe paid by any other article posted in Britain.[40] The postal treaty, signed December 15, 1848, and ratified the following January, expressed a confidence that international postal communication by steam could be structured in mutually beneficial ways and at rates that permitted greater correspondence. Yet achieving a postal treaty was one thing. Establishing lines of American steamers under its terms was another.

The Politics of Steam

The development of American mail steamers turned as much on the vagaries of domestic politics, and often very local politics, as it did on great national questions. Shipping agents in New York did not have to think deeply about transit rates and postal schedules to know that high costs and insufficient facilities would hinder their access to foreign markets. Traders in Charleston, or Philadelphia, or Boston did not have to consider the dynamics of an American steam marine to see that a line originating from their city promised a boon to local commerce. Such sentiments would lead Americans from across the country to send scores of petitions and memorials to Congress in the mid-nineteenth century seeking federal support for various lines of mail steamships. In the early 1840s, however, political support for such lines emerged most strongly under threat of war.

Americans evinced a striking sense of territorial insecurity even at the moment when the nation's continental imperialism reached its apex. The United States acquired Texas in 1845, settled with Britain title over the disputed territory of Oregon in 1846, and wrested California and much of what became the Southwest in war with Mexico in 1848. Still, many Americans perceived more weakness than strength. Alexis de Tocqueville wrote in 1840 that "Americans have no neighbors and thus no great wars, financial crises, devastations, or conquests to dread." If this statement had ever been true, Americans certainly did not see things that way in the decade that followed. The ambition to annex Cuba was matched by a perception that American states were likewise vulnerable to foreign invasion. In particular, the growth of the British steam marine, and to a lesser degree the French one, produced considerable anxiety, particularly in the South. Thomas Butler King, a Massachusetts native and for many years a Whig representative from Georgia, proclaimed his nation unprepared for war in an 1841 plea for the creation of a squadron of defensive coastal steamers. King pointed to undefended harbors, crumbling fortifications, and coastal defenses lacking cannons, especially along the country's broad southeastern coast from South Carolina to Alabama. New technology and domestic unpreparedness combined to produce a uniquely southern fear: "Any of our unprotected harbors might be entered by fleets of armed steamers," King wrote, "loaded with black troops from the West Indies, to annoy and plunder the country." In such an invasion, he warned, tens of thousands of regimented black soldiers would arrive by British mail steamers detailed for naval service, ships that would draw on coal depots in the West Indies and Halifax and at the same time sever American communications and trade routes from Florida to the Gulf of Mexico. The fact that these warships provided conduits to international information meant that the force could gather before the federal government would know a conflict had begun. Even in peacetime, King contended that the organization of Britain's mail steamers posed a threat, for the ships were commanded by Royal Navy officers who were gaining intimate familiarity with the American coast. Knowledge of its harbors and shoals, its winds and currents, offered the British strategic information that might prove decisive in a future war. As the size of the British fleet grew with government subsidies, Americans, and especially southerners like King, felt increasingly threatened.[41]

When Congress failed to respond to King's plea with funds for new war steamers, the representative took a different tack, proposing a combination of new naval construction and the establishment of lines of private mail steamships. His plan revisited the longstanding policy of funding the navy, however modestly, while leaving American merchant shipping to private investors. Instead, King sought for the United States "to render the transmission of the mail, passengers, and freight subservient to the extension of her naval establishment," much as Britain had already done. Congressional support for new steamers rested on these ships being privately maintained as contracted mail vessels in peacetime and as converted warships in the event of a national emergency. Already, many Americans conflated the new merchant marine and the developing naval fleet and considered Britain's maritime dominance at least as much a product of its mail subsidies as its already substantial navy.[42] "*Speed*!!!!" scribbled King in his notes, emphasizing what he saw as perhaps the greatest advantage of steam communication. He also played down the martial aspects of the plan. "They bind our country together," he wrote. "These are the *triumphs* of peace."[43] Language like this makes it difficult to tell to what extent members of Congress weighed the defensive and communications aspects of the steamer program. Without question, however, American merchants, who produced the bulk of correspondence, memorials, petitions, and news items on the subject, clearly expressed a greater interest in the enhanced flow of postal communication.

Despite the House Committee on Naval Affairs' urgent calls for great steamships to protect the country, the navy itself reacted uneasily to proposals for first-class war steamers. Seasoned officers questioned the value of the kinds of ships Congress proposed. Their objection was not to steamers as such—many officers in fact favored the construction of smaller steamers for raiding enemy commerce—but to large steamers, which seemed to them to offer too few advantages. They were costly, consumed "ruinous" quantities of coal, and possessed exposed machinery that prevented close engagements with the enemy.[44] Still, some officers saw potential advantages. If mail steam lines employed both junior officers as well as steam engineers, the new lines could help familiarize many in the service with the complexities of the new machines. Matthew Perry, for example, recognized these advantages. Beginning in 1848, Perry served as steamship inspector in New York; in this capacity, he scrutinized the construction of new mail steamers to ensure their suitability for naval purposes, until embarking on his naval expedition to Japan in 1853.[45]

King sat on the House Committee on Naval Affairs, and his proposal complemented that of another southerner, Alabama's Henry Hilliard of the House Committee on the Post Office and Post Roads. Hilliard echoed the American merchants and newspapers and advocated a more aggressive policy on steam communication.[46] Steam mail service offered apparently boundless opportunities for trade. "The rapid and certain transmission of intelligence is of the highest importance to a commercial people," he declared, "and instead of relying upon the steamships of Great Britain for the transportation of our mails, we should enter at once upon an enterprise to which we are invited by the most powerful considerations connected with our relations to the world, and which can no longer be neglected if we would keep pace with the movements of an enlightened age." Both King and Hilliard believed that the 1845 act creating the Bremen Line was only the beginning of what should become a vast system of subsidized mail steam lines, and they sought to build support for more.[47]

At first, they had impressive success. What Hilliard had described as his "enlightened age" brought a new round of mail steamships in 1847, when Congress authorized letting contracts for three additional lines. The terms of the act help illuminate the mixed motives of the new policy; while the 1845 act that funded Edward Mills's underwhelming Bremen Line had entrusted contracting for steam communication with the postmaster general, the new act granted this authority to the secretary of the navy. Both northern and southern states received potentially profitable lines: the successful proprietor of the Dramatic Line of sailing vessels, Edward Knight Collins, secured a line from New York to Liverpool, while A. G. Sloo obtained one from New York to New Orleans, with stops at Savannah, Havana, possibly Charleston, and Chagres in Panama. A third line, later organized as William Aspinwall's Pacific Mail Steamship Company, would be contracted by the secretary of the navy and connect with Sloo's line at Panama and carry mail up the Pacific coast to Oregon.[48]

It is telling that both King and Hilliard were southerners. Though historians frequently characterize the agrarian South, supported on the backs of slave labor, as less inclined toward technological innovation than the North, advocates of both sections believed that steam communication offered many advantages. King himself had long advocated for southern steamers; as early as 1841, he had corresponded with other southerners to propose steam lines that would couple carrying the mail with coastal defense.[49] Southerners like King were also connected to the maritime Caribbean economy. King's brother, Andrew, for instance, left the United States to manage a sugar refinery in Cuba, one of many southerners who looked to the Caribbean for commercial opportunities. While there's no evidence that the reason Thomas Butler King sought more steam mail connections to Cuba was to help his brother, the two suffered from the same kinds of informational obstructions that hobbled American commerce elsewhere: in 1842 the brothers quarreled over a mutual perception that the other kept in insufficient contact, though it turned out that owing to poor mail service, letters they had posted were simply never delivered.[50]

More generally, swift and direct southern steam lines promised to cut out middle men from more distant ports, reduce the time it took to ship cotton and especially the uncertainty connected with shipping it, and cut shipping costs to levels that would permit local banks to be able to finance them at lower rates than foreign ones. In pursuit of these steam lines, after Edward Collins introduced his subsidized service between New York and Liverpool, southerners returned to Congress seeking their own connections to Europe. In 1851, for example, the mayor, city council, and leading merchants of Baltimore asked Congress to support a direct line from that city and the nearby Virginia port of Norfolk to England, with the explicit aim of supporting southern commerce and providing a countervailing infrastructure to northern-dominated shipping.[51] William Barney and his associates fought for many years in the 1850s to establish a steam line between New Orleans and Bordeaux that would lessen southern dependence on northern merchants. "A direct trade between New Orleans and Bordeaux would save to the producers of our exports large sums of money," Barney wrote, "and to the consumers in the Southern States of foreign goods about 20 per cent., that being about the difference of cost of the same articles in New York and Southern cities."[52]

Interstate rivalries in the South also shaped the dynamics of that section's support for steam lines. The commercial ports of Savannah and Charleston, for example, were coastal neighbors and fierce rivals for foreign trade. When one city obtained a foreign steam line or when mail for one city was detained in the other, city merchants and politicians were quick to lobby the federal government for additional mail facilities.[53] When the postmaster general declined in 1845 to support a steamboat connection between Charleston and Savannah, routing the mails through inland Augusta instead, merchants of Savannah were outraged. By sea, the two coastal cities lay less than a hundred miles apart and the road between them delivered the mail in about a day. The proposed steamboat would have cut that time in half. Routed through Augusta, slower travel over 150 miles threatened to lengthen the delivery time by a day or longer, giving Charleston a crucial commercial advantage in receiving the latest news from New York and Europe. Boosters of the steam line in Savannah hinted ominously that if Cave Johnson refused to change course, there could be "no remedy but to wake up Mr. Johnson with a small speck of nullification"—presumably of some federal postal laws in Georgia—all to express the seriousness of postal routes conveyed by steam.[54]

Cities like Savannah desired mail steamers for advantages in capturing additional foreign trade, but the ships enticed Americans with diplomatic advantages as well. When the postmaster general, Nathan K. Hall, reported in 1851 that his department had endorsed the formation of a new line of steamers from New Orleans to the Mexican port of Vera Cruz, he noted that enhanced communication was only one benefit of the program. The United States and Mexico were recovering from a controversial war that had resulted in Mexico's loss of some half of its territory, a fact that could hardly be ignored in the relations between the two countries. Steady correspondence by steam, Hall stated, "would also, it is believed, be productive of great political and commercial advantages; would abate national antipathies and prejudices; promote and increase friendly views and relations between the people of the two countries, and unite more closely by mutual benefits the two great Republics of the western Hemisphere." This sentiment linking better international relations via faster and cheaper postal communication perfectly presaged arguments later advanced about telegraphs, radio, and most recently, the internet.[55]

Mail steamers had a use in domestic politics too. If western expansion and the debate over the extension of slavery exacerbated sectional tensions, then the global communication and international trade that the mail steam lines would facilitate could serve, advocates claimed, as a mechanism for bringing the sections closer together. Duff Green, a Whig editor, industrialist, and political adviser from Missouri, argued that any expansion of southern trade would, in turn, cultivate a greater demand for agricultural staples from the central and northwestern states to the benefit of all sections. As agent for the Georgia Exporting Company in 1850, Green advocated a direct southern steam line to Europe from Savannah in an attempt to advance southern commerce.[56] William Caldwell Templeton, a trader in cotton and sugar and a river steamboat proprietor from New Orleans, hoped to see New Orleans obtain the Mexican specie trade then flowing to England, which he believed could help the city become a southern complement to New York as the great American commercial mart. In Templeton's view, British steamers had established trade circuits that did not properly belong to them, trade circuits in which New Orleans, not London, was the "natural depot."[57]

Mail steamers hardly interested the South alone, however. Proposed steam lines for northern port cities promised direct connections to Europe and Asia, and these cities, who competed with one another for trade, capital, and influence, quickly embraced any plan that favored them. When Ambrose W. Thompson, a Philadelphia shipping entrepreneur, memorialized Congress in 1850 to subsidize an ambitious network of steamers connecting California and China in the Pacific and Philadelphia, Norfolk, and Antwerp in the Atlantic, the

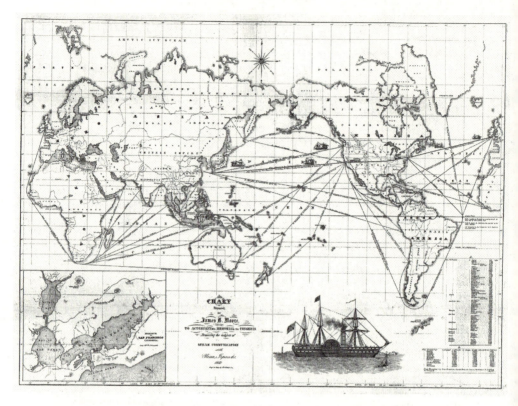

James Moore commissioned this engraved world map as part of his campaign for federal subsidies for an American steam line to the Far East. Moore was among dozens of Americans sending petitions and memorials to Congress in the 1840s and 1850s for similar lines to carry mail, freight, and passengers. Congress's policy of steam subsidies helped make coal and its global availability for American vessels a political concern for the first time. *Chart Prepared by James B. Moore, to Accompany His Memorial to Congress Respecting the Subject of Steam Communication with China, Japan &c., 1850* (Cincinnati, OH: Hugo Gollmer, 1850), courtesy David Rumsey Map Collection.

legislature of Pennsylvania was quick to favor the measure and request its congressional delegation do the same.[58]

Although southerners played a considerable role in advocating for mail steam lines and although, at least initially, the proposed lines would mostly have benefited the North, it is clear that political support for mail steam lines did not fall into simple partisan or sectional categories. In 1852, for example, Congress endorsed Edward Collins's request for additional subsidies for his line between New York and Liverpool because of support by merchants from around the country. In March and April 1852, Collins had organized a campaign stretching from Boston, New York, and Philadelphia to Baltimore, Charleston, and Mobile. His preprinted petitions bore the signatures of ship owners, trading firms, marine insurers, city aldermen, clerks, and comptrollers. Nearly 350 residents of Charleston agreed that "the Collins Line is not a local, but a National interest." It was a bulwark of American independence against British domination of transatlantic communication through subsidies to Cunard steamers. A massive petition from New York collected over 1,000 signatories—both individuals and commercial firms—while supporters from other cities as distant as Detroit and Portland, Maine, totaled over 1,350.[59] In the House of Representatives, the key vote passed narrowly eighty-nine to eighty-seven, only because sixty-five northerners were joined by twenty-two representatives from southern and border states, along with two from California.[60] Yet as the captains and engineers of this and other steam lines tried to turn funding, both government and private, into functioning networks of communication, they found a range of obstacles in their path. Chief among them was the difficulty of securing coal.

The Challenges of Coal

When Gansevoort Melville crossed the ocean to take a post in the American minister's office in London in 1845, he experienced firsthand the travails of early transatlantic steam travel. Upon finally nearing Liverpool after an exhausting voyage, he complained to his mother that "our passage has been long owing to contrary winds, bad coal & deficient power in the engine." It was a characteristic complaint. In the 1840s and 1850s, even as engineers figured out how to apply steam power to long-distance ships, the lack of good fuel routinely hindered voyages. It was a problem shared by nearly all early shipping lines.[61]

Like his father, a former shipmaster, William Wheelwright was drawn to the sea. Born in 1798 in Newburyport, Massachusetts, Wheelwright spent his early childhood in school, including a few years at Andover. After the War of 1812, he began his working life as a cabin boy on a ship sailing for the West Indies. By 1817, he was a captain. In 1823, he found himself in Argentina, victim of a shipwreck off the coast of Buenos Aires, and committed to building a life in South America. Eventually settling in the port of Guayaquil, he established a merchant house and became the U.S. consul. After a brief visit to Newburyport in 1829, he returned to Guayaquil with a new wife, discovered that his business had collapsed in his absence, and settled instead in Chile's commercial center, Valparaiso. There, he began a series of commercial and industrial ventures, most significantly the establishment of a steamship line in 1835.[62]

Wheelwright's object was accelerating communication along South America's Pacific coast and connecting the region to the western coasts of Mexico and the United States. "Owing to the present irregularity of advices," Wheelwright explained in 1838 to investors in London, "vessels are often indefinitely detained at the different ports of the Coast; and from the same cause no changes in markets can be beneficially and mutually acted upon." In Wheelwright's view, the unpredictable arrivals and departures of sailing vessels hindered otherwise attractive business opportunities. And if the opportunity to keep a finger on the pulse of Latin American commerce were not sufficient to induce his British backers to support a steam mail service, Wheelwright also appealed to their fiscal sense. By the late 1830s, British lenders had lent millions of pounds to South American governments, though few countries appeared likely to repay them any time soon. Political unrest and scant infrastructure hampered the development of stable state institutions. Wheelwright argued that an effective transportation and communication network based on steam power would rid the continent of these constraints. "The effect of it would be," he insisted, "to strengthen the executive authorities, to promote the industry of the people, and to contribute to an improved state of public and private credit." Steamships, however, consumed vast quantities of coal, and securing coal in South America proved far more difficult than Wheelwright or his backers had first anticipated.[63]

Wheelwright had good reason to believe that adequate coal supplies existed to support his steamer project. He had assurances, in fact, from Robert FitzRoy, captain of the *Beagle* during Charles Darwin's voyage around the world between 1831 and 1836. "In my own mind," wrote FitzRoy to Wheelwright in 1838, "there is no doubt whatever of the existence of coal in abundance at various places on the western coasts of South America." Moreover, continued FitzRoy, "its quality is sufficiently good to make it available for steam-vessels."[64] Wheelwright himself had investigated reports of coal in the Chilean port town of Talcahuano in 1834. As he planned his steamship service, he anticipated obtaining inexpensive supplies from these or other Chilean mines nearby. If necessary, he believed imports from Britain or Australia could provide additional stocks. He was soon forced to reconsider these plans, however. Returning to Lima after meeting with investors in London in early September 1840, he expected the imminent arrival of two vessels he had commissioned in England. One of them, the *Peru* arrived in November, but the unanticipated absence of coal in both the Peruvian port of Callao and the Chilean port of Valparaiso, however, threatened to doom his steamship service before it even began.[65]

Wheelwright began frantically searching for more fuel. Rumors of coal from the nearby island of San Lorenzo failed to pan out, while anthracite from the Cordillera proved too distant to supply the port. Samples purporting to be coal from Piura province turned out to be mineral pitch, useless for steaming. With few options for fuel remaining and his charter from the government of Chile expired, he prepared to accept defeat. He boarded the *Peru* and left Callao for Valparaiso. There, to his surprise, he discovered the *Portsea* had arrived with 600 tons of coal. Wheelwright was elated, but only until he discovered that the coal not only failed to generate steam—"little better than sulphur" he called it—but that it damaged his ship's boilers as well. This episode, as Wheelwright recounted to his directors in London, "has brought this beautiful enterprise, commenced under the most brilliant circumstances, upon the verge of ruin." When the *Peru* left Valparaiso for Callao, Wheelwright expected that owing to a lack of fuel, it would not return again to Valparaiso. It did in fact return, but only because the ship's captain encountered a shipment of wood in the port, which he promptly purchased and consumed.[66]

For two and a half months in early 1841, Wheelwright and Captain Peacock of the *Peru* canvassed the Chilean coast from the Maule River to the island of Chiloé. They made their way to Talcahuano, where Wheelwright had collected coal samples seven years earlier. There, they deposited mining equipment, rounded up forty men living nearby, and began mining. According to one observer, coal there "was found to give abundance of steam, although yielding a large amount of residuum, and about 20 per cent greater consumption than the best Welsh coals, requiring consequently more space in the ship and greater labour in working." By 1843, Wheelwright's miners had excavated almost 5,000 tons; barely a decade later, that number stood at 30,000. All told, Wheelwright estimated that his difficulties in supplying coal cost his company some £23,000, or $121,000. Although Chile undoubtedly contained substantial deposits of coal, turning natural abundance into practical resources was far from easy.[67]

Over the two decades that followed, other new steam lines repeatedly faced similar difficulties in securing coal, including those lines directly subsidized by the U.S. government. The mail steamer *Oregon* belonged to William Aspinwall's Pacific Mail Steamship Company, the second vessel of the new line funded with federal subsidies and government mail contracts. Aspinwall built the ship to connect San Francisco with Panama, where mules, and eventually a railroad, would carry passengers, mail, and specie across the isthmus to a second steam line to New York. On the *Oregon*'s first voyage from New York to California during the winter of 1849, however, the crew discovered that it took more than new laws to build a new steam communications network. Operating conditions were terrible. The ship's engine room reached temperatures of at least 132 degrees Fahrenheit and within days of leaving port, according to the engineer's log, the berths for engineers and firemen were "to[o] hot" for human habitation. After more than two months of stultifying heat, the firemen finally changed their quarters, but by then, they proved too weak to continue feeding the boiler. The furnace-like temperatures of the engine room encouraged, or perhaps attracted, further problems, and the engineer's log serves as a chronicle of the travails of a mid-nineteenth-century steamer. On January 12, the log reports that "Peter Gurney Fireman and Alfred H. Bentham Stoker Absconded." On March 15, the engineer "took a Bottle containing Liquor from John Galaghar 2nd Asst Engineer, who has rendered himself unqualified to do his duty by drinking[.] I have no confidence in the man. [T]o his room is the best place for him." On March 16, "Ed Dezel Fireman made coal passer he not being able to stand the fires[.] John Eddy made Fireman in his place." On April 2, "John Diddy violently assaulted Fireman[.] Lawrence Willis & Fenton Bowes deserted." More desertions, discharges, and shackling in irons followed, as the ship steadily hemorrhaged engine workers and the ship's commanding officers struggled to replace them.[68]

Labor in the debilitating heat of the bowels of the ship was only one of the challenges faced by the new steamer. Conflict and exhaustion while tending the engine were compounded by a difficulty in keeping the engine running at all. Coal that the crew purchased at Rio de Janeiro after only a month at sea came aboard waterlogged from rain. Less than three months later, coal from San Blas proved half sand. When the sand melted and fused into glass, it clogged the boiler's grate bars and choked the flow of oxygen. "Having bad Coal and little or no Draught [it is] almost impossible to raise steam," recorded the ship's engineer in May. By July, the ship was reduced to burning wood. The rest of the year witnessed continued difficulties in fueling: more wet coal, more impurities of ash and clinker, more coal that could not maintain steam. By mid-July 1850, after the ship had entered its regular mail route between Panama and San Francisco, the engineer recorded that its latest load of coal proved "very bad having lost its strength by being exposed to the weather[;] makes a great smoak [sic] but very little flame." The following September, the engineer reported "shocking bad coal one third of it Dust."[69]

Ensuring that the *Oregon* and her sister ships had coal was a central problem for William Aspinwall and the company's management. From Panama to Oregon, the Pacific Mail struggled to fuel its ships. At first, the line purchased Vancouver coal from the Hudson's Bay Company and also experimented with Cowlitz River coal from the Oregon Territory.[70] To meet the line's anticipated needs, Sir George Simpson, the powerful governor of the Hudson's Bay Company, dispatched a mining team from London to begin coal extraction on Vancouver Island. Yet Canadian coal was slow to arrive and stocks at Pacific Mail depots in San Diego and San Blas remained low. The company purchased some coal shipped from distant England, and agents received instructions never to loan or supply fuel to any other passing steamer. From his desk in New York, William Aspinwall fretted about even the appearance of insufficient fuel. "If at any time you have to send a steamer off short of coal," he wrote his agent in San Francisco, "do not let it be known that we are short at San Diego, & San Blas." Instead, he offered instruction in deception to his ship commanders: if they leave "the coast under steam they will find their sails will secure them a fair passage in case of need, & the less said about any disappointment the better."[71]

Despite the obstacles, the Pacific Mail survived its early years precisely because it could draw on a global, if fragile and expensive, market for coal supplies. The flood of ships and passengers to San Francisco tantalized by the prospect of gold alleviated some of the early plight of the company, as many ships arrived laden with coal as ballast. The Pacific Mail's carefully orchestrated fuel shipments from the East Coast during the summer of 1849, along with orders from Vancouver and Liverpool, also took some pressure off.[72] By 1850, the company was placing coal orders around the Pacific rim, from Valparaiso to Sydney, though Aspinwall believed these coals were not likely to perform as well as the prized Welsh varieties.[73] After over a year of negotiations with the Hudson's Bay Company, Aspinwall finally discovered that Vancouver coal proved poor for generating steam.[74] Whatever the source, though, the prices of coal remained incredibly high—as much as the astronomical $40 per ton, more than ten times the rates prevailing in the mid-Atlantic. Still, the company opted to pay rather than risk falling low on supplies.[75]

These challenges hindered the management of the ships. When it came to operating steam engines themselves, sufficient supplies mattered little if engineers, among whom there was rapid turnover, consumed coal wastefully or damaged engines, both problems affecting the Pacific Mail in the summer of 1849. In response, the company appointed a new, superintending engineer to establish steam allowances to regularize practices and regulate coal consumption aboard the company's several ships—"an object of the first importance," according to Aspinwall. The new superintendent further worked to incorporate new, experienced engineers from the navy into the company at whatever pay they needed to keep the ships running. "Whilst economy with high wages ruling is very desirable," wrote a partner of William Aspinwall of the chief engineers, "no means will be spared by the Company to have an ample force on hand." Yet even new hires and new regulations could not ensure that vessels had enough fuel. When the Pacific Mail Steamer *Golden Gate* began its service between Panama and San Francisco in 1851, the naval engineer Charles Stuart considered the ship the fastest in the Pacific. With coal so expensive, however, its engineer could not afford to run at high speeds. Instead, it ambled along, propelled by coals of mostly poor quality, some deteriorated from weather exposure, some scrounged from the detritus of other ships.[76]

New steamship lines along the southern Atlantic coast and the Caribbean hardly fared much better than the Pacific Mail. If coal was already an international commodity in the 1840s, it was one that largely lacked infrastructures of credit, warehousing, and distribution. Until steamship lines hired agents and established depots of their own, ship commanders rarely knew what kind of coal to expect when they arrived at a new port. Even by the late 1840s, steam coal merchants in Maryland and Pennsylvania had not yet developed markets for their wares beyond the narrow regions of the eastern seaboard. The captain of the United States Mail Steam Ship Company vessel *Falcon* discovered that coal dealers in New Orleans could supply American coal only at a very high cost, while the highest quality Welsh coal in Havana sold for no more than $4 a ton. But coaling in Havana proved a more complicated proposition when the city began enforcing a quarantine against a global cholera epidemic in 1848, impeding normal port activities. Even after costly delays, however, there was no guarantee of the quality of coal actually available in Havana markets. After one U.S. mail vessel passed the quarantine period there and could finally purchase fuel, the results were disappointing—the engineer of the line's *Isthmus* called the coal he found there "the poorest stuff he ever burnt." Perhaps the company could turn to Key West? Perhaps establish a dedicated depot en route in Savannah? The firm tried sending coal from Philadelphia to Chagres but abandoned the project after discovering that unloading it there proved too difficult. From New York, the company's management struggled to find economical ways to supply its line. Only by 1850 would it work out reliable coaling arrangements with merchants and commission agents in its various ports of call. Even then, however, its management remained fixated on the issue of economy, as prices remained high and supplies low. There was barely enough coal to sustain ships—and then only so long as they didn't run too quickly.[77]

Sometimes, the problem with coaling was as much a matter of law as it was of supplies or expense. A short-lived rival to the Pacific Mail, the New York and San Francisco Steamship Company, discovered that shipping restrictions imposed by Mexico complicated the process of refueling their ships between Panama and California. "The labor and delay of landing coal and other supplies and again reshipping them is immense at Acapulco," the company's agent wrote the secretary of the navy in 1852. Realizing that avoiding the delays of transshipment required sanction from the Mexican government, the company pleaded with the secretary to persuade the country's minister to the United States to permit them simply to refuel from a store ship anchored in Acapulco harbor instead of forcing them to go into the port itself. The appeal does not appear to have helped the company, and it soon went out of business.[78]

Increasing coal consumption in naval vessels also introduced new challenges to the American navy. At the beginning of the 1850s, the navy had only seven steamers in commission—two steam frigates, the *Mississippi* and the *Saranac*; the lake ship *Michigan*; the former merchant ship and army transport *Massachusetts*; and three smaller steamers, the *Engineer* and *General Taylor*, which were assigned to harbor duty, and the *Union*, which was a receiving ship.[79] By February 1861, the navy operated twenty-six steamers in all its squadrons from the Atlantic coast to Africa to Japan.[80] With the growth in the steam force, coal consumption grew as well. In 1843, the department estimated its six steamers would consume fewer than 14,000 tons of coal in a year; in 1860, it estimated it needed over 50,000 tons.[81] From the adoption of war steamers to the end of 1858, the navy purchased some 136,500 tons for use aboard its ships and another 115,250 tons for its navy yards.[82] Matthew Perry's expedition to Japan alone cost over $500,000 to fuel with domestic coal, over $50,000 more in commissions to Howland and Aspinwall for providing it, plus an additional $51,112.10 worth of fuel purchased overseas.[83] Not surprisingly, coal purchased near the United States or Britain tended to be cheapest, at times under $5 a ton. In the East Indies, distant ports in South America and Africa, and around the Pacific, coal could exceed $20 or even $30 a ton.[84]

Though their histories are usually told separately, the development of the mail steamers and navy's steamers were tightly connected, and not only because they both led the government to deal with coal in new ways. Beyond functioning as bearers of information and cultivators of trade, the subsidized mail steamers also helped teach a generation of young naval officers about the new problems posed by industrial fuel. In 1850, a young lieutenant, David Dixon Porter, commanded the mail steamer *Georgia*. The ship was among George Law's steamers cruising between New York, Charleston, Havana, and Chagres, comprising the eastern half of the route that, together with Aspinwall's Pacific Mail, connected New York with California. Porter had enthusiastically accepted command of the vessel, believing that "this service is one in which officers have better opportunities to gain experience than perhaps in any other position in the regular service."[85] Yet he quickly encountered the challenges and limitations of steam. Porter regularly reported to the department that while his ship was fast and its machinery reliable, he often had to rely on auxiliary sails and could not push the ship's engines too hard because coal supplies were so hard to come by and rapidly exhausted.[86] On the run between New York to Havana, the *Georgia* benefited from full bunkers of high quality coal. Continuing from Havana to Chagres, the entrance to the Isthmus of Panama, Porter was forced to reduce the power of his engines by fully 25 percent. He maintained the same reduction on the return trip to New York, all to save fuel. Operating the *Georgia*'s engines economically engrossed the ship's commander, who managed only after paying "great attention" to the ship's engines to steam some 240 miles a day in smooth weather with just 32 tons of coal. On an average day the ship needed 50.[87]

Before commanding the *Georgia* for the Law line, Porter had been dispatched by the navy to ferry the Pacific Mail's new steamer *Panama* from the Cunard dock in Jersey City through the Straits of Magellan to its new route hugging the Pacific coast between San Francisco and Panama City, making it the third ship of the Pacific Mail.[88] Porter was hardly alone in this mail steamer service. Between 1848 and the mid-1850s, nearly two dozen naval officers and engineers were detailed to command or operate the private vessels. They were mostly lieutenants and passed midshipmen, though many would end their careers after the Civil War as captains, commodores, and admirals—Robert W. Shufeldt, Fabius Stanly, and Charles Stuart Boggs, for example, in addition to Porter. But their years of service aboard the several mail steam lines was comparatively brief. By the end of the 1850s, all federal subsidies for mail steamers had expired. Some officers and engineers returned to naval duty, while others left the service altogether. When the Civil War came, many would join either Union or Confederate fleets; other officers would command the mail steamers themselves, detailed in the emergency for war duty as they had been initially designed.[89]

Given the collapse of the Collins Line (two of its massive ships were lost at sea) and expiration of the contracts for the other lines in the late 1850s, it might appear that even without the Civil War, this policy experiment in federal steam communication subsidies had run its course. Yet even up to the eve of the war in 1861, memorials for new mail steam lines and modified routes continued to arrive in Congress, congressional committees that dealt with the post office and post roads continued to favorably report bills endorsing new subsidies, and both national politicians and the commercial public continued to debate new proposals.[90] If anything, growing sectionalism made it increasingly difficult to undertake these projects not because they were undesired but because it could not be ensured that their advantages would be distributed evenly.

After the Civil War, the political economy of international communication would again change. Countries would rely more heavily on new submerged telegraph lines, and international postal conventions would eliminate the threats of discriminatory actions by the nation whose ships carried the mail. These normalizations and modernizations of international communications would make the American mail subsidies less important than they had seemed in the 1840s and 1850s, though some subsidies were in fact reinstituted, like those to the Pacific Mail for mail carriage to Japan and China. But the antebellum steamers had another consequence. They brought the subject of coal before Washington in a way it had never been brought before. If Congress sought to connect the United States to the rest of the world by steam, it had to think about how to sustain that infrastructure. Where could coal come from? Who would provide it? How would it go from its source in urban eastern markets to potentially distant sites of consumption? In the later nineteenth century, some Americans claimed that the increased use of steam power demanded that the United States secure coaling stations overseas. Americans of the antebellum period came to other conclusions and experimented with a range of ways to support the fuel needs of steam vessels. What united the various approaches was a consistent concern with the idea of economy.

CHAPTER TWO

ENGINEERING ECONOMY

The ocean pales where'er I sweep,
    To hear my strength rejoice,
And the monsters of the briny deep
    Cower, trembling at my voice.
I carry the wealth and the lord of earth,
    The thoughts of his godlike mind;
The wind lags after my flying forth,
    The lightning is left behind.

George W. Cutter, "The Song of Steam"

As Americans in the Pacific and Caribbean quickly discovered, the challenges of limited fuel resources quickly shattered the fantasy that steam power would annihilate time and space. This tension between imaginable networks of communication and transportation and the practical limitations that confronted them persisted throughout the nineteenth century. Daniel Webster could declare of steam power in 1828 that "no visible limit yet appears, beyond which its progress is seen to be impossible," but even then, limits were, in fact, plainly visible. It was one thing to imagine a transpacific steamship service, to petition Congress, to draw up a business prospectus; it was quite another to ensure the availability of abundant quantities of coal—of precise varieties of coal—all at reasonable prices halfway around the world. It was more challenging still to commit the national defense to machines never before tried by war. All these challenges demanded careful attention to anything that might facilitate powering ships by steam power. Nineteenth-century Americans had a word for managing this attention to progress amid scarcity of time, money, and resources: "economy."[1]

Economy did not mean efficiency. The two words, similar in connotation by the turn of the twentieth century, once expressed two very different concepts. In the nineteenth century, as Timothy Mitchell notes, economy "referred to a process, not a thing."[2] Economy evoked proper management, responsible government, and a frugality—but not parsimony—with money or resources. Economy could describe the regulation of the household, as in the phrase "domestic economy," or the polity, as in "political economy." Economy was a moral value, an obligation to family and country. "The man who is economical," wrote Lydia Maria Child in her bestselling guide to home management, "is laying up for himself the permanent power of being useful and generous." As Child suggested, this economical man was inherently forward looking, husbanding resources in the present to ensure sufficiency in the future.[3] A responsible public official steered the ship of state in a similar way.

In contrast, in the early nineteenth century, "efficiency" was much closer in meaning to the related word "efficacy." Both words expressed an ability to cause some desired consequence. The words so closely shared a meaning that Webster's 1841 dictionary defined them nearly synonymously: efficiency was "the act of producing effects," "effectual agency," and the "power of producing the effect intended." Efficacy was the "power to produce effects" and "production of the effect intended."[4] This sense of efficiency had roots that stretched back to antiquity and the notion of "efficient causes," what Aristotle defined as "the source of the first beginning of change or rest."[5] Within the sciences, efficiency likewise expressed a notion of effective causality. Davies Gilbert, serving in 1827 as president of the Royal Society, defined "efficiency" as a physical quantity: what was done to a machine to cause it to operate. How the machine reacted in response he labeled "duty." An operator expended efficiency on a machine and in return, a machine performed duty.[6] Five years later, this definition was adopted by the prolific polymath William Whewell, who employed it as a now-forgotten means for explicating the science of mechanics.[7]

In the middle of the nineteenth century, "efficiency" was just beginning to take on its modern connotations. Among engineers, the word evolved from meaning an action administered on a machine (as employed by Gilbert and Whewell) to a property of that machine—a number measuring the actual performance of a machine against its ideal performance. This usage was developed most significantly by W. J. M. Rankine, a Scottish engineer and central figure in the development of thermodynamics. In 1858, Rankine, building on several years of earlier investigations, defined a machine's efficiency as "a fraction expressing the ratio of the useful work to the whole work performed." For Rankine, efficiency expressed how much work a machine could perform "in producing the effect for which the machine is designed"—pumping water, driving a paddle wheel—divided by all the work the machine performed, useful work as well as work lost to friction, heat dissipation, or other impediments. By this measure, a "perfect" machine was one that wasted no work, whose total work was entirely "useful," making the efficiency fraction simply one, or "unity." As a corollary, this definition implied the responsibility of the machine-building engineer, which was "to bring their efficiency as near to unity as possible."[8] Gradually, this usage slipped from engineering into wider circulation. By 1911, Frederick

Winslow Taylor could use efficiency in its fully modern sense, describing a worker's "highest state of efficiency" as "when he is turning out his largest daily output."[9]

In the 1840s and 1850s, the pursuit of economy expressed a more expansive concept than efficiency in either its earlier or later usages. Thinking about efficiency meant thinking about machines, either what powered them or how they operated. Thinking about economy connected those machines to wider networks of fuel and broader methods of operation. The economy of fuel implied attention not merely to prudent means but desired ends. "It is not the *saving* only of fuel which merits attention," instructed the Scottish engineer Robertson Buchanan in his 1815 *Treatise on the Economy of Fuel*, "but its *safe, easy*, and *healthful* application to the various purposes of life."[10] Economy meant ideas, judgment, and attention to the complex relationships that linked people to the world around them. If for Rankine, the engineer's responsibility was building steam engines that operated closer to a calculable ideal, for the French engineer Sadi Carnot, achieving "the considerations of convenience and economy" with steam engines required the cultivation of "the man called to direct"—the wise engineer trained to evaluate the factors of expense, materials, design, constraints of space, and safety of operation in particular ways for particular purposes.[11]

One could pursue economy in any realm, and economy affected everything. Discussing the increasing adoption of anthracite coal as a domestic and manufacturing fuel, the American chemist Walter R. Johnson noted that "the consequences of such changes, if judiciously made, will doubtless be the diminution of expense, the saving of labor, the gaining of comfort, and the economizing of space and time."[12] Economy could also frame the perception of limits. For Columbia College professor James Renwick, transoceanic steam navigation was both possible and useful, but "in point of economy," it could "never compete with sails" and would likely only be used for passenger travel or naval purposes.[13] In these terms, achieving economy of fuel encompassed all aspects of what historians would later call a socio-technical system.[14]

As the construction of naval and mail steamers increased during the antebellum period, the economy of fuel became a subject for the national government. These projects introduced new demands on resources, budgets, and bureaucratic organization. They raised new questions about the role and responsibility of government in providing material means for achieving policy ends. Along the way, the adoption of steam power led the federal government to rely on new forms of technical expertise. This technical expertise addressed fuel economy in primarily three forms: first, through chemical and physical investigations into different varieties of fuel and their combustion; second, through engineering experimentation and ship design; third, though geological and diplomatic expeditions to investigate fuel supplies in distant lands. This chapter explores the first two, and the following chapter considers the third. In all areas, the pursuit of technical knowledge both influenced political actions and was, in turn, influenced by them.

The Calculus of Combustion

Economy of fuel began with adequate supplies. As mail steamer commanders and line proprietors quickly discovered, ensuring sufficient coal presented one of the fundamental challenges to establishing global, or even simply coastal, communications networks. The problem was not that the United States lacked mineral deposits—in the eighteenth century, Americans had become aware of tremendous strata of coal near the Appalachians and further west. Enterprising operators in Virginia had begun commercial mining in the coalfields surrounding Richmond in the 1740s and in the western, mountainous portion of the state during the first two decades of the nineteenth century. Jefferson mentioned both in *Notes on the State of Virginia*, repeating the widespread belief in the vastness of western deposits—it was thought "that the whole tract between the Laurel mountain, Missisipi [*sic*], and Ohio, yields coal." In neighboring Pennsylvania, accounts of the "Pittsburgh seam" date back at least to the French and Indian War, and local coal consumption began there no later than the 1780s.[15]

In the decades that followed, political campaigns for internal improvements and economic development blossomed, as did a desire to harness the capacity of the state. Between 1823 and 1850, twenty-two states commissioned surveys to better understand regional geological structures. Most importantly for state legislatures, these surveys sought to locate, identify, and map commercially valuable minerals, coal notably among them. North Carolina's pioneering state survey, begun in 1823 under Denison Olmsted, was the first to characterize the state's Deep River coal formation. Larger and more sophisticated surveys followed, especially in Virginia and Pennsylvania (both initially undertaken in 1836), the former under William Barton Rogers and the latter by his brother Henry Darwin Rogers. Both of the Rogers brothers devoted considerable efforts to describing the coalfields of their respective states (what William called "our great western coal region" and what Henry described as "the enormous series of coal measures"), while geologists in Maryland, Ohio, Indiana, and elsewhere mapped extensive coalfields in those states as well. In the minds of scientists, legislators, and aspiring industrialists, there was little doubt that the United States possessed enormous deposits of coal.[16]

Still, even late into the 1830s, there was reason to doubt that American steamships could ever compete with British ones on account of the inferior quality of coal for steaming purposes. When a London newspaper in 1829 criticized the prospects of American steam navigation because of the limited extent of American coal and its suitability for steaming purposes, U.S. newspapers reprinting the article swiftly pointed out the vast extent of the country's coalfields. No paper, however, could respond to the charges of poor quality—no one in fact knew whether the quality of the coal was good or bad—and on that subject they remained conspicuously silent.[17] This question of quality haunted plans for ocean steam navigation. When the British *Sirius* and *Great Western* raced across the Atlantic in April 1838 in the first transatlantic steamship competition, the smaller *Sirius* arrived in New York nearly depleted of fuel. The larger and more carefully outfitted *Great Western* still had nearly a third of its coal remaining (203 of 660 tons), seemingly easing the fears of those who had fretted over the ability of any steamship to carry enough fuel to make it across the ocean. Returning home, however, remained a problem, for there was still no adequate American variety of steaming coal to fuel the vessels. Nearly two months after the ships had successfully reached New York, editors at the *Albion* worried that the expense of shipping British coal to America would still doom transatlantic steam navigation, calling the absence of American coal suited for steaming "the only difficulty in the way of this enterprise."[18]

One way to address this difficulty was to locate a superior variety of American coal. The *New York Herald* mocked those who threw up their hands and declared "that nature has interposed an effectual barrier to prevent the United States from competing with Great Britain in steam navigation, owing the scarcity and inferior quality of our bituminous coals." True, American bituminous coals consumed valuable space aboard ships, fouled decks, and were known to release distinctive plumes of billowing smoke, thus revealing the presence of American warships as much as seventy miles away, but according to the *Herald*, skeptics had not considered the introduction of vast quantities of American anthracite. Still, simply pointing to anthracite was an expression of hope, not a solution.[19]

Even as American mining companies, geologists, and chemists uncovered new varieties of domestic coals, the difficulty of identifying the ideal steaming fuel persisted through the Civil War. Engineers understood that different industrial processes called for different kinds of coal, the precise chemical compositions of which favored different uses. Weighing these compositions against price and availability, steam engine operators selected bituminous coal or anthracite or sometimes hardwood or pine. "Each of these has its peculiar manner of burning," instructed a popular engineering manual, "and hence the furnaces or fire-places in which they are used must differ in form and arrangement."[20] This peculiarity meant that between the 1820s and 1850s, research into steaming fuels required careful attention to specific varieties of fuel from specific places. Unlike wheat or hogs, high-precision fuel woods and coals were not easily commodified across different states, fields, or strata. Wood from apple trees, American chestnuts, or Jersey pines all burned in particular ways that rooted them to particular geographies, just as coal from the Lehigh Valley would forever burn differently from specimens mined along the Schuylkill or in faraway Newcastle. Coal for copper smelting could not contain large quantities of sulfur or iron. Cannel coal suited steam engines but not iron making. Broad Mountain white-ash anthracite coal of the Lehigh Valley was ideal for making iron but Buck Mountain coal, also of the valley, was better for steam generation. Especially for steel making or transatlantic steaming, the choice of coal varieties was critically important, and investigating the properties of fuels revealed the inextricable connection between nature, politics, and the market.[21]

At first, American investigators looked to Europe, where experiments on the economy of different fuels had begun in the late eighteenth century. In Paris, the French Ministry of Finance had asked Antoine Lavoisier in 1779 to examine various domestic fuels and determine their heating capacities when their price in the marketplace was taken into account. Turning to Paris's most common fuels, Lavoisier selected a local coal, coke, charcoal, beech, and oak. Despite his coal samples exhibiting roughly double the heating effects of wood, Lavoisier found that the taxes, fees, and transportation costs levied on coal made that fuel more expensive per unit of heat it provided, a fact of political economy that the chemist considered absurd in a kingdom of forests "chers et rares" and where the more abundant fuel, found in accessible riverside mines, was made more expensive by the state.[22]

In Munich, Benjamin Thompson undertook more elaborate experiments almost twenty years later. As part of his ongoing investigations into the practical applications of heat, the American-born Thompson (ennobled Count Rumford in Bavaria in 1792) considered understanding the properties of fuels and combustion essential for social betterment. "The great waste of fuel in all countries must be apparent to the most cursory observer," he noted in an essay of 1797. Focused on lessening this waste, especially in the furnaces of the poor, Thompson concocted novel mixtures of fuel that generated greater heat, devised innovative fireplace and kitchen designs to better conserve wood and coal, and manipulated the conditions of combustion in boilers to achieve maximal effect. Thompson also investigated the combustion of different fuels. In one experiment, he employed a specially designed calorimeter to determine the heat produced by burning different varieties of wood (elm, oak, ash—twelve species in all) in a variety of preparations. In another, he determined how much combustible charcoal he could produce from various species. As did Lavoisier, Thompson focused on practical improvements.[23]

Lavoisier's and Thompson's research influenced investigations on the other side of the Atlantic. In the United States, the Philadelphian Marcus Bull followed their research by analyzing the combustion of forty-six species of American trees in a series of experiments in the 1820s. Like Thompson, Bull justified his research by pointing to its social utility, noting the lengthy American winter, particularly for those too impoverished to ensure an adequate supply of fuel. His work constituted a contribution to what he called "an improvement in the domestic economy of society." Bull's results showed that eleven kinds of oak each burned differently, as did cedar, chestnut, poplar, and swamp whortleberry. Bull also discovered that various woods and coals of equal weights produced roughly similar quantities of heat, a warning to those consumers who purchased fuels by standard volume measures such as a cord. Due to the wide variance in density of different woods and coals, equal volumes of different fuels could produce a considerable range of heat.[24]

Lavoisier, Rumford, and Bull pursued their fuel studies as applications of science for social betterment. The proliferation of railroads made consideration of the fuel question vital to the success of highly capitalized corporations while simultaneously stimulating research into the problem for steamers. Even into the late 1840s, coal use on American railroads remained rare, unlike in England, where locomotives burned coke (a coal product). At first, American railroads followed the English example, but then they quickly adopted cheap and abundant pinewood. There were exceptions, however. "Strange to say we commenced with anthracite and at a time when people hardly thought it was stuff that would burn at all in anything," wrote Benjamin Henry Latrobe II of the Baltimore and Ohio in 1845. Fifteen years earlier, the railroad had begun using special anthracite-burning engines designed by New York's Peter Cooper, an experiment adopted by few other lines. But while the B&O continued consuming anthracite in these older engines, its experiments on different fuels arranged by Latrobe in the late 1830s revealed that burning Maryland's Cumberland coal—a bituminous variety—both saved money and more effectively evaporated water. In subsequent years, all of the B&O's new engines used wood or a mixture of wood and Cumberland coal.[25]

Latrobe's observations about the perceived obstacles to burning anthracite stemmed from how it combusted. In the most commonly used engines, anthracite ignited slowly; when finally burning, it generated so much heat that it ruined boilers. Furthermore, its ash fused into damaging clinker, and hard chunks blown out with the steam damaged copper engine components, leaving railroad mechanics struggling to prevent leaks from the joints of the boiler's iron tubes. The challenges posed by anthracite notwithstanding, wood had its own problems, ranging from its bulkiness to its relative weakness in generating fire to the frequency with which its sparks, ejected from the smokestack, tended to ignite the farms and forests through which locomotives rolled.[26]

Still, anthracite's abundance in Pennsylvania encouraged railroads there to continue experimenting with it. Some small, coal-carrying roads running from the anthracite fields of eastern Pennsylvania were able to make use of the locally abundant fuel by employing specially designed boilers (as had stationary steam engines and some river and sound steamers), but the much larger Reading Railroad struggled to do so. The anthracite engines of smaller roads had to perform less strenuous work than their giant neighbor, and engineers for the Reading discovered that small-road operations simply could not scale up. To accommodate its existing infrastructure, the Reading tried manufacturing patent fuels from anthracite coal dust, but it knew a better solution would somehow employ the coal directly.[27] During the late 1840s and early 1850s, the Reading pursued a series of investigations into anthracite fuel, adopting specially designed coal-burning engines and carefully analyzing their behavior.[28] These investigations yielded positive results in a short period of time, success that was aided by reductions in coal prices due to increased national production. In 1846, the Reading burned 66,000 cords of firewood to haul 1,188,258 tons of anthracite coal to market. That wood cost the railroad over $200,000, compared with barely $1,000 for the sporadic use of anthracite as a fuel, making the line's fuel budget the largest single expense—over 30 percent—of its Transportation Department.[29] After experiments and engine innovations, within a decade, wood use declined by nearly two-thirds, to a mere 23,274 cords, while consumption of anthracite fuel rose to over $100,000 for more than 50,000 tons of coal. Over the following decades, this transformation took place in various forms on lines across the United States, and by the 1880s, some 90 percent of American railroads burned coal.[30]

Despite many similarities between railroads and steamships, there was never any prospect of transoceanic lines consuming wood, as steamers needed the more energy-dense fuel to travel for weeks without stopping. Successful ocean steamers meant coal. For ocean steam navigation, there were three qualities in particular that the coal needed to possess. As articulated by Maryland chemist James Higgins, steam coal required "quickness of combustion, continuance of combustion, and steady combustion." Unfortunately, as late as the mid-1850s, neither chemists nor engineers knew of a variety of coal that exhibited all three attributes simultaneously. Most bituminous coals possessed considerable quantities of bitumen, the sticky, flammable substance that accelerated ignition but burned so quickly that fires required continual refueling. Anthracite coals contained little or no bitumen, slowing their ignition but lengthening their combustion once alight. This characteristic of chemical composition had real consequences. As steamship firemen often discovered, unless they burned anthracite in specially designed engines, shoveling additional anthracite into a firebox "deadened" fires, lowering fire temperatures and rates of combustion until the new batch of coal could fully ignite and leading to uneven engine performance.[31]

Higgins represented the scientific boosters of state surveys and highlighted the connections—real and rhetorical—between science, economic promotion, and security. He argued that western Maryland's Cumberland coals possessed the perfect amount of bitumen—just enough to ignite quickly but too little to consume a fire quickly. At stake was national defense. "The policy of the world at present is for steam navigation," wrote Higgins, "not only for commercial, but also for warlike purposes." War steamers in particular needed coals that could reliably enable the ship to engage with—or escape from—a potential adversary. "A minute's delay may prove disastrous," he concluded, while "the increased revolution of the paddlewheels for a few times will frequently insure success." This exhortation was steam engineering booster boilerplate; Higgins had a product to push. "Our national flag may float gloriously over the sea," he continued, "or be stricken from the mast, as the ship which bears it is well or ill supplied with fuel, and these ships should always use the Cumberland coal."[32] These arguments, by a state-supported scientist advocating the economic interests of his state, were part of a larger effort in Maryland to leverage naval coal consumption to capture growing foreign markets for steamship fuel. This effort had begun in 1842, when the navy commissioned Walter R. Johnson, a professor of chemistry and physics at the University of Pennsylvania, to analyze American coals to identify the ideal naval steaming fuel. It was a project designed to utilize the needs of national defense to launch research that might yield a broader social and economic benefit.

Johnson was an institutionalist in search of an institution, a scientist seeking to apply the insights of science not merely for public betterment but state-sponsored public betterment. In 1838, he had sketched a plan to use James Smithson's unexpected bequest to the country to create a great American scientific body for research for the national welfare. That same year he advised Congress on the prospects of establishing a national foundry in Washington to forge naval cannon (a project dependent on the nearby coal mines in western Maryland). In 1843, Johnson joined a navy commission to investigate the causes of explosions in steam boilers. In 1845 he investigated the public water supply for Boston. His most significant technical contributions, however, came from a series of experiments on the comparative qualities of different kinds of coal, a subject he long believed had never received the attention its importance in the industrializing world deserved. In contrast to textiles or metals, "the material which furnishes *motive power*," he lamented in 1850, "is either wholly overlooked, or soon forgotten."[33]

Given the fuel needs of the navy and prospective commercial steamers, as well as those of growing industrial and commercial interests, Johnson believed that coal was a problem for the federal government. "The Government of the United States," he wrote, "though not possessing this direct interest of proprietorship in mines, has still such a stake in the value of their resources, and the prosperity of citizens more immediately concerned in making them available, that the least which could reasonably be expected of it, is, to aid in some measure in ascertaining their true value." To this end, Johnson's research program followed the kind of public-private partnership that characterized a great deal of governance in mid-nineteenth-century America. Johnson had approached the navy in June 1841, offering his scientific services, and the department accepted. In early 1842, the navy issued a call to American coal mine owners and coal dealers to supply the chemist with samples for comparative analysis, an analysis that not only would aid the navy in evaluating different fuels but also promised to help coal companies themselves learn to what purposes their products were ideally suited. Soon, coal samples reached Johnson from mines in Pennsylvania, Maryland, Virginia, Indiana, and Nova Scotia, while an international dealer in New York supplied a range of British specimens. Johnson, essentially a contracted scientist, performed his research in the facilities of the Washington Navy Yard. Receiving the final report, navy secretary John Y. Mason indicated the value of Johnson's experiments beyond their contribution to naval service by referencing "the large and growing interests which the United States possess in their vast coal mines, scarcely yet developed, and the numerous national and domestic uses to which the article of coal is applied."[34]

Johnson's research reinforced the notion that with coal, geography mattered. After testing samples from the range of coalfields, Johnson ranked them by ten characteristics. For ocean steaming, the most important was "evaporative power under equal bulks," or the weight of steam produced by a cubic foot of coal. Stark differences separated economical coals from uneconomical ones; the most powerful produced nearly 5¾ times as much steam per volume as simple pinewood, while the worst coal produced only 3½ times as much. This difference could mean making it across the Atlantic or not. To the delight of Maryland's coal industry, the outstanding sample by this measure was a bituminous coal specimen from Cumberland, "taken from a vein 9 feet some inches in thickness, on the eastern slope of Dan's mountain, about 40 feet below the surface of the earth, on a stream known by the name of Clary's run, two miles south of the national road."[35] Johnson's results suggested the value of similar coals mined nearby, which could improve the economic prospects of the coal region. Another Cumberland coal sample rounded out his top five, along with, unsurprisingly, three anthracite coals from eastern Pennsylvania.[36] "For Maryland this ministerial step has a considerable amount of interest," noted the *Baltimore Sun*, adding that "we think we may venture to predict an immense advantage to her, to be derived through one of her staples, but very partially developed as yet, as the result of Professor Johnson's experiments."[37]

Johnson's report had immediate consequences for both producers and consumers of coal. Following its publication, the navy began issuing proposals for contracts to supply Cumberland coal to its new ocean steamers, including the

Mississippi, Susquehanna, and *Saranac.* At least one coal producer published a promotional brochure based on Johnson's results, advertising the consistently high performance of its product. Consumers of coal similarly saw the value of his research. After Johnson exhausted his research funds, over sixty prominent citizens of Massachusetts, including numerous railroad and manufacturing executives, petitioned Congress in 1850 to renew its support of the investigations, citing newly uncovered coalfields, the proliferation of railroads and steamships, and burgeoning industry, all of which had contributed to a doubling of American coal consumption in just seven years.[38]

Operators of Pennsylvania's anthracite mines, however, refused to cede what might become a lucrative market to their southern neighbor. They railed against interpreting Johnson's report as evidence for the superiority of Cumberland coal over anthracite for steaming, dismissing Maryland coal as having merely performed "an inappreciable shade above the Anthracite—a mere shade, amounting to exactly nothing in practice."[39] The frustration of anthracite operators reflected the fact that they did not see themselves as engaged in mere domestic competition with Cumberland. While the quantity of coal used for steam navigation represented only a small fraction of total American coal consumption, capturing a major steamship contract—or even better, a naval one—was the first necessary step toward entering a burgeoning global marketplace—a marketplace rapidly becoming a British domain.[40] Between 1830 and 1845, British coal exports came to dominate international markets. Their exports to Prussia increased by 1214 percent; to the East Indies and Ceylon by 2025 percent; to Denmark by 1800 percent; and to the United States by 287 percent. By the mid-1840s, Britain exported nearly 650,000 tons of coal annually to France alone.[41]

American coal producers had good reason to worry. By the end of the 1840s, they watched as the Royal Navy tried to cement Britain's growing global dominance of coal export markets with the development of a research program into the steaming qualities of various domestic and foreign coals far larger than Walter Johnson's American program. The British experiments, conducted for the Royal Navy by Sir Henry de la Beche and Lyon Playfair at the Museum of Practical Geology, again highlighted the geographic particularity of fuel quality. Geographic origins mattered. De la Beche and Playfair tested Myndd Newydd and Pentrefelin coals from Wales, Dalkeith Jewel and Grangemouth coals from Scotland; Slievardagh coal from Ireland, coal from Borneo, Formosa, Patagonia, and Vancouver, and six kinds of manufactured patent fuels—133 varieties of fuel in all.[42] The experimenters performed chemical analysis on each of these coals, surveyed their mechanical structure, and analyzed their behavior in actual steam engines under various conditions, what de la Beche and Playfair described as research of "rather a practical than a scientific character."[43] Like other chemists before them, the pair observed that ideal naval fuels should possess a range of characteristics: they should ignite quickly, boil large quantities of water into steam, generate no position-betraying smoke, hold together without crumbling and yet be dense enough to stow compactly aboard ship, and be chemically free from sulfur and not prone to spontaneous combustion. And like their competitors across the Atlantic, the researchers found that no single coal exhibited all of these characteristics. Anthracite, for example, packed a lot of energy but ignited slowly. It held together without pulverizing in storage, but since it did not fuse together while burning, it risked tumbling inside the furnace with the inevitable pitches of the ship. It was smokeless, but its intense heat rapidly oxidized the iron of grate bars and boilers.[44] Still, four years of research provided a guide for both purchasers in the Royal Navy as well as coal dealers working in both domestic and international markets. While other researchers in Britain, like the natural philosopher William Thompson and the engineer W. J. M. Rankine, pursued a more theoretical and fundamental understanding of the nature of energy, de la Beche and Playfair attended to the materials at hand to support Britain's global commercial and naval predominance.[45]

Americans abroad were among the consumers of British coal exports. Both U.S. naval vessels and merchant ships depended on it when cruising on faraway stations. In the Mediterranean, American consuls supplied British coal to American ships, as they did for the steamer *Mississippi* during its cruise there in 1849. Yet some officers, along with domestic coal merchants, worried about a false economy. They questioned whether the fees, duties, and costs of transportation—not to mention the presumed greater efficiency of American coals established by Walter Johnson—really made American coals more costly. And even if the costs of American and English coals were simply equal, wondered navy captain Charles W. Morgan upon taking charge of the Mediterranean squadron in 1849, would it not make sense to support American industry? American coal burned cleaner, he argued in a brief for sending Cumberland coal overseas, and "the Government would be giving large and valuable orders to our own citizens which would otherwise be supplied by foreigners."[46] Anthracite merchants in Pennsylvania thought the same about their coal, imagining that if they could claim even a small portion of this global coal trade, they would earn fabulous profits. All they needed was a little help from the government.

Philadelphia anthracite merchants believed they could break into the global market with their high-grade coals by appealing to the need for national defense. Some time around 1845, they nominated Benjamin H. Springer, himself a coal dealer and former president of the Coal Mining Association of Schuylkill County's board of trade, to visit Washington and lobby the navy to adopt their higher grade, more expensive anthracite fuels. If the lobbying succeeded, coal

mines would see profits and commission dealers would receive income, but neither of those would matter as much as the fact that American naval vessels
overseas would be advertising the products of American coal country to foreign
navies and steamship lines. The merchants sought to turn the navy into a floating promotion of their wares. "The trade urged me, as I was acquainted in
Washington," recalled Springer years later, "to get the appointment with a view
to that more than anything else."[47]

In Washington, Springer argued that naval operations were too important
and coal characteristics too inscrutable to the inexperienced to rely on the old
practice of simply purchasing from the lowest bidder. This method, which had
been the navy's modus operandi since the first naval steamer *Fulton* had been
built in 1815, had been used to obtain all manner of naval materiel. Coal, Springer
argued, was different. "The properties of coal are so various that a person who
is not thoroughly acquainted with it may purchase a bad article and endanger
the ship and all on board," he explained. "The received opinion of persons not
acquainted with the subject is that all coals are alike; but there is as much difference between different coals as there is between the best hickory and the worst
pine wood." After the failure of his initial efforts, Springer returned to Washington during every session of Congress through 1850. Millard Fillmore's navy
secretary, William A. Graham, advised him that if Congress would only grant
the department more flexibility on coal purchases, Graham would appoint a
special agent to manage the business. Speaking for the measure in the Senate,
Pennsylvania senator James Cooper, himself a resident of his state's anthracite
country, supported the plan by discrediting the bituminous competition, claiming "it is impossible to purchase the coal and wood without getting the worst
article in the market, and very often at higher prices than it would be necessary
to pay for good articles." Cooper surely exaggerated, but his remarks suggested
the ways the anthracite interests sought to expand their market at the expense
of bituminous coal dealers, especially coal dealers from Cumberland. Springer
finally succeeded in September 1850, when Congress granted the secretary the
"power to discriminate and purchase" whatever fuel best suited the public
service.[48]

The new law allowed the secretary of the navy to appoint two agents, one for
anthracite coal in Philadelphia and another for bituminous in Baltimore. After
having lobbied for the creation of the post for over half a decade, Benjamin
Springer secured the anthracite agency for himself. Almost immediately,
Pennsylvania's Senator Cooper began pushing the value of using anthracite.
At Cooper's prodding, Springer dispatched questionnaires to leading engine
manufacturers and figures in the coal industry, the responses to which confirmed Springer's belief that anthracite offered a superior fuel for ocean steamers and government use. He began trying to persuade the navy to abandon its

preference for bituminous steaming coal—which had been department policy since Walter Johnson's research program in the early 1840s—and adopt anthracite instead. Already, several naval steamers had begun experimenting with it.[49] Believing that the Pennsylvania fuel possessed both economic as well as technical advantages over bituminous, Springer asked the secretary to allow a comparative test to be conducted, ceteris paribus, a plan that was approved and overseen by the navy's engineer-in-chief, Charles B. Stuart, at the Brooklyn Navy Yard. Though Springer represented the proposed evaluation of the two fuels in the dispassionate language of scientific objectivity ("the trial can be made by the same men," he had explained, "and under the same boilers; and it is fair to infer that a full and impartial result will be attained"), the political and economic consequences of the investigation were clear in Maryland. Maryland coal dealers had viewed the creation of the two anthracite and bituminous agencies in 1851 as validating the value of their state's product and as a defeat for Pennsylvania forces bent on snatching the lucrative naval contracts from Cumberland bituminous. When news of the navy's new experiments became widely known just a year later, Maryland's general assembly hastily instructed its congressional delegation to discover what could possibly have happened to cause the navy to reconsider its reliance on Cumberland coal.[50]

Charles Stuart's steaming tests pitted Cumberland against Pennsylvania anthracite coals. Walter Johnson had found Cumberland a superior steam generator for its density; this time, Stuart found no such thing. Stuart attributed his results, "not in accordance with theories heretofore received," to Johnson's experimental design, different from any conditions a steamer actually encountered at sea. Johnson had only tested small quantities of coal (usually less than half a ton per trial and never close to even a single ton), burned coal at less than half the rate of actual steamers, and used a boiler unlike any in use aboard ships.[51] In contrast, Stuart had ton upon ton of both coals for his experiments, and even in a pumping engine designed for bituminous coal, he found that anthracite coal enjoyed what he called an "economical superiority" about two-thirds greater than bituminous. This result meant that a ship could steam two-thirds farther on the same weight of fuel. Anthracite had the additional virtues of greater density (so captains could store even more coal aboard ship) and of burning without smoke. After the success of the navy yard tests, Stuart recommended the adoption of anthracite fuel aboard all naval vessels with iron boilers. After one of those ships, the steamer *Fulton* (the third to carry that name), had burned anthracite for several days, her engineer exclaimed that her "engine worked as well as any I ever saw, but the boilers exceeded my calculations." Little soot, constant steam pressure, no need to force a draft—he predicted the ship "will do *more* service at *less expense*, than any steamer government will have in five years."[52]

In the 1840s and 1850s, combustion experiments helped define the character of various coals for commercial and naval purposes. But it was not the only way Americans considered making sense of the new challenges of steam power. Another possibility was that instead of merely arbitrating between commercial mines, the government could itself purchase coal lands for future naval purposes. This was the approach favored by Charles Miner, a former Pennsylvanian representative, editor, and promoter who had helped first open the great anthracite fields of the state's Wyoming Valley during the War of 1812. Four decades later, Miner came to regret the capitalist frenzy in anthracite country he had helped unleash. "The Anthracite Coal Lands are being absorbed by wealth and monopolized by speculators," he grumbled in 1852. Though he confessed that he "sometimes thought it was almost to be wished that the use of Anthracite, so limited in Quantity, so invaluable for naval purposes, should be excluded from common use, wherever a substitute could be found," Miner knew that such a proposal was impossible. Instead, he urged the navy to secure its own thousand or fifteen hundred acres of anthracite land in Wyoming Valley. With this reserve, he explained, the security of the nation could never be threatened by "monopoly purchasers" or be forced to "submit to their terms." Of course, Miner touted anthracite coal as a better fuel than bituminous, but the significance of his proposal was its integration of antimonopoly sentiment with the specter of a failure in war preparedness. While nothing came of the proposal in the 1850s, the establishment of naval fuel reserves would be pursued for both coal and oil lands after the turn of the twentieth century.[53]

Unlike Miner, most anthracite operators in Pennsylvania were content merely to siphon the trade from Maryland. Through the 1850s, they continued boosting their product. This pressure had little immediate effect, and until the end of the decade, the navy retained both its bituminous and anthracite agents and continued to purchase coal from both Maryland and Pennsylvania. Still, Pennsylvania anthracite producers did not stop lobbying the department or sending samples for analysis—chief engineer Benjamin Isherwood conducted one influential comparative analysis in 1859—and during the Civil War, the rapidly growing Union navy would overwhelmingly consume anthracite in its steamers.[54]

Even before the war, however, Isherwood was as interested in designing engines to suit available coal as he was in analyzing coal to suit available engines. In this interest he was not alone. Since the invention of the steam engine itself, inventors had tinkered with it to improve economy and often pursued alternatives to steam that would hopefully replace it. In the 1840s and 1850s, some of these inventors turned to the federal government at precisely the moment when the navy and mail service were becoming dependent on coal. The question was, who would design the new engines?[55]

Political Engineering

Choices about engine design were political choices. In 1859, a special House committee, chaired by a rising Republican from Ohio, John Sherman, investigated a series of charges against the Navy Department alleging corruption, graft, and gross incompetence. Among the claims: that the navy had allowed Daniel B. Martin, one of its own chief engineers with patent interests in a particular boiler design, to sit on a board selecting engine manufacturers for five new steam sloops recently authorized by Congress. When the board only approved contractors incorporating Martin's design—at a higher cost than competing proposals— critics cried foul. As Sherman's committee discovered, however, when it came to proving corruption, plausible inference was not the same as dispositive evidence.[56]

As they investigated, the members of the Sherman's committee received considerable technical educations. They considered the relative merits of horizontal and vertical tubular boilers, the strength of propeller shafts, and the effects of excessive propeller revolutions. They reviewed the operations of high and low pressure engines, the advantages of varying cylinder diameters and lengths of stroke, and the limits to the structural integrity of the longitudinal bulkhead. Determining whether Martin's judgment was shaped by financial gain or engineering expertise would require members of Congress to think like engineers. How did they evaluate the merits of competing experimental designs? How did the government balance its interests in security with cost and administrative capacity and without showing favoritism to politically connected contractors? Answering these questions took the committee deep into the weeds of steam engineering and technical design.[57]

At the heart of the investigation was a basic question that would be revived over a century later by historians of technology: did artifacts have politics? Did design choices instantiate particular relationships and empower certain groups, like the nascent brotherhood of professional naval engineers who jealously guarded their claims to expertise? Did they weaken others, like the independent inventors who believed they could not have their own innovations fairly examined? Moreover, were the designs of government steam engines the result of abstract engineering principles or the temptations of power, political connections, and greed? How would Congress, itself bitterly divided in the 1850s along partisan lines, ultimately adjudicate these complex, technical questions? Congress faced the particular questions of Martin's boiler after two decades of government experience pursuing fuel economy in the engines of naval and mail steamers. At first, the Navy Department alone had handled this subject, but inventors and patent holders increasingly turned to Congress to press their innovations. Though these appeals for congressional support only occasionally resulted in legislation, they kept the issue of engineering fuel economy before legislators, who generally debated more over the proper means of attaining econ-omy than the desirability of the ends.[58]

The role of Congress in naval ship construction has usually been understood to have involved appropriations for new shipbuilding programs, the designs of which remained the obligation of the navy itself.[59] But throughout the 1850s, members of Congress received proposals for a range of technical innovations in naval steam engines, nearly all of which promised reductions in fuel consump-tion. Both chambers of Congress devoted time in committees and floor debates to wrangling over the merits of novel condensers or boilers and considering even more radical proposals for propulsion innovations and the question of whether Congress should legislate their adoption. Some proposals resulted in appropri-ations or other legislation; others merely led members of Congress to debate the relationship between technological innovation and government action. Taken as a whole, these episodes reveal the ways fossil-fueled steam technology looked in the 1850s rather than in hindsight decades later. Unlike subsequent historians and naval analysts, engineers and politicians of the 1850s did not see the absence of American coaling stations around the world as the limiting con-straint on embracing steam power. Instead, they looked to a range of technical innovations to exploit the advantages of new machines within the constraints of national policy. In the antebellum period, Americans preferred to seek tech-nical innovations, not foreign coaling stations. Which technical innovations would work was a different matter, however. The English engineer Josiah Parkes claimed that in studying the problems of fuel consumption and engine perfor-mance "any person endowed with common powers of observation and experi-mental tact" was "as capable of discovering the position of an engine, in the scale of economy, as if he were gifted with the genius of a Newton."[60] But in practice, engineering could not be so easily cleaved from politics.

The inventors who approached Congress arrived at the importance of engi-neering for fuel economy by a variety of paths. In the mid-1840s, Thomas Ew-bank found inspiration at a New York fish market. Delayed there while shut-tling to Harlem, Ewbank began drawing what he called "these natural propellers" arrayed before him—the tails of porgee, salmon, cod, mackerel, and flounder. Later, he would add sketches of the curves and angles of porpoises and seals, the webbed feet of cormorants and geese, the legs of frogs and wings of bats. Accord-ing to Ewbank, nature held lessons for contemporary engineers. "In the tails and fins of fishes," he wrote, "in wings of birds and insects, and especially in the palmipeds, she has nowhere sanctioned a rectangular propeller."[61] According to Ewbank, as with all steam innovations to save fuel, the objective was doing more with less. Redesigning paddle wheels according to the lessons of nature would shave twelve to twenty-four hours from a transoceanic voyage "without any in-crease of power."[62]

Ewbank, then serving as commissioner of patents, was not solely motivated by the forms he found in the fish market but also by the transformation in international communication he witnessed from his homes in New York and Washington. "Engineers and naval constructors, animated with the ambition of Olympian competitors, are preparing for a series of Atlantic chariot races," he declared in one essay.[63] In another, he exclaimed that "oceanic steamers are too essential links to the system of cheap and free postage—domestic and international,—to be allowed to pursue undisturbed their present average passages."[64] For Ewbank, the pursuit of speed through improved design benefited the nation and demanded government attention. Ewbank lobbied Congress to appropriate $10,000 for additional experiments, pointing to the navy's ability to leverage its size and technical sophistication to promote mechanical innovation for both public and private purposes. In this pursuit he was joined by navy officials. "Private individuals cannot well make the experiments but the Government interest in Steam Navigation is already sufficiently large to warrant the resolution of these problems," wrote Charles B. Stuart.[65] Like many similar appeals, Ewbank's was rejected by the Senate's Committee on Naval Affairs, but not before William Seward declared on the floor of the Senate that people of the future would look back on the inefficient paddle wheels of his present day and "wonder at their gross unmechanical action."[66] Even if Congress failed to appropriate funds for further study, it began regularly debating the importance of economizing mechanical designs for public benefit.

A greater challenge for achieving fuel economy came from the innovation that made Ewbank's research on paddle wheels increasingly outmoded—the introduction of screw propellers. The navy's *Princeton*, designed by John Ericsson and launched in 1843, featured this new propulsion system, and more propeller ships followed over the coming decade. But by the early 1850s, engineers found that the thrust these propellers produced also generated enormous friction and taxed ship engines. Screw ships had to steam slower than their paddle-wheeled counterparts. Engineers variously applied discs, collars, and grooved rings in futile attempts to reduce friction and conserve fuel. An invention by George Parry, a peculiarly shaped circular casing of rollers, finally appeared to solve the problem. Parry noted the foremost advantage he offered for rotating screw propellers—"securing additional Speed, efficiency, and safety combined with a great saving in *Fuel* and *Oil*." A navy board examined the device in 1855, finding it reduced coal consumption by 35 percent and shaved twenty minutes from a three-and-a-half-hour voyage. When navy chief engineer J. W. King sought to test Parry's thrust bearing aboard the *Wabash*, he found it so superior to the ordinary one (which rapidly overheated) that he abandoned the test and simply continued using Parry's device. At least fifteen firms and engineers from around Philadelphia and New York similarly reported superior results aboard their ships. Meanwhile, the expanding scope of American commercial interests added to its prospective value. "In China, the East Indies, or any part of the Pacific Ocean, or Coast of Brazil where Coal costs $20 per ton," wrote Parry, "this would effect a saving of $58.20 per day." Parry estimated that a naval frigate steaming there would save over $20,000 in a typical three-year cruise, not to mention avoid the reduced physical work of coaling, decrease the space needed for stowing fuel, and eliminate time lost in potentially hazardous ports of call.[67]

Stephen Mallory, chairman of the Senate's Committee on Naval Affairs, was so impressed with the device that he felt he could only express his committee's thoughts "by presenting to the inspection of each member of the Senate a working model of the 'Anti-friction Box'" along with an account of its myriad advantages to the navy. This rolling mechanism offered a better way to relieve the massive friction of new screw propellers, explained Mallory, producing "greater speed, with *saving in fuel*, together with a diminished consumption of *oil* used in lubricating the thrust-bearing." Mallory, however, advised against Congress mandating that the navy adopt the contrivance, but only because he was sure that if all the testimony Parry offered proved accurate, the navy would surely do so on its own. A year later, he added that government support should come from naval adoption rather than an outright purchase of the patent, principally because government patent rights could preclude the device's use in the general economy.[68]

Congress proved more forthcoming with support for a particular invention when the request for action came from within the navy itself. In 1850, Congress funded the navy to experiment with variously designed steam condensers. For years, ships attempting ocean voyages generated steam by boiling salt water, the saline residues of which fouled boilers. Condensers purified water, keeping engines running smoothly and with an accompanying savings in coal. Following the congressional appropriation, a naval scientific commission examined twenty-nine condenser designs and found four excellent, but each in different ways. Faced with mixed conclusions, the navy secretary, William Graham, proposed brokering an agreement between the patent-holding parties, thus allowing the navy to combine the most desirable features of each condenser into a single device. This negotiation need not have involved Congress, but following the release of the navy commission's equivocal report, subsequent, more conclusive experiments found that overwhelmingly just one condenser, invented by Joseph Pirsson, alone fully satisfied government needs. Of its value, according to one engineer who adopted it, "no better evidence is required than the fact that a much greater volume of steam can be produced by the same amount of fuel than when salt water is used." According to an account in the *New York Herald*, the device could shave two full days off the transatlantic route between New York and Liverpool. Uncertain of how to proceed without incurring criticism from competing patent holders, the secretary turned to Congress. As an amendment to the annual naval appropriations bill, Graham asked Congress to require that the navy adopt Pirsson's condenser alone.[69]

As senators debated this appropriation bill in August 1852, they faced questions of how to deal with technological change. Should Congress specify the details of engine designs? Did the navy secretary not already possess sufficient authority to chose between competing designs? Did the Senate have the expertise needed for such judgments? Party affiliation and ideology was hardly a sure guide. Some, like Lewis Cass (a Democrat) and John Davis (a Whig), objected that designs and inventions properly remained a matter for the navy. Supporters of having Congress mandate the adoption of Pirsson's condenser appealed to the urgent need to save fuel and money. New Jersey Senator Robert Stockton, a Democrat and himself a retired commodore and advocate of the naval adoption of new steam technology, presented the endorsements of nearly twenty engineers, engine builders, steamship line proprietors, and naval officers, along with the unified voice of the Senate's Naval Affairs Committee, all favoring requiring Pirsson's patent for naval use. After recounting the condenser's merits, Stockton exclaimed that "nothing remains for me to do but to make a long, scientific discussion on the subject of marine engines, and the use of coal, to show the absolute necessity that something should be done to reduce the expense of your steam navy," an expense Stockton estimated could be lowered by the use of Pirsson's condenser by as much as $200,000 a year.[70] In the end, Stockton's arguments carried the day, and with Pirsson's name removed (on principle) from the amendment, the Senate voted to empower the navy secretary to adopt "any steam-condenser which may be found best calculated for the purpose"—a criterion met by Pirsson's condenser and in language sufficiently prescriptive to allow the navy secretary to chose Pirsson's design over competing ones.[71]

Pirsson's condenser, which promised to save the government coal, was just one innovation amid a flurry of experimentation in both Britain and the United States to improve the economy of steam engines. But coal and steam power had hardly begun to transform oceanic transportation when mechanics and entrepreneurs began experimenting with alternatives. In 1849, in a project championed by Missouri senator Thomas Hart Benton, Congress appropriated $20,000 to Charles Grafton Page, a patent examiner and chemistry professor at Washington's Columbian College, to pursue experiments on "electromagnetic power as a mechanical agent for the purposes of navigation and locomotion." Though it quickly became apparent that the expense of a viable electromagnetic engine would be far greater than that of existing steam engines, Page hoped the public would evaluate his work not merely by the relative costs of zinc and coal but by what the *National Intelligencer* reported as "the cost of human life, the sacrifice of millions of property, and risk of many millions more"—the entire existing sociotechnical system for producing coal and sustaining the infrastructure for steam power.[72]

American scientists and engineers enthusiastically greeted Page's initial exhibitions of his engine. His engine attracted particular attention at an 1850 demonstration in New Haven attended by many of the leading figures of American science. Joseph Henry, America's expert on electricity and magnetism, proclaimed his interest, while another member of the so-called American Lazzaroni, Benjamin Pierce, "felt astonishment and great delight." The elder statesman of American science, Benjamin Silliman, was impressed by how far Page's research had progressed in so short a time. Two men who had spent years examining coal and steam, Walter R. Johnson and William Barton Rogers, both discussed the new engine's cost relative to steam, with Johnson concluding that he anticipated that the two sources of power would find complementary uses. "Where there were serious objections to the use of steam power," he was reported as saying, "this power would come in very well."[73]

Interest in an electromagnetic engine next reached Washington. Benton, the leading spokesman of the West in Congress, saw the project as both a boon to his state of Missouri, as it could provide a way to efficiently excavate untapped deposits of zinc, and his section as a whole, as it could power locomotives across the wide expanse of western North America on the way to increased trade with the Far East. But Benton was particularly interested in the nautical uses to which the engine might be put. Though Page designed his engine to power an experimental locomotive, Benton provided an exhaustive list of reasons that favored the electromagnetic engine over the steam engine at sea. The navy of the future, explained Benton, would find it "saving room in the vessel, the engine and battery requiring but little space, and the fuel very compact compared to coal—doing away with chimneys, smoke-stacks, and their cumbrous fixtures—instantaneous communicability of the full power, so important in changing course and avoiding collision—capacity to run a blockade, making no noise and showing no light, except at pleasure—simplicity in the construction of vessels—diminution of insurance from absence of danger from explosions and conflagrations, and less danger from collisions."[74] The electromagnetic engine would eliminate the constraints imposed by coal and conventional steam engines and herald a new dawn of safety, savings, and security.

Still, Page's efforts to construct an experimental electromagnetic locomotive succumbed to technical obstacles, and he depleted his political and financial resources. His most efficient battery, a design adopted from a cell built by the British chemist William Grove, required zinc but also copious quantities of platinum. The battery itself proved exceedingly fragile and difficult to operate. With little to show for his efforts, his initial appropriation quickly ran out. When Benton pushed his Senate colleagues for a second round of funding, twice as large as the first, they balked, and Page instead futilely tried supporting his work on his own. Lacking adequate resources, assistants, and technical expertise, Page saw his trial locomotive barely travel a few miles before its batteries quickly fell apart.[75]

No challenge to the limits of fuel economy, however, elicited as much anticipation and subsequent sense of failure as John Ericsson's hot air or "caloric" engine. Before the development of thermodynamics in the 1850s, Ericsson's caloric engine represented one of many attempts to devise a source of motive power superior to steam in cost, convenience, and economy of fuel. Unlike Page's electromagnetic engine, these many and varied attempts relied on the same basic principles of steam engines—using a fluid to propel an oscillating piston—but they substituted various agents for steam. Since the late eighteenth century, mechanics had experimented with engines propelled by substances as varied as alcohol, ether, mercury, and carbonic acid. The U.S. Navy investigated a carbon bisulphide engine in the late 1850s, a design that would periodically resurface for decades afterward. Ericsson, however, focused on pistons powered solely by atmospheric air, a substance universally (and freely) available. Employing air meant no need for frequent replenishment with fresh water, a challenge at sea or in arid terrain. Most importantly, Ericsson promised a vehicle that would consume a mere fraction of the coal as a comparable steam engine, just enough to put the caloric engine in motion and keep it moving as it slowly lost heat.[76]

Ericsson's efforts to champion caloric engines spanned two decades. After leaving his native Sweden for London in 1826 to pursue a career in engineering, he spent six years crafting various machines to improve the fuel economy of steam engines through new designs or added apparatuses. None proved satisfactory. By 1833, convinced that heat—"caloric"—was a physical quantity that could produce effects without changing itself, Ericsson constructed his first caloric engine. This five horsepower model included the key elements, what he called "regenerators," that Ericsson would employ in later versions, including his largest experiment aboard the ship bearing his name in 1853. Regenerators recaptured the caloric of heated air that had already been used to raise a piston, held it, and then imparted it to a fresh blast of air to raise the piston still more times. To begin the cycle, the engine called only for a small quantity of coal, the substance whose relative scarcity and expense lay behind the project. Ericsson insisted his engine was not quite perpetual motion—some heat would indeed be lost and need occasionally to be replenished—but the design promised fantastic savings of fuel. Ericsson's nineteenth-century biographer characterized the inventor's ambition as "to remove farther into the future the inevitable period when the world's coal supply will be exhausted."[77]

After several more years in England, in 1839, Ericsson sailed for America. For the next dozen years, he labored on a range of projects, including an ill-fated navy propeller steamer, the *Princeton*, but he also continued his research on caloric engines.[78] In the decade after 1840, he constructed eight new prototypes of progressively larger size and expense. In 1851, Ericsson built a ninth model: it cost $17,000 more than all his previous engines combined and was capable of running for three or more hours without refueling. Its complex network of heat-retaining wire mesh effectively recycled waste heat but could not yet produce enough power to compete economically with steam. By late 1851, Ericsson was ready to seek investors for a full-sized prototype, to run aboard a specially designed ship. Financially underwritten by some $500,000 from New York merchants and bankers and constructed at a breakneck pace, the *Ericsson* launched in New York harbor in September 1852, beginning its trial voyage on January 11, 1853.[79] Ericsson's creation was nothing if not original. Examining the ship before its launch, the navy's former engineer in chief Charles Haswell pronounced it "the strangest ship out of the port."[80]

For expectant observers, the *Ericsson* evoked more than simple wonder at its design, for it was a machine whose operation blurred the line between the living and the inert. It was "the breathing ship," according to a party of early passengers, "an immense breathing monster," and a vessel "with lungs, respiratory organs, and every visible sign of vitality." The New York press nearly universally fawned over it. The *Tribune* trumpeted that "the age of Steam is closed; the age of Caloric opens." The *Express* emphasized the engine's ultimate advantages: "Economy in fuel, economy in space, economy in manual labor, and economy in the expense of machinery." Turning to the great reduction of dangerous engine-room jobs to as few as a fifth of what steam required, the paper added that "there is what perhaps ought to be valued more than all the rest, economy in human life." The *Times* proclaimed that "no mechanical event since the time of Fulton has promised so well for the interest of mankind." As for the unprecedented mobility the engine offered, the paper noted that "the vessel will be able to carry her coals for the longest trips out and back, even should the voyage be extended beyond the customary route of our packet steamers." In contrast, "steamships can carry a supply sufficient only for a single trip." The only sour note came from *Scientific American* editor Orson Munn, who had snuck aboard uninvited. Munn leveled his criticism more at his credulous colleagues in the press than the inventor, calling Ericsson "more modest in lauding the merits of his invention, than the few un-scientific croakers who blunderingly call the invention a new motive power."[81]

Despite Munn's grousing, when Ericsson took the ship to sea for a voyage to Washington, the vessel was a roaring success. It kept good time in bad weather along the coast, and navy commander Joshua Sands, along for the voyage, expressed his surprise at the coolness of the ship's fire rooms and the ability of a single tender to keep the ship supplied with coal. As word of the voyage reached New Orleans, the *Times-Picayune* opined that once Ericsson engines would be seen on the continent's inland waterways, the labor needed to operate steamboats would fall by as much as 80 percent and the cost of fuel would drop even more. "New Orleans will then be better able to compete with the East and North than she now is," the paper wrote, "for freights will fall enormously, and boats will increase enormously, and the river will thus be enabled to compete to some advantage with railroads."[82] The ship similarly captured the imaginations of politicians eager to apply the innovation to the same challenges steam vessels faced in the realm of international trade and in shoring up sectional economies. At a banquet in February, just as the *Ericsson* was making its way from New York to Washington, Alexander Stephens of Georgia—later the Confederate vice president—toasted his hosts and a gathering of political dignitaries with a request to remember the need for mail steam packets for the south. "Steamers," he exclaimed, "no, not steamers, for they were behind the times—but an Ericsson motor or two."[83]

Once anchored in Alexandria, Virginia, Ericsson and his ship were met by a delegation headed by President Fillmore and his successor, who had just arrived, Franklin Pierce. Accompanying them was a party of over a hundred—the sitting cabinet, the heads of naval bureaus, four commodores, distinguished younger officers like Charles Wilkes and Matthew Maury, and three members of the House Committee on Naval Affairs. Mail steamer champion Thomas Butler King was there, as was editor and power broker Francis P. Blair, former speaker of the house Robert Winthrop, the visiting William Thackeray, and literary light (and former diplomat) Washington Irving. "The Ericsson appeared to justify all that had been said in her praise," Irving wrote his sister, "and promises to produce a great change in navigation." Irving may have watched as the two presidents, Ericsson, the secretary of state, Edward Everett, and the navy secretary, John P. Kennedy (who had organized the demonstration) illustrated the engine's power by sitting atop one of her pistons as it rhythmically "breathed" up and down. The enthusiastic Kennedy anticipated contracting with Ericsson to build a caloric frigate for the government, a recommendation he passed along to the House Committee on Naval Affairs.[84]

The committee's chairman, a pro-navy Democrat from Tennessee named Frederick Stanton, embraced the proposal. His committee endorsed it, too, but it soon met a roadblock of parliamentary dysfunction. That year, partisan deadlock in the House had ground the normal mechanisms of the legislative process to a halt. Stanton found himself stymied in his attempts to persuade the full chamber to even consider a bill recommended by his committee to appropriate

$2.5 million toward building eight new vessels that used either steam power or Ericsson's new hot air engine. Abandoning his efforts to force the House to consider the full bill, Stanton tried to raise the proposal again in late February, the day after the public demonstration in Alexandria. With the thirty-second Congress just days from ending, Stanton attempted to secure an amendment to the regular naval appropriation. This proposal called for six ships, at least two frigates of which would be built by Ericsson with his novel power system. Ericsson, Stanton assured his colleagues, promised "that they will acquire a speed of ten miles an hour, and burn only eight tons of coal per day" and guaranteed a plan whose technical innovations, whether through steam or hot air, "will secure economy in the expenditures of the Navy Department." Still, though the proposal had garnered considerable support, opponents engaged in still more parliamentary maneuvers, with the chair ultimately ruling that an amendment for new construction was out of order, as it did not appropriate funds for any already existing authorization, and declaring that the naval appropriation must be limited only to the repair of existing vessels. Stanton protested in exasperation that there was no law authorizing the repair of vessels either and demanded the matter be appealed to the rest of the chamber. After several minutes of canvassing, Stanton lost by a single vote, sixty-one to sixty.[85]

This legislative defeat began the end of the caloric engine's seemingly inevitable triumph over steam. Congress never funded the ships. Ericsson, meanwhile, returned to New York to improve the design and increase the power of the engines. As part of this work, he continued planning the construction of caloric ships for the navy. On April 27, 1854, on a trial run off Sandy Hook, Ericsson reported reaching a record eleven miles an hour without even pushing the engine to its fullest, consuming coal at close to the promised rate of eight tons per day. But despite an otherwise calm day, a sudden tornado struck the ship, dunking its starboard side and causing a rush of seawater to flood into her portholes. Minutes later, the ship was entirely underwater. A distraught Ericsson conceded that even after raising the ship, repairing her caloric engine would be too costly, so he consented to replacing it with more conventional steam power.[86]

Whether caloric engines ever really offered an alternative to steam remains a complicated question. The engines occupied too much space aboard the *Ericsson* to leave room for other essential features like cargo or armaments. At the size required, the machinery also reached higher temperatures than most nineteenth-century materials could handle for long periods of time. Still, smaller caloric engines became popular in the years that followed 1854. The inventor's biographer notes that Ericsson sold a thousand engines in two years and as many as three thousand over the years that followed. They found employment powering small yachts, pumping water, and driving sewing machines. One promotional manual of 1860—itself printed by a press powered by a caloric engine—prominently advertised that the engine consumed only a third the coal as a comparable steam engine, and numerous testimonials affirmed the value of its simple operation and savings of fuel. By this time, however, Ericsson had abandoned his efforts to persuade the government to adopt his invention. His promise to the government of economy through a radical engineering innovation remained unfulfilled.[87]

Which brings us back to the Sherman committee of 1859. Just as an assessment of the value of Ericsson's caloric engine remained elusive, so too an authoritative technical resolution to the best design of steamship boilers remained out of reach. Here, partisan politics clouded definitive conclusions. Three of its five members, two Democrats and one Know-Nothing, voted essentially to acknowledge mismanagement and errors of judgment in the Navy Department but absolved anyone with authority of any actual responsibility. According to the majority, the Brooklyn Navy Yard indeed exhibited "glaring abuses" but they had grown slowly over so long a period of time, no one administration could be held accountable. The anthracite coal agent, they concluded, had become a worthless sinecure, but no one in the navy was at fault and, in any event, the navy always got the best coal at a reasonable price anyway. There was no evidence of corruption in the awarding of engine contracts, only the zeal of the secretary to maintain "the good of the public and the interests of the service." On the other hand, Sherman and his fellow Republican David Ritchie came to different conclusions, blaming navy secretary Isaac Toucey directly for appointing a coal agent with no knowledge of the business, for abuses of patronage in the navy yards, for supposedly granting contracts based on party membership, and especially for allowing navy engineer Daniel Martin to sit on boards of engineers when he held patent interests in the matters under consideration, a failure for which they demanded congressional censure. Congress took no action during the remainder of the thirty-fifth Congress, which ended a week after the reports were released, but a year later, Sherman forced the issue again and won passage of five resolutions, each condemning the management of Isaac Toucey's navy.[88]

Was Martin's patented vertical boiler design inferior to unpatented horizontal boilers? This question is only answerable in specific contexts. Every part of an antebellum steamship was an evolving element of what were perhaps the most sophisticated technological systems of their day. Particular innovations like Martin's boiler were superior when the boiler was boiling salt water, as it allowed the easy removal of saline incrustations that accreted inside boilers, but not when it was boiling fresh water, which was becoming increasingly common in the late 1850s with the use of surface condensers.[89]

Twenty years later, the navy engineer Benjamin Isherwood would note that Martin's vertical boilers consumed coal more economically than horizontal alternatives of the same dimensions, suggesting that their commercial rejection by engine builders in both the United States and Britain was a result of manufacturers' incentive structure, not the inferiority of the design. Marine engineering firms, Isherwood noted, typically built their engines for fixed fees to produce ships of stated horsepower or speed. Martin's vertical boilers were more expensive to build and weighed more than other designs. Yet they consumed coal more efficiently and could thus be more economical for consumers in the long run. Since the manufacturers never paid for coal, they rarely paid attention to this cost.[90]

The problem also reflected fundamentally different ways of conceiving of the process of engineering itself. Edward Dickerson, a New York patent lawyer and partner in the engineering firm Sickels and Dickerson (and informal consultant to navy secretary Isaac Toucey), explained the philosophy of steam engine design through what he called the "two theories upon which engines are built," exemplified by the country saw mill and the precision marine engine. "The one is to make the simplest possible form of a machine," he explained, "without regard to its efficiency. The other is to make a machine that will develop the highest possible power from the steam, and then to make that as simple as it can be made without detriment to its efficiency." With an abundance of fuel, the country steam engine could afford to be inefficient. The steamship at sea could not. But what the navy lacked in fuel it compensated for with labor, for it could afford to dedicate a crew to maintaining the marine engines to a degree not possible in the old country saw mill. That, at least, was the theory, and Dickerson was among those engineers who believed the navy had so far failed to see the difference, the consequence of which was wasteful engines, weak ships, and a considerable waste of precious coal. "Heretofore we have been making for the man-of-war the same engine which was adapted to the country saw mill, to get the engine into as few pieces as possible and then to attain as much efficiency as possible with that simplicity. In other words," he explained, "we have been making the engine for the engineer, instead of making the engineer for the engine."[91]

During and after the Civil War, Dickerson would engage in a public and acrimonious fight with engineer in chief Benjamin Isherwood. Historians have not remembered Dickerson kindly, in large part for his aggressive attacks on the integrity of Isherwood and the naval administration. He has also been criticized for a series of failed projects like the engines for the navy's *Pensacola* that went over budget and under specifications using engine designs of baffling complexity. Isherwood, in contrast, has been characterized as an engineering visionary, having undertaken influential experiments on coal and steam engines, not to mention having successfully designed numerous vessels. Yet Dickerson's testimony to the Sherman committee reveals a great deal about his philosophy, which was no less innovative than Isherwood's, even if the two men could not understand or value each other. Dickerson conceded the complexity of his designs but justified them in the name of efficiency—a term he repeatedly employed in its modern connotation with reference to measurable characteristics of steam engines—and claimed that in the long term, experience would make it possible to simplify them. It was an approach to engineering that for all of Dickerson's failures would become increasingly common in the decades that followed.[92]

Until then, the pursuit of economy remained the prevailing American approach to addressing the new challenges created by steam power. Between combustion experiments and new engineering innovations (or attempted innovations), Americans tried to alleviate the constraints imposed by coal. Rather than rethink their expectations of ocean travel, Americans sought economy, hoping to reap all the advantages in speed and power that steam offered while somehow retaining the freedom to travel long distances at low costs more characteristic of sailing vessels.

Still, through the 1850s, even with improvements in engine and boiler economy and a greater understanding among engineers of the properties of different varieties of coal, American steamers struggled when operating far from domestic ports. "At foreign stations we have to buy coal from merchants and other persons who have shipped it there for sale," explained John Lenthall, chief of the Bureau of Construction, Equipment, and Repairs, the governmental department that was responsible for coal purchases, "and we must buy such as the market affords. We can have no assurance that we can obtain the best coal." Lenthall believed that the superiority of Pennsylvania anthracite demanded that the navy continue shipping it to foreign stations.[93] Others saw a different future, hoping to develop coal resources in distant lands themselves.

CHAPTER THREE

THE ECONOMY OF TIME AND SPACE

By our recent acquisitions on the Pacific, Asia has suddenly become our neighbor, with a placid, intervening ocean, inviting our steamships upon the track of a commerce greater than that of all Europe combined."

Robert J. Walker, *Report of the Secretary of the Treasury*, December 9, 1848

After returning in September 1845 from circumnavigating the globe, Captain John Percival of the USS *Constitution* did what many frustrated public servants before him had done: he asked Congress to be paid. The trouble was a question of law. While preparing for sea in early 1844, Percival had asked President Tyler if he might employ a naturalist for his coming voyage to the East Indies. Tyler agreed, as did the acting secretary of the navy, Lewis Warrington. But Warrington cautioned that there was no provision in naval statutes to raise the number of officers serving the ship without congressional approval. The captain, however, believed he found a clever solution. Percival's choice for the post, John Chandler, was also a clergyman. Percival could thus appoint him to serve as the ship's chaplain—at the ample annual salary of $1,200—while assigning him additional scientific pursuits once at sea.

Though Percival refrained from disclosing this appointment until leaving the United States, it is unlikely that either the auditor at the Treasury Department or members of Congress would have much cared had Chandler not fallen ill en route to Rio and, once there, been discharged from the ship. Finding himself once again in need of a naturalist, Percival hired a native Pennsylvanian residing in Brazil, a Dr. J. C. Reinhardt. Reinhardt was fortunately skilled as a natural historian but, unlike the man he replaced, not as a minister of the gospel. Percival appointed him naturalist anyway (at the lower pay of a passed midshipman), then set sail for the Far East. Upon returning to Boston a year later, the captain found that the Treasury had rejected his claims for reimbursement for the pay of both men, thus leading to his appeal to Congress.[1]

Part of Percival's troubles in paying his naturalists derived from the ambiguous purposes of the cruise itself. The Tyler administration had presented it not as a scientific voyage, like Charles Wilkes's recent United States South Seas Exploring Expedition around the Pacific, but instead as a trade mission. His task was to promote American commercial opportunities around the Indian Ocean rim, and his well-publicized instructions from navy secretary David Henshaw had directed him to visit lands remote from the usual currents of American exchange like Mozambique, Madagascar, and Cochin China. His objectives included encouraging American trade, fostering amity among nations, and gathering intelligence on "the people, resources and commerce" along his journey.[2]

Though merchants in Boston and other commercial cities applauded the mission, this expansive charge and exotic itinerary made it unpopular among many members of Congress. Whig Congressmen were already alienated from Tyler, the "accidental" president, because his positions opposed much of his nominal party's platform. Democrats disliked him for having bolted to the Whigs in 1835. When news of the *Constitution*'s expedition first broke, members of both parties lampooned it by circulating a satirical letter mocking Henshaw's instructions. Casting Tyler as "King Jonathan," the letter purported to direct the fictional Percival to contact Tyler-connected merchants in Bombay and "eat as much curry & rice with them as may suit your digestive organs," sail in adverse weather, and in an admonition critical of the lengthy and unfocused itinerary, remember that "the best way of keeping out of harms way is to remain in Port the shortest possible time."[3]

When Percival returned seeking reimbursement for his naturalists, partisan politics intervened. Whigs generally supported relief (under the theory that if anyone was at fault, it was Tyler for having approved the appointment), while Democrats were split. Among the Democratic opposition, Indiana's Jesse Bright warned against those "who would be glad to ship themselves on board the public ships as naturalists and men of science, and thus go a bug hunting or possum catching, if they can travel free or at the public expense."[4] The leader of the charge to prevent reimbursement, Ohio's William Allen, railed against Percival's apparent violation of law and claimed that a naturalist at sea was so patently useless a post that it must have been cover for greater corruption still. According to Allen, Percival's actions constituted a "plain" and "palpable" constitutional violation and a "flagrant abuse in that arm of the public service."[5]

As Allen suspected, there was indeed more to this story, just not what he had imagined. After Allen finished leveling his accusations, David Yulee of Florida, himself a Democrat and an influential member of the Committee on Naval Affairs, subsequently rose to reveal the secret purpose of the cruise, until then kept even from most members of Congress. "The true object of the expedition," explained Yulee, "although ostensibly to sail around the globe, was to secure to the United States the benefit of the coal mines in the island of Borneo, situated in the Asiatic seas." A trade mission it was, but one specifically outfitted to meet the needs of American merchants seeking to compete in the region with British steam vessels. Percival had sought a naturalist to evaluate coal deposits recently discovered on and around Borneo, near enough to the commercial marts of Canton and Shanghai to supply prospective lines of American steamers across the Pacific. Yulee's unexpected revelation effectively ended debate and ensured that Percival would receive his money. The naturalists had been appointed for a vital, but sensitive, public purpose. But the disclosure also illuminated for the first time the measures that agents of the United States were beginning to take to ensure the availability of coal in distant lands.[6]

Few Americans in the mid-nineteenth century knew much about Borneo, Canton, or Shanghai, and fewer still had actually seen the other side of the world. But Americans imagined how to travel to these places long before they knew anything about what they would find when they got there. However distant in miles or culture, steam power made these once distant places at least *seem* closer or made them seem like they would be closer in the very near future. This sense derived from a changing perception of space and time. As historians have often noted, in the mid-nineteenth century, whether in the context of the postal service, railroads, or steamships, Americans pointed to the "annihilation of space and time" with a messianic zeal. Early railroad riders described how the spaces and places between destinations faded into a blur, while the need to manage railway schedules helped eliminate local time in favor of "railway" or "standard" time.[7] But few Americans actually involved in building the new networks of transportation and communication employed the language of annihilation. For them, the *economy* of space and especially of time, not their annihilation, remained the guiding conceptual metaphor. According to Francis Lieber's 1845 *Encyclopaedia Americana*, the value of machinery in achieving this economy of time was "too apparent to require illustration."[8]

The measurement of the economy of time and space, however, was always relative—faster communication was no intrinsic virtue but faster communication than the British was. This competitive economy became even more pressing after the annexation of California, a point noted by Robert J. Walker, the secretary of the treasury, at the close of the Mexican War. With the new Pacific territory, combined with transit through Panama from the East Coast and Gulf of Mexico ports, "we would be much nearer to the west coast of America, as well as Asia, than any European power," wrote Walker, "and with the best steamships in adequate number, with the greater certainty of the voyage, of the period of arrival and departure, and economy of time and saving of interest, and with diminished cost of carriage, we would ultimately supply the western coast of America, as well as Asia, with our products and manufactures on better terms than any European nation." Pursuing the economy of space and time encouraged Americans to reimagine the geography of global trade made possible by steam power, and this focus on new places led to interest in coal on distant shores. This chapter explores antebellum American efforts to secure coal in east

Asia, putting particular emphasis on the ways geological knowledge undergirded these diplomatic efforts.[9]

"A Beautiful Fact"

The American interest in coal from Borneo had come about entirely by accident. In 1836, an American merchant house in Canton, Olyphant and Company, began outfitting the brig *Himmaleh* to begin commerce with the sultan of Brunei. Olyphant sought a revival of the pepper trade, which had flourished in the late eighteenth century but suffered a more recent decline. At the same time that Olyphant planned this voyage, the British and Foreign Bible Society, a missionary organization operating from Batavia and Manila, was seeking further access to southeast Asia. Olyphant and the Bible Society agreed to work together, a collaboration eased by the strong Quaker faith of the trading house's founder, David W. C. Olyphant (his company was but one of two western houses in Asia that avoided the opium trade). The bible society's Far Eastern representative, George Tradescant Lay, who joined the *Himmaleh* in Macao, had experience in natural history, having served as naturalist aboard the HMS *Blossom* during its Pacific voyage in the late 1820s. By May 10, 1837, Lay and the *Himmaleh* reached Brunei.[10]

It was here that Lay stumbled on something unexpected. While a guest in the sultan's palace, Lay received a sample of local coal brought to the court for his perusal. Pressing his hosts for its origin, he could determine only that it came from "Kianggi," though to his chagrin "no one could point out the spot, nor had any definite idea of the extent and limits of this *Kianggi*." Eventually, a court official claimed to know the place and offered to supply the *Himmaleh*, but there the matter rested until nearly the end of Lay's visit to Borneo. Returning from a final trek outside the city, Lay and a companion paused for refreshments beside a fresh spring. While his partner drank, Lay continued to explore. "I struck my hammer upon what seemed to be a vein of sandstone," he later recalled, "but to my very great delight, I discovered that it was the very thing I had so often sought for in vain, the coal of 'Kianggi.'" Lay called the discovery "a beautiful fact."[11]

Lay published his memoirs of the voyage in New York two years later, and knowledge of his discovery spread quickly in both Britain and the United States (Salmon Chase, later Lincoln's first treasury secretary, reported reading the account before bed in 1841).[12] In 1842, the British governor of Bengal appointed agents to begin testing Borneo's coal for steaming purposes. The agents constituted the Committee for Investigating the Coal and Mineral Resources of India, which found that Borneo's coal appeared to be an outstanding steamship fuel. This announcement inspired the editors of the English-language *Singapore Free Press* to applaud the news. "There is no quarter in the East where a Coal

Depot would be more valuable or is more urgently required," they wrote. No other part of the world offered the British greater promise of commercial gain and yet presented greater obstacles to the flow of goods and information. British steamers in the region then depended on coal from Burdwan in Bengal or even more distant England. Penang and Singapore occasionally exhausted their coal stocks altogether. British traders in Singapore believed that "a mine at Borneo would serve to keep those two Stations well supplied and thereby greatly facilitate our Steam communication with China." With communication would come trade, and with trade, wealth and power.[13]

But George Lay and merchants in Singapore were not the only British subjects interested in Brunei. As news of coal in Borneo percolated through the trading houses and consulates of southeast Asia, James Brooke was just beginning his career in the region. Brooke, an adventurer, admirer of Singapore founder Sir Stamford Raffles, and heir to a substantial fortune, had grand visions for Borneo, its people, and its resources. With proper aid from Parliament, he anticipated subduing regional piracy, quashing the seemingly endless contests for power within the Brunei court, and developing the resources and commerce of southeast Asia. Brooke began his enterprise on Borneo pledging a different kind of empire, seeking what he called "a pure spot in the troubled ocean of colonial politics." Should any local resource draw investment from Europe, Brooke intended a share of that capital to enrich the local court and encourage political stability and economic growth—and Borneo had no shortage of resources. Brunei and Sarawak, together comprising the northwest coast of the island, possessed a bounty of potential commodities, from the pepper sought by Olyphant to even more exotic coconuts, birds' nests, and tortoise shells. Elsewhere on the island, one might find gutta percha, bees' wax, vegetable wax, and betel nuts, as well as oils, camphor, and ebony wood. And then there was the coal.[14]

Brooke had learned of Lay's coal discoveries, and since his own arrival in Brunei in 1838, additional coal outcrops had been identified around the city. By March 1843, Brooke had concluded that the British government ought to secure an outright monopoly of the coal and establish a naval station nearby.[15] Nevertheless, he maintained his skepticism that coal was in fact the most valuable offering of the region, and for almost two years, he took little decisive action. "The truth is," Brooke confided to a friend on New Year's Eve 1844, that British officials "are pottering about coal and neglecting far greater objects." These greater objects included suppressing piracy, stabilizing the court, and securing British supremacy in commerce. "Coal there is," Brooke conceded, "the country is a coal country, but when gentlemen are sent to make specific reports, it is not known that great difficulty exists in finding this coal, and that the search, in a wild country, will occupy months, or else the report will be imperfect."[16]

Traces of coal, even coal suitable for steam engines, did not immediately translate into the availability of the fuel.

Brooke changed his mind on the central importance of coal in the region three months later, after one more discovery persuaded him to reconsider the strategic and commercial value of the island. Brooke had learned that while exploring the island of Labuan, located at the entrance to Brunei Bay, a British navy lieutenant named Leopold George Heath had encountered a large coal seam. The outcrop was abundant and easily mined. Captain Rodney Mundy of the Royal Navy, a friend and admirer of Brooke, related that engineers aboard the British steamer *Nemesis* "report it to be the best coal for steaming purposes which they have met with in India." It was also easy to burn and deposited only an inconsequential residue of ash. Samples of Labuan coal soon made their way to Britain, along with news of the discovery. Chemical experiments performed at London's Museum of Practical Geology by Dr. Lyon Playfair confirmed its high carbon and low ash content. Geologist Henry de la Beche, Playfair's colleague, advised "that the coal of Labuan should be systematically and carefully worked" to protect the deposits for future mining.[17] Brooke at last conceded that Borneo contained more coal than even he had originally anticipated. "I now begin to think it really may become a prize some future day to our steamers," he wrote.[18]

"Labuan" is a Malay word for "anchorage." The island is some eleven miles long, roughly forty square miles in area, an isosceles triangle with a bite taken out of the base and tapering to a point in the north. It was near all the major commercial ports in southeast Asia: 650 miles from Manila, 707 from Singapore, 984 from Siam, 1,009 from Hong Kong. Owing to the coal discovery and reports of its steaming qualities, the island came to figure into great power struggles in southeast Asia. "Should there ever be another war," wrote Mundy, "the command of this coal district will be of vast importance; and in the mean time, the quickly increasing numbers of steamers in the neighboring seas will probably draw their supplies from there."[19]

As the island was uninhabited, strategically located, and now known to contain large quantities of coal, British expatriates began discussing the prospects of a formal colony. A new settlement on Labuan would "almost perfect the chain of posts that connects, by means of steam navigation, Southampton with Victoria in Hong Kong," reported the *Singapore Free Press*. This network already linked England to China via coaling bases in Malta, Alexandria, Suez, Aden, the Ceylon port of Galle, Singapore, and finally Hong Kong. According to the paper, "in a very few years we may expect to see the world fairly *belted* by the steam navy of England."[20]

Americans had other ideas. They were also paying attention to the discovery of coal on Labuan, but given their weaker diplomatic and commercial posi-

In this sketch, Leopold George Heath illustrated his 1845 discovery of coal on the small island of Labuan, off the coast of Borneo. Heath was then a twenty-eight-year-old Royal Navy lieutenant serving in the East Indies station aboard the HMS *Iris*; he would later serve as the island's colonial governor. Labuan coal would attract Americans aboard the USS *Constitution*, sent by the Tyler administration to investigate coal deposits in southeast Asia for prospective American steam lines. Rodney Mundy, *Narrative of Events in Borneo and Celebes: Down to the Occupation of Labuan* (London: John Murray, 1848), 2, plate facing 348.

tion, could not follow Brooke and entertain thoughts of a coal monopoly. When David Henshaw, the secretary of the navy, learned of the coal, he assigned John Percival and the *Constitution* the task of gaining access for the United States to support American steamers in the region and across the Pacific. Though Henshaw's public instructions to Percival acknowledged that the ship's ultimate route depended on conditions of climate, weather, health, and politics far beyond the limits of advance planning, he insisted in his secret instructions that the ship visit Borneo. "It is represented that this island possesses Coal mines of great richness," he explained to Percival, "both for quantity and quality. Your enquiries will therefore be especially directed to this subject, of finding coal that can be readily procured for the use of sea steamers; and if deposits be found, easily accessible, for supplying steamers or other vessels. You are authorized to purchase a right to such mine, for the United States, of the Government which owns it, at a reasonable compensation."[21] Ensuring access to the island for future commercial vessels, not control of it as a colonial outpost, was to guide Percival's mission.

By the time the ship approached Borneo in March 1845, the object of the *Constitution*'s visit there was known to the crew. One crewmember, Henry George Thomas, recorded that "we had reports that there was abundant coal in the area and we hoped that an agreement between the Sultan and our country could be reached" over mining it. Thomas and others had also heard rumors of Brunei's association with regional piracy, though the crew's "extra defenses" proved unnecessary when the Sultan warmly welcomed the ship's expedition with a nine-gun salute.[22]

Negotiations took place almost immediately. Percival, for his part, had long suffered from gout and found his condition deteriorating as the ship approached Brunei. Too ill to play diplomat, he sent Lieutenant William C. Chaplin to the royal court in his place. Also absent was James Brooke, then away visiting Singapore. At the palace, Chaplin introduced his party as representatives of the sultan of America, boasting of his nation's maritime strength and extensive trade. Offering the sultan (of Brunei) samples of American goods, he announced his desire to open a regular commerce with the country, promising abundant revenue from trade duties and the gift (from the American sultan) of American manufacturing. The sultan, Omar Ali, acknowledged the offer but explained that only weeks before, he had given the English "the exclusive right of trade in Borneo Proper and now he could do nothing for America." Chaplin protested that such a policy ran counter to the usual arrangements for international commerce. He insisted that exclusive rights limited the development of industry, the arts, and agriculture. The prohibition of trade was, in fact, a violation of the natural order. As he informed Washington, he explained to the court that "the Divine Hand for a wise purpose had not deposited the fruits of the earth equally and alike upon every country and climate, and that when it was too late, the Sultan might have cause to regret so ruinous a policy." Omar Ali remained unmoved, explaining that with the absence of James Brooke, the new English rajah, they could conduct no substantive business.[23]

Nevertheless, Chaplin persisted with his negotiations. When he broached the subject of coal, he was rebuffed a second time. According to Chaplin's report, the sultan explained that barely three weeks previously, an English steamer brought "a special agent of the Queen of England who had purchased the exclusive right to all the coal" in the sultan's "dominions." At first, this response tempted Chaplin to conclude that the court was merely bargaining for better terms in the negotiation. Yet as he recalled events and observations from the preceding month, viewed with a newfound clarity, he concluded otherwise. Chaplin was aware, for instance, of James Brooke's entreaties to the British crown to incorporate his influence in Brunei into the formal British empire. The lieutenant also realized the significance of a November announcement from the Royal Navy creating a "special agent" to Borneo. The agent, Captain Charles

Bethune, had reached Singapore while the Americans were recuperating there from weeks of shipboard illness. Bethune suddenly left his sailing ship for a steamer and unexpectedly altered the course of that ship for what was, to the Americans, an unknown destination. All this "at a time when we had reason to know that he was aware of the destination and object of *this* Ship." To Chaplin, the conclusion was unmistakable. Bethune had hurried to Brunei to conclude a commercial treaty before the Americans could arrive. The British had not only bested the Americans in securing trade with Borneo but in "that which is still more important," according to Chaplin, namely, "the use of the immense mines of coal supposed to exist, in this part of the Island, which in course of time must render incalculable benefits to commerce, when Steam, already an important auxiliary, becomes a chief agent in the Commerce of the world." But Chaplin's appeals led nowhere and the American party soon left the island having failed in their central objective.[24]

Before hauling anchor, Percival allowed J. C. Reinhardt, the naturalist he had employed in Rio, to make a brief, if futile, reconnaissance of coal on Labuan. The island lay twelve miles from the ship, and upon landing Reinhardt began a hurried three hours of observing and collecting specimens. Impenetrable vegetation prevented him from trekking more than half a mile into the forest, but even close to shore he recorded local topography, noted an unknown species of black squirrel, and disemboweled a sea snake for preservation. But he saw no trace of coal, which he reasoned must be found on the other side of the island. Since access to fuel, not scientific observation, had been the central motive for the mission, Reinhardt concluded that little could have been gained had he found some in any event. Labuan was indeed "an interesting place to a naturalist," but "a survey of the coal field here could have been of no benefit to our country" now that Brooke had apparently secured British control. Without the prospect of further negotiations, "it could only have been of interest to science to have remained there," Reinhardt concluded. And he was not employed to worry about science.[25]

After Brooke returned from Singapore, the court in Brunei reported that the Americans had proposed protecting their government, acquiring exclusive privileges for mining the region's coal, and securing a monopoly on the Borneo trade. Brooke doubted whether in fact the Americans had made the final stipulation but worried nevertheless that the opportunity for British supremacy in the region was fast disappearing. "The Americans act," he observed, "while the English are deliberating about straws." To his uncle, Brooke fretted that the American arrival "proves that while one nation is deliberating another can act."[26]

But American actions amounted to little, as Americans had fallen victim both to local politics as well as miscommunication. Muda Hassim, an influential minister in the royal court had, in fact, "pledged to forbear from negotiation with other powers, pending his negotiation with the English to repress piracy and to cede Labuan," yet Chaplin's conclusions notwithstanding, these negotiations had not yet been completed. Poor communication may have played a role in the collapse of the American mission. "It is probable," Brooke grumbled, that "the demand for exclusive trade has been erroneously understood," owing to "the badness of their interpreter (who was formerly my drunken servant)." Brooke, however, was grateful for the misunderstanding. He believed that had the Americans been better prepared, the court would readily have consented to their request, much to the detriment of his imperial project. He brooded about the shifting alliances of the Brunei court, only one faction of which he supported. "Even now they twit our party with the Americans doing at once, what the English cannot do[;] they are blamed for repulsing the Americans and for preferring our friendship," he groused. Brooke considered what he saw as Percival's near success as evidence of his own inability to control the situation. He feared that he was trapped within the snares of local politics. "I can see no direct and immediate line of conduct, which can extricate our friends, and in the mean time we are in a wretched, inefficient steamer, which could, on occasion, neither fight or run away." In short—a steamer without coal. In contrast, "The Americans would act first and inquire afterwards—and they are right."[27]

Nevertheless, the danger for Brooke, at least for the moment, had passed. The *Constitution* left after its officers were rebuffed at the sultan's court, and Reinhardt returned from his brief reconnaissance of Labuan. A year and a half of negotiations later, in November 1846, Brooke received instructions from Viscount Palmerston to formally seize Labuan for England. Sultan Omar Ali and Captain Rodney Mundy signed the treaty ceding Labuan on December 18.[28]

As the British government continued its expansion of steam communication in southeast Asia, the coal fields of Labuan grew in significance. By the late 1840s, mining there and on the nearby coast of Borneo was well under way. "The European governments have during late years made careful researches to ascertain the distribution of coal fields," wrote the journalist Horace St. John, noting discoveries of coal in India's Tennasserim provinces (in present-day southern Myanmar), along the Malay Peninsula, on Sumatra, Borneo, and many other less familiar locations. Only in Labuan, however, were large enough deposits, suited for steaming purposes, discovered by Europeans. According to Hugh Low, by then Labuan's colonial secretary, coal "will prove of the greatest value to our increasing steam communication with the East. It has been tried by various government steamers, the engineers of which pronounce it to be of the finest quality, superior to that imported to Singapore from England." Further, he added, "one of the principal reasons which has caused our government to form the settlement at Labuh-an is the value that this mineral will prove both in time of peace and in case of war." By the 1870s, the mines were producing around 5,000 tons a year, before a series of accidents sharply curtailed output.[29]

Despite the British colonization of Brunei and Labuan, these same concerns over the value of the region's coal in peace and war led Americans to make a second bid for access. The agent was Joseph Balestier, a son-in-law of Boston's Paul Revere and a man already well informed about Borneo's coal. When appointed U.S. consul to Singapore back in 1836, he had aided Olyphant and Company in preparing Captain Frasier and the brig *Himmaleh* for their commercial and missionary voyage to Borneo. His association with Olyphant would have made him one of the first to hear of George Lay's discoveries. Almost ten years later and still in Singapore, he almost certainly knew of the *Constitution*'s secret instructions to secure a coal supply for American vessels as well, for he was still serving as consul when the ship arrived en route to Borneo, and he hosted Percival at his plantation some three miles from the city's central business district.[30]

In August 1849, Secretary of State John Clayton contacted Balestier to undertake a series of diplomatic missions in southeast Asia. Appointing him a special agent of the United States "to Cochin China and other portions of South Eastern Asia," Clayton included in his instructions a request that Balestier visit the sultan in Brunei. Two circumstances drew Clayton's attention to Borneo. First, British naval expeditions to crush piracy around the China sea aroused American expectations of expanded, safe commerce in the region. Second, Clayton recounted the object that attracted the government to Borneo five years earlier, "the abundant deposits of fossil coal, suitable for the purposes of Steam Navigation, at Labuan, Sarrawack [*sic*], and in other districts on the coast of that Island." In the intervening years, of course, the United States had also gained a direct outlet to the Pacific with the annexation of California. If Percival had failed to win a coal concession by arriving too late for an outright grant by the Sultan but too early to broker a deal with the British, Clayton hoped that now, with a British company actively mining coal on the island, he could secure "treaties of amity and commerce" between the United States and Brunei with the sanction of both Brooke and the Omar Ali.[31]

At Macao, Balestier met the USS *Plymouth* at the end of December, and two months later he boarded the ship to begin his mission. After a failed diplomatic venture in Cochin China, Balestier and the *Plymouth* made their way to Borneo. Balestier's general mission was to extend diplomatic recognition to Brooke's government in Sarawak as well as to secure the commercial treaty with the sultan that Percival and Chaplin had failed to secure five years before. Although Brooke himself was once again not present on the island, Balestier obtained both objectives (though Americans were barred from trading for Brunei's antimony, a mineral Brooke kept as a monopoly for exclusively British consumption). From

Brunei, Balestier and the *Plymouth* sailed to Labuan, where they continued their negotiations. "Mr. Balestier's object in coming to this place," recorded George Welsh, a young American officer aboard the *Plymouth*, "was to make inquiries concerning coal, its price, and at what price it would be furnished American steamers, besides to form a sort of treaty with the Sultan of Borneo." Balestier, Welsh, and the rest of the crew discovered that the British had by then already set up a company to mine coal "of a very superior quality," coal that was near the surface and evidently abundant. After a brief negotiation, the mining company agreed to sell Americans Labuan coal for $6 a ton. A successful negotiation it was, but the company also sold the same coal to British vessels for only $4.50 a ton. Coming in second had its consequences.[32]

Still, the deal gave a greater opening for American trade in east Asia, especially for textiles. This trade, Balestier hoped, would displace piracy and make Borneo once again a source of lucrative commerce in tropical agriculture and forestry. These were articles "to which may now be added bituminous coal of the very best quality in the greatest abundance," Balestier wrote home to Washington, "in exchange for which a new avenue will be opened for the export of our cotton clothes and other commodities thus creating a new & valuable trade to us and giving importance to the treaty just concluded."[33] Although U.S. trade with China remained a small fraction of its total trade in the nineteenth century (its trade was mostly with Europe), American exports to China were comprised overwhelmingly of cotton goods. In 1850, the $1,203,000 worth of cotton cloth delivered to China represented fully 81 percent of all American exports there.[34] At the same time, in the mid-nineteenth century, China constituted the largest single foreign market for finished American textiles. In 1845, this market meant that nearly 35 percent of American exports of finished textiles landed in China. Although that figure dropped to about 10 percent in the mid-1850s, it rebounded again by 1860, a year that saw a more than doubling of value exported from just fifteen years earlier.[35] These statistics, which went beyond crude calculations of China's vast (and presumed cotton-needy) population, contributed to the allure of the Chinese market for American textiles and the appeal of obtaining coal nearby.

Balestier's mission was to help further this American cotton trade with the Far East. Securing access to Borneo coal was a step toward developing the transportation and communication infrastructure this trade might require. In the State Department, however, interest in Borneo's coal cooled with efforts to find alternative supplies in eastern Asia. After the death of President Zachary Taylor on July 9, 1850, the incoming Fillmore administration shook up the Cabinet. John Clayton left the State Department, replaced by Daniel Webster, who returned to a post he had previously filled under John Tyler. Webster did not have a lot of enthusiasm for Balestier's mission, noting (unfairly) that the endeavor "has not, thus far, produced any important result, and does not seem to promise much for the future." With that the new secretary of state ended the mission.[36] Webster, however, soon had other ideas about finding coal in the Far East, and he soon turned his attention to Japan.

Commodore Perry's Pacific

Americans had many reasons for voyaging to Japan, closed to most foreign contact since the 1630s. Though Americans had, in fact, had sporadic contacts with the country for several decades—there had been some limited trade during the Napoleonic Wars and a series of tense encounters between Japanese officials and American merchant and naval vessels in the 1830s and 1840s—rapid changes in the United States whetted the appetite for a more regular relationship. The cession of California—and the rush for gold there in 1849—strengthened old dreams of increased commerce with the Far East, especially with China, and some Americans eagerly anticipated manifest destiny continuing beyond their continent's shores. For those for whom Mammon was an insufficient motivation, there was also the vast United States whaling fleet, at least three hundred ships a year plying the length and breadth of the Pacific. These ships were, of course, in pursuit of highly valuable whale oil, but there was no doubting the emotional resonance of stories of sailors shipwrecked on Japan, who were then trapped and tortured. Other Americans espoused a duty to civilize, to bring liberty and Christianity to benighted heathens. For Americans who gave the matter any thought, there was little question that American influence would improve the world. "The islander will cease to go naked, the Chinaman will give up his chop sticks, and the Asiatic Russian his train oil," wrote the navy's oceanographer Matthew Maury, "the moment they shall find that they can exchange the productions of their climate and labor for that which is more pleasing to the taste and fancy."[37]

By the late 1840s, the development of steam power and the policy of federal steamship subsidies made east Asia seem closer than ever. In 1848, the legislative architect of the steam subsidy system, Thomas Butler King, called for a new line between California, Japan, and China.[38] Soon, Congress began receiving proposals, of which Philadelphian Ambrose W. Thompson's quickly emerged as the most promising (the House Naval Affairs Committee endorsed it as "a most favorable contract for the government").[39] By May 1851, both the State and Navy departments were considering an official expedition to the Far East to secure coal supplies and refueling arrangements for the future line. These objectives focused attention on Japan. Webster believed that Japan possessed vast quantities of coal, or, as he called it, "that great necessary of commerce." Like Lieutenant William Chaplin in Brunei, Webster believed that no nation had the right to withhold necessities from those who needed them. "The interests of commerce, and even those of humanity," he wrote, required Japanese coal, "a gift of Providence, deposited, by the Creator of all things in the depths of the Japanese Islands, for the benefit of the human family." Webster, navy secretary William Graham, and President Fillmore began planning the mission, and after losing confidence in their first choice to lead it, Captain John Aulick, they selected the fifty-seven-year-old Matthew Calbraith Perry.[40]

Perry already had considerable experience with the new technology of steamships. Since the 1830s, he had pushed the navy to develop steam vessels and helped design some early ships, and he oversaw the *Mississippi* during her construction in Philadelphia between 1839 and 1841. During the war with Mexico, he commanded the *Mississippi* and eventually the whole of the Gulf Squadron.[41] After the war, navy secretary John Y. Mason appointed Perry inspector of the mail steamers under construction in New York. For the next three years, Perry commuted from his home in Tarrytown to the bustling machine works and shipyards at the southern tip of Manhattan, where he worked with constructors, naval officers, engine builders, and ship owners. Perry concerned himself above all with ensuring that the mail steamers met naval specifications, permitting their conversion into naval vessels in an emergency (a policy he supported publicly but in private believed of limited value in war). In his role as steamboat inspector, the tensions between commercial exigency and naval-grade construction occasionally flared up, as when the proprietors of George Law's line allowed their ships *Ohio* and *Georgia* to sail repeatedly without protective copper sheathing on their hulls, thus making them susceptible to destructive shipworms. But Perry also concerned himself with how these ships would receive the highest quality fuel. In one instance, he recommended that the department employ a newly invented water-measuring device to test the evaporative power of various coals, which would allow it to avoid "the purchase of coal of an inferior quality." The device could be used, additionally, "in ascertaining the most economical description of Boiler for the use of Government Steamers." Perry also joined a commission to investigate a novel condenser that saved fuel by converting salt water into fresh for use in ship engines.[42]

Once selected to lead the expedition to Japan, Perry resolutely sought to keep his principal interests hidden. He chose instead to make the public face of his efforts a remonstrance against the mistreatment of shipwrecked American sailors and whaling crews who increasingly washed up on Japanese shores. Perry further sought to resolve a series of petty diplomatic indignities inflicted on the handful of Americans who had recently attempted to breach the gates of the secluded islands. But about coal and steamships he preferred to say as little as possible. "The real object of the expedition should be concealed from public view," Perry wrote. Yet when senators learned in March 1852 of the hitherto secret mission and demanded details, supporters privy to it found themselves in the lurch. One senator whose state stood to benefit most from a successful mission, California's William Gwin, needed to signal the importance of the expedition without yet revealing its full purpose. This challenge led to comical paralipsis on the floor of the Senate. "Suppose that part of the instructions given to the commander of the fleet was to explore these seas for the purpose of finding an island in which coal might be obtained, in order that we might have a depot to supply vessels" Gwin posited to his colleagues. "Is it proper to make it known to all the world, that some other power may go and take possession of it?"[43]

Unsurprisingly, the purpose of the expedition did become known to the world. And even as Perry deemphasized the centrality of coal, the very process of outfitting the expedition dragged the navy deeper into domestic coal politics. Unlike in the case of the voyage of the *Constitution*, Perry's fleet would include several steamers, including the frigates *Mississippi* and *Susquehanna*, and all ships would require constant supplies of fuel. The question, as always, was from where. As news of the expedition leaked to the public (and Congress), Maryland and Pennsylvania coal dealers traded in rumors that the navy sought to outfit the entire mission with foreign coal, which, if true, would land a serious blow to their industry. If Daniel Webster could declare that Japanese coal existed "for the benefit of the human family," would pursuing it come at the expense of the specifically American family? More precisely, for Maryland and Pennsylvania coal dealers, would it come at the expense of American miners' families? Confronted with these concerns and the potential political backlash of the dealers, William Graham assured the industry that he "never intended to do any act that would militate against the American coal interest." Instead, he decided to circumvent the recently established coal agent system and appoint instead a "special agent" to handle all coal purchases for the expedition. Graham's choice for this post was the politically connected New York firm of Howland and Aspinwall (whose partner, William Aspinwall, already operated the navy-funded Pacific Mail Steamship line between Panama and California). The firm would receive double the 5 percent commission that was standard for the regular agents, with the stated expectation that this commission would be used to pay the existing agents, Benjamin Springer in Philadelphia and John Jameson in Baltimore, for examining all the coal before it shipped. Graham defended the arrangement as a mere "experiment only," promising that bids from other dealers would be considered in the future.[44]

There was only one problem. Graham's arrangement with Howland and Aspinwall specified that the regular agents would be paid only for examining *American* coal. And while Howland and Aspinwall did indeed ship some domestic bituminous and anthracite, the greater part of the coal they provided was in fact purchased in England. This arrangement aggravated American merchants (while also depriving the two agents of thousands of dollars of lost commission fees). In the Senate, Pennsylvania's James Cooper protested that "the coal trade [had] reached a point that puts it comparatively beyond the reach of destructive competition" but only "if the Government would extend to it such incidental encouragement as it has in its power to do." Baffled that the navy would spend what was likely to become hundreds of thousands of dollars on English bituminous coal instead of supporting the domestic industry, Cooper insisted that American (that is, Pennsylvanian) anthracite bested English bituminous on cost, quality for steaming, and durability against the elements—the last point especially relevant with the likelihood that the Japan expedition would need to deposit large supplies months or years in advance for later consumption. For Cooper and his constituents, naval adoption of anthracite fuel was an essential element of the growth of the industry, and the government's plans to purchase foreign coal presented an unexpected obstacle.[45]

More generally, Americans debated the controversial relationship between coal and the foreign policy aims of the expedition. The Washington correspondent to Pennsylvania's largest paper, the penny-press *Public Ledger*, explained Webster's idea in Jeffersonian language: "Deposits of coal were intended by Nature and Nature's God for all mankind," and no people could thus prevent its exchange "for purposes benefitting the whole human race."[46] This portrayal produced a firestorm of criticism, particularly from Democratic and antislavery newspapers. In Cleveland, the *Plain Dealer* titled one article the "Invasion of Japan," calling the expedition "a swaggering show, intended to bully the Japanese."[47] The *Boston Commonwealth* complained that "the coal, as we take it, is not to be carried from this country and deposited there, but dug on the island, so that the enterprise is as honest as it would be for the Emperor of Russia to seize Mauch Chunk and appropriate it to himself."[48] Ohio's influential *Anti-Slavery Bugle* described the purpose of the "warlike expedition" as "stealing one of the Islands of Japan, for a coaling station for our Pacific steamers, in their passage to and from China."[49] As Perry had hoped, more supportive papers denied the centrality of coal, promoting the expedition as one principally to restore dignity to shipwrecked sailors, suggesting the political wisdom of the Fillmore administration's emphasizing this aspect of the mission.[50] Still, the *New York Herald* could not resist noting that the expedition was manifestly to benefit the business interests of California and would be "of much service to the future lines of vessels to be established between our new State on the Pacific and China and the East Indies" by yielding havens for transpacific steamers.[51]

As he planned for his fuel supply, Perry also mapped out his route around the world. The limited fuel capacity of steamships in the 1850s constrained his choices, as well as those of prospective Pacific steamship line operators, a point that generated considerable confusion at the time and since. Americans sought coal and coal depots in east Asia to accommodate the great length of the voyage across the ocean, but this length varied according to the route a ship traveled. Supporters wrote casually of the "great circle" or northern route, a northwest path from San Francisco that brushed the Aleutian Islands before returning to lower latitudes and reaching China. It was undoubtedly (and indeed, by definition) the shortest path across the ocean. An early booster of a mission to Japan, San Francisco's most important newspaper, the *Alta California*, cited the judgment of "gentlemen best acquainted with Pacific navigation," who believed that only the northern route would suffice and that a port at the southern extremity of Japan" that had been supplied with coal from northern Formosa "would be indispensable as a depot" in such a route.[52] The significance of this great circle route to the expedition has been repeated by historians ever since.[53]

However, a great circle path from San Francisco to Shanghai would not pass through the southern islands of Japan but instead perfectly intersect the *northern* island of Hokkaido. The great circle was important to planning prospective transpacific steamship lines, but not in the way most accounts describe. The northern route was indeed shorter than a voyage across the central Pacific through Hawaii by about eight hundred miles, or three days' sailing. Matthew Maury, the navy's prolific oceanographer and mapmaker, made this point in 1848, in an analysis of Pacific steam travel that helped shape the terms of the debate.[54] But despite steam booster rhetoric, steamships still relied a great deal on winds and currents, which meant that voyages east and west could not always (or even typically) follow the same paths. A more plausible transpacific steamship service would follow the great circle route, but only on the return trip from the Far East to the United States, carried along by the Kuro Siwo (Kuroshio) current. Heading west, however, steamers would have been wiser to use auxiliary sails and catch the powerful trade winds blowing across the central Pacific, even though this route was longer. "By this detour," explained a young lieutenant, Daniel Ammen, "fair winds would be obtained over the whole distance to Shanghai and favorable points be found along the route for depots of coal &c at the Sandwich and also at the Bonin Islands."[55]

This prospective Pacific loop—westward to Shanghai via Hawaii then returning eastward by the great circle and the northern Pacific rim—explains Perry's focus during his expedition on the Bonin Islands, which Maury never mentioned, as a prospective American coaling station. These islands, southeast of the larger Japanese archipelago, lay nearly along another great circle route— this one from Hawaii. While the central Pacific route was longer than the northern great circle route by the Aleutians, according to Ammen, "the detour brings us in the region of almost invariable winds, tending yet to shorten the voyage, economize fuel and save the wear and tear of machinery." In the 1850s,

steam alone could not annihilate time and space across the Pacific Ocean. Instead, it would need to work within the constraints of costs, geographical position, diplomatic possibilities, and the natural features of wind and currents.[56]

The struggles Perry found in fueling his expedition itself make clear why
developing sources of coal in the Far East was such a pressing concern. Perry
left Norfolk for Madeira November 24, 1852. He had thought as carefully about
how to get to Japan as he had about what to do upon arriving. His ship, the
Mississippi, had been specially outfitted to hold 600 tons of coal, some 150 tons
beyond its original design, and reconfigured to ensure it could travel farther
without refueling. Perry anticipated an initial coaling at Madeira, then subsequent stops at the Cape of Good Hope, Mauritius, and Singapore.[57]

Along the way, however, Perry realized his ship was consuming more coal
than he had expected. Unseasonably foul winds and weather delayed the voyage from Madeira south along the African coast, and as the ship consumed its
limited fuel supply, Perry ordered an unplanned stop at St. Helena for an emergency resupply before continuing on to the cape. Steam engines may have
helped ocean vessels travel along sea routes less constrained by currents of wind
or water, but they did not free them from the constraints of economy: the 440
tons Perry purchased in Madeira cost $9.30 a ton—expensive compared with the
domestic market but roughly average for a major transatlantic port. In St. Helena, however, the 124 tons loaded aboard the *Mississippi* sold for $25.20 a ton—
about a quarter of the coal but for three-quarters of the cost.[58]

That Perry could coal at all at Capetown and Mauritius was the result of judicious planning. Before leaving Norfolk, Perry had arranged with the shipping firm of Howland and Aspinwall to dispatch two ships from New York
loaded with Pennsylvanian anthracite to sail ahead of the *Mississippi*. Both ships
arrived at their destinations only days before Perry. Perry believed that without
them, purchasing coal for the *Mississippi* and *Susquehanna*, as well as the *Pow-
hatan* and *Allegheny* that were following them, would have involved "great difficulty." The arrangement with Howland and Aspinwall was a success, and Perry
noted that future commanders of steam vessels should likewise send cargoes of
coal ahead of themselves to ensure an adequate supply once they arrived at their
ports of call along the way to their destinations.[59]

Yet even after his arrangements with Howland and Aspinwall, Perry remained concerned about his fuel supply. He projected that his three steamers
would devour some 90 tons of coal a day, creating, in Perry's words, a "serious
risk in depending too much on probable purchases abroad." He supposed Aspinwall could send out some 10,000 or 12,000 tons for the mission, but even then
he believed he would need to make special arrangements with the British
Peninsular and Oriental Steam Navigation Company to sell him coal from its
network of depots. Moreover, he desired the detail of a dedicated coaling

vessel and wanted a special reserve supply of 1,500 tons to be established at Honolulu.[60]

Perry's difficulties obtaining coal persisted in Singapore. The British port had become a major coaling depot for mail steamers, facilitating "a constant postal communication, by means of the English and one or more Dutch steamers, with Hong Kong, Penang, Batavia, Shanghai, Calcutta, Madras, Bengal, Bombay, Ceylon, the Mauritius, Cape of Good Hope, and, by the Red Sea, with Europe and America." This mail communication kept Europeans and Americans in constant contact wherever they might be conducting business. Yet despite the operation of these mail steamers, "there was not a pound of coal . . . to be purchased at Singapore." Perry lamented that "there was reason to fear that the Mississippi would be deprived of her necessary supplies." Perry was learning, in fact, the cost of Percival's failure to secure a favorable coal concession in Borneo before the British. By the time of Perry's arrival in Singapore, the Labuan mines were producing a substantial 1,000 tons a month, but the Peninsular and Oriental maintained a lock on the supply, consuming it entirely in the company's own vessels.[61]

Good fortune assisted Perry again, however, as Peninsular and Oriental coal supplies had fallen low in Hong Kong. Although it had enough coal for both ports in Singapore, the company lacked an available ship to transport it. Perry and the company reached a deal. Perry would coal the *Mississippi* at Singapore in exchange for returning the same amount upon a later visit to Hong Kong.[62] Incidents like this one convinced Perry and his officers to ration their coal consumption and jealously guard what supplies they were able to amass in various Asian ports.

That guarding was not always successful, as suggested by the frustration Perry encountered in trying to store coal in Shanghai. Perry left coal he had purchased there under the protection of a storekeeper, J. S. Amory, whom he forbade to release any amount without his written permission. Perry was concerned, in particular, that the French or Russian navies might try to take advantage of his actions in Japan before he himself could complete his negotiations. And indeed when a Russian vessel arrived in Shanghai in November 1853, the vice admiral aboard, Evfimii Vasilievich Putiatin, approached Amory for a loan of 20 tons of coal. Amory, under Perry's orders, refused. The Russians, however, maintained an agent in Shanghai who served in a second role as the American vice consul. The agent pressured Amory to release the coal, which he eventually did. When he learned of the transaction, Perry was predictably outraged, blaming both Amory and the consular system, which he described as "fraught with much evil." He would not fire Amory, as he was concerned such a move might offend the Russians, but his hopes of using the little coal he collected in Asia to his advantage had clearly failed.[63]

For the duration of the mission, Perry and his ships scrounged for coal supplies all around southeast Asia. Some coal continued to arrive from the United States, while Perry purchased other stocks from British or Asian suppliers. His supplies were precarious and dealers unreliable. "At no time" Perry observed, "have we had more than fifty days of steaming for the three steamers." Moreover, Perry limited travel so as to make "provident and economical use" of the supplies they had, and so the voyage itself was constrained by this exigency. At their usual pace, his ships consumed between 28 and 32 tons of coal a day.[64] When the *New York Tribune* reported that two of Perry's ships had been consigned to Chinese ports and a third released from the expedition for the use of the new commissioner, it blamed the weakening of the fleet on the exorbitant costs of coal. "Our correspondent in alluding to this fact states . . . that the cost of coal for a day's steaming of a single ship in those waters is *eight hundred dollars*. It is in this way that patient Uncle Sam is fleeced."[65] This was the logistical situation in the Pacific faced by the American steamers in the 1850s. As Perry was forced to rely on coal supplied by others—whether shipped from America or England or mined by western companies in concessions in Asia—he began planning the development of coal supplies for American ships himself.

These plans involved geological expeditions to identify and, if possible, secure coal fields for American steamers. There were two expeditions of note. The first was to what Americans then called the Lew Chew Islands, now known as the Ryukyu group, home of Okinawa. In January 1854, Perry dispatched his geologist, the Reverend George Jones, to explore reports of coal on the island. Four men accompanied him with other research portfolios: Dr. Daniel Green was tasked with studying disease and agriculture, Dr. Charles Fahs with studying botany, Dr. James Morrow with studying botany and agriculture, and Wilhelm Heine with recording the expedition as its official artist. Two enlisted crew came along as well to manage food and supplies. Adding to the party, some thirty residents of the island also joined them as they began their trek. Their mission, as Jones affirmed in his report to Perry, "was to examine some indications of coal at Shah bay," coal Perry hoped might support American steamers.[66]

The group trekked north from Napha in early February, recording geological, botanical, agricultural, and medical observations. At Farnigi, some fifty-five miles north of Napha, the party encountered their first indications of coal. They found traces of conglomerate, a jumble of diverse rock fragments fused into a single mass. Conglomerate passed into course sandstone, and coarse gave way to fine. For seven miles, sandstone alternated with slate, until the group encountered "some outcroppings of the black bituminous slate," the kind, Jones noted, "usually accompanying coal." Three miles further and they reached Shah Nehatu, or Shah anchorage. There they found an even larger deposit of the bituminous slate. The village of Shah itself sat atop a small island in the bay. As the Americans encircled it in their boat, they observed additional outcrops. Slate, however, even bituminous slate, was not coal, but was often found near it, frequently forming the ceilings of coal mines in America. Nevertheless, Jones acknowledged that the evidence for coal on Lew Chew remained circumstantial: "I wish to guard against too sanguine or certain expectations," he explained in his report to Perry. The slate would not burn and, as Jones noted, "for steam navigation, it would be useless." Even so, the expedition had suggestive evidence that coal might still be found around Shah.[67]

The second and more extensive coal expedition occurred five months later. While visiting the Japanese port of Simoda in June 1854, Perry instructed Captain Joel Abbot of the *Macedonian* to detach from the fleet and sail to Formosa. Perry gave Abbot two objectives: to inquire on the island for American sailors thought shipwrecked nearby and to explore for coal. Perry offered specific instructions. According to the letter of dispatch, Abbot was to ascertain "the productiveness of the mines; the quality of the coal for steaming purposes; its cost per ton of 2,240 pounds at the mines; the convenience and cost of shipping, &c., &c." Geological exploration was again delegated to George Jones, while Abbot was instructed to inquire whether coal might be purchased there directly. If so, Perry sent along the storeship *Supply* to collect as much as 300 tons—but only if it was relatively inexpensive. If the price was high (Perry quoted $20 a ton), Abbot was to buy less.[68]

Jones presented his expedition for coal in Formosa in the style of a travel narrative. As best he could, he obscured the fact that he was not exploring virgin land for coal outcrops but searching for coal mines already in use. He first tried to gather information about coal from the residents of Keelung, the port where the *Macedonian* dropped anchor on Tuesday, July 11. To his frustration, they refused to share any information. "Nearly all that we have learned about the coal in this region has, therefore," Jones noted, "been by pushing and persevering investigations, in the face of constant attempts of the inhabitants to mislead us or to blind us as to the facts." When Jones persisted with his pushing and persevering, he met little additional resistance, though his behavior along the way suggested that the local recalcitrance was a reasonable response.[69]

A few hours after reaching Keelung, Jones shuttled ashore with the ship's purser, two midshipmen, and the master's mate, "determined to commence our explorations before the authorities could suspect our object and throw difficulties in the way." After collecting an interpreter from Amoy, the group was led to a house in town with a large pile of some 10 or 12 tons of coal. The owner offered to sell it to the visitors for a few dollars a ton. The coal, reasoned Jones, was probably mined nearby, and the group "set out on an exploratory walk" to a valley in the east. Residents of Keelung tried discouraging the Americans, insisting they would find nothing. The party, collecting scraps of coal from their path, ignored them and continued on, followed as far as mile out of town, where a handful of coal piles again encouraged the Americans. Only then did their discouragers return home. Alone, Jones and his group continued along a path that cut through another valley heading south. Along the way, "with the help of some country people," they were led further to where Jones triumphantly declared that "to our great pleasure, [we] discovered some mines." With nighttime approaching and lacking lanterns or other tools, the party returned to the *Macedonian*, "gratified with our first day's work."[70]

The next day, Jones, Abbot, and the purser, Richard Allison, returned to shore for the coal that Jones had located at the house in Keelung the day before. This time, however, the owner explained he could not, in fact, sell the coal to the Americans. He did not explain why, but Jones surmised that "the mandarins had interfered," and in his opinion, "the man seemed almost afraid to speak to us." The ruling mandarin, described by Jones as the "hip-toy" Le-chu-ou, met the party, explained the coal *could* in fact be sold, but that it came from an island some one hundred miles away. Confused, and skeptical, the crew returned again to the ship.[71]

The following morning, Jones, the midshipman Williams, and four sailors (armed, Jones noted), returned to the island to explore the mines they had encountered two days before. The party first found three separate entrances to the mine, each about thirty inches wide and four feet tall. A short crawl inside brought the Americans to the coal seam. Horizontal drifts stretching along the length of the seam for what Jones estimated to be about 120 feet indicated the extent of the workings of the mine. The report noted that the miners there evidently used only a sharp pick to remove coal, and carted the dislodged pieces away in baskets. Jones believed this method led to substantial waste, as a certain amount of coal would be pulverized into an unusable dust. Jones, however, "found no difficulty in getting it out in large pieces, of which, as specimens, we brought away as much as we could carry." Further exploration and negotiations continued over the next few days, as Jones tried convincing the hip-toy that "if we can find coal here of a suitable quality for our steamers, it will be greatly to the profit of your country making you rich and prosperous," only to be told again that the coal came from too far away and that he had no control over it (adding that the residents there were cannibals and that he had to steal coal from them anyway). Though Jones employed a narrative of investigation and discovery, he was hardly "discovering" coal in any meaningful sense; he sought to identify mines already in use and plan for their extension to serve needs of Western commercial steamers.[72]

Monday, July 17, brought another expedition. One of the two midshipmen who had accompanied Jones on the first day of exploration, Kidder Randolph

Breese, penned an entry in a friend's journal explaining with both humor and irritation the events of his day. "I started this morning at half past four in the cutter for the famous mines of Formosa," he began, "from which so much benefit to the whole world (some few speculating Amer[ican] merchants) is to be derived, and for which I, poor fellow, was turned out of my ship to incommode some and benefit others." Breese spends most of the rest of the entry complaining about being forced into the coal exploration business and his repeated failures to secure breakfast.[73]

This time, Jones, the reluctant Breese, and purser Allison were joined by two disguised Chinese guides ("who to serve Mammon forgot their master," according to Breese) and an interpreter. When the party, disguised guides and all, set off for the mines Monday morning, they expected substantial "discoveries" and they were not disappointed. The previous day, Jones had been suspicious of Le-chu-ou's claim that the town's coal came from one hundred miles away. With the help of the guides, they found it was closer to a mere three miles. Slipping by a channel in the bay and turning past the "Sphinx head" promontory, the party came upon the coal mines in a location that "was also everything that could be desired." After they learned of the mine, Le-chu-ou permitted them to purchase coal there; they bought about 12 tons at $3 a ton. Another midshipman, John Sproston, who learned of these events from his friend Kidder Breese, noted that the exploration was successful, "a great source of satisfaction to all, as it has placed beyond a doubt the fact of the existence of extensive veins of coal upon the Island, of easy access, and from all appearances of good quality." He added an observation that placed the coal expeditions on Formosa into a larger context of fuel supplies and markets in the Far East. "When we consider," he observed, "that not three hundred miles from this port (Shanghai) coal is selling for $60 a ton, it is truly astonishing that more notice has not been taken of the existence of it here, and [a] depot for the useful article established." Perry would later cite the knowledge gained on this exploration as material support for steam power and for a future American colonization of the region as a means of competing with Great Britain.[74]

Should Americans succeed in securing the coal from any of these mines, Perry also needed to plan for depots to refuel prospective steamers. On this subject, Perry boasted to Washington that with respect to the "bountiful Island of Lew Chew" he could "at any moment secure an entire control over it, without the shedding of a drop of blood." He had already secured land there for a coal shed, deposited a supply of coal, and arranged for an additional shipment. As for the Bonin Islands, Perry purchased land on Port Lloyd for a coal depot to supply ships en route from Hawaii to China. In these efforts, Perry believed he was simply following the policy of the United States; "it is only necessary of the gov't to say the word," he wrote John P. Kennedy at the Navy Department, "and Lew Chew, can be brought quickly under the American flag, or its protection—and the same with respect to the Bonins."[75]

President Pierce approved of Perry's establishment of a coal depot at Port Lloyd (which had no indigenous population of its own) but immediately rejected the prospect of claiming one of the Lew Chew islands for the United States. Instead, he urged Perry to continue negotiations with Japan for access to at least one port at which Americans could rest and coal. While this instruction sailed from Washington to meet Perry, the commodore had, in fact, already accomplished this goal. On March 31, 1854, Perry and three Japanese commissioners signed a treaty, the Convention of Kanagawa, the second article of which ensured that the ports of Simoda (Shimoda) in the south and Hakodadi (Hakodate) in the north would open to American vessels for coal and other supplies. The eighth article established that Americans should purchase coal only though official Japanese government agents. It was but a basis for negotiations to come, but for the first time, Americans believed they had secured access to coal for future steam lines in east Asia.[76]

Perry carried the findings of his coal expeditions, as well as his agreement with Japan to provide coal to American steamers, back to the United States. In a paper read before a packed crowd at the American Geographical and Statistical Society in New York, he observed the prospects for "a flourishing trade" with Japan and "the boundless elements of trade" offered by China. Of all the products of Asia, however, Perry noted the "one mineral that calls for special remark." Coal, Perry observed, had become "the most valuable to commerce of all the minerals since the introduction of steam in aid of navigation." As he had learned while traveling to Japan, the availability of coal prescribed the limits of steam communication. It meant the difference between a successful voyage and being stuck in port. Perry encouraged the members of his American audience by informing them that coal could be found in China, as well as in Japan, Formosa, and Borneo.[77]

Like many of his contemporaries, Perry wrote enthusiastically about America's future in world affairs in the Pacific and beyond. "It may be looked upon as mere speculation in me," he apologized to his audience in New York, "but I have been long a believer in the doctrine of the 'manifest destiny' of this great nation, still in its youth." His youthful nation, no less than a youthful person, was "destined at some *indefinite* time to attain a full and vigorous manhood." But Perry's account of history was cyclical in nature, and like a man, the United States would inevitably weaken with age. Like empires past, it would rise to economic power and then fall. Glory was transient, he concluded, and it was also destiny for the United States, "like all earthly governments, to fall into decadence, to decline in power, and at last, to fall asunder, by the consequences of its own vices and misdoings; thus making room for some new empire now scarcely in embryo." But in 1856, decline seemed a long way off.[78]

In the meantime, by 1860, Americans had established a clear pattern of responses to the opportunities and challenges presented by coal. Steam lines, whether with direct government subsidies or not, could aid American commerce only if they could economically reach their destinations. To protect that commerce, naval steam vessels similarly needed coal. Combustion experiments helped identify which varieties of coal best suited these voyages, while engineering experimentation sought improved designs or alternative mechanisms to reduce steam engines' voracious appetites for fuel. Finally, as steam helped Americans renew their desire for commerce with east Asia, the government employed geological and diplomatic missions to obtain access to foreign ports and coal supplies to create markets for American vessels. Yet notwithstanding Perry's roundly rejected suggestion to seize one of the Lew Chew islands for the United States, few Americans at the time imagined that steam power somehow demanded the acquisition of fortified foreign coaling stations. Instead, the relationship between new technology and American foreign relations was more rooted in existing American approaches to diplomacy: assist Americans in making commercial arrangements whenever possible but otherwise solve problems by employing native technical ingenuity, whether chemical, engineering, or geological. But American responses to coal and steam power were not always related to their uses in commerce and defense. Coal also soon came to be seen as a tool for confronting other challenges, including America's most divisive, slavery.

CHAPTER FOUR

THE SLAVERY SOLUTION

All creation is a mine, and every man, a miner.

Abraham Lincoln, "First Lecture on Discoveries and Inventions"

On August 14, 1862, Abraham Lincoln welcomed a delegation of black representatives to the White House. It was the first time black men had visited the executive mansion not as servants or laborers but as guests. The committee of five were doyens of the capital's black community, educated, politically astute, and socially connected, chosen by members of the city's elite black institutions to receive the president's proposal to resettle volunteers somewhere outside the United States. According to an account of the meeting published by the *New York Tribune*, Lincoln spoke of the suffering experienced by both races and the impossibility of true equality within the boundaries of a single nation. "For the sake of your race you should sacrifice something of your present comfort," he implored, asserting that given the relations between the races, "it is better for us both, therefore, to be separated."[1]

But where to go? Lincoln mentioned Liberia, but equivocated that while "in a certain sense it is a success," it was not ideal as a colony. Liberia lay far from the only land most prospective colonists had ever known, and after forty years of emigration, only some twelve thousand black Americans had ever elected to settle there. Instead, the president spoke vaguely of a place in Central America. This location was closer to the United States than Africa, situated along "a great line of travel," rich with natural resources and endowed with a climate, Lincoln told his guests, "suited to your physical condition." Most importantly, it had coal. Lincoln put great emphasis on this colony's coalfields. "Why I attach so much importance to coal is," the president explained, "it will afford an opportunity to the inhabitants for immediate employment till they get ready to settle permanently in their homes." Agriculture took time; coal mining offered instant work. "Coal land is the best thing I know of with which to commence an enterprise," he argued. Within a month, the government had contracted with Ambrose W. Thompson, a man claiming grants of over one million acres in Chiriquí, the westernmost province of Panama, bordering Costa Rica. While the prospect of colonization remained controversial within the Union's free black community, by the spring of 1863, some fourteen thousand men, women, and children had volunteered to set sail.[2]

Historians have long been interested in Lincoln's efforts to colonize free blacks. Even though colonization failed—the Central American project was ultimately abandoned and another one, to Île à Vache off the coast of Haiti, ended in disaster—it has long been understood as providing insights into the political possibilities and racial anxieties of the mid-nineteenth century. Explanations for Lincoln's own interest have ranged widely. Historians have portrayed the president as a reluctant advocate of colonization, pursuing it only in the heartfelt wish to improve the condition of American blacks. Alternatively, Lincoln has been criticized for using colonization as a kind of psychological defense mechanism, a crutch to help him cope with the challenges of imagining the United States after slavery with a large, free black population. Another interpretation portrays Lincoln's pursuit of colonization as a canny political strategy to thread a needle between abolitionists and conservatives along the road to the Final Emancipation Proclamation.[3]

Most accounts, however, have offered only passing reference to the coal that for most of the life of the project, rested at the center of the plan. Historians have tended to portray the Chiriquí coal mine as an excuse for a colonization scheme, when, at least in the mind of its chief promoter, it was more a colonization scheme pursued as an excuse to open a coal mine. Undoubtedly, during the Civil War, coal and colonization became intertwined. In the mid-nineteenth century, steam power created new demands for coal in places far distant from domestic markets. This demand was a central reason why so many Americans saw Chiriquí and its supposed coal reserves as a valuable property. But steam power also encouraged new ideas about how machines and government policy together could become vehicles for social transformation. For Lincoln, coal and steam offered a technological fix to address the slavery question, peacefully, profitably, and permanently.

As a practical program, however, it was a total failure. In the end, its most fruitful outcome was that it bought time for more northerners to come around to supporting abolition as a military necessity. No colonists ever sailed for Chiriquí and no miners ever extracted commercial quantities of coal there. Failures, however, can be illuminating. The experience of war led Americans to think differently about the security demands of fossil energy. For two decades, Americans had debated how steam power changed the country's engagement with the world and how the naval and commercial need for coal required seeking out new sources of fuel. It was a principal reason Chiriquí attracted Lincoln's attention in the first place. The war revealed, however, how well the United States could supply itself with coal, even at a time of great conflict. The episode also revealed how the demand for coal created new diplomatic constraints for the

United States. Working out the diplomacy of coal showed how steam power made distant lands more dependent on one another and also how countries with desirable fuel or harbors gained new leverage in international relations. The story begins not with the war, however, but the mail steamer debate a decade before.

Development in the Valley of the Moon

In 1850, Ambrose W. Thompson was thinking a great deal about steamships but little yet about slavery, emancipation, or colonization. His efforts over the next decade, however, demonstrate that his colonization project during the Civil War was as much an attempt to industrialize energy as it was a promised solution to racial strife. It is a story of private speculation passing into the international politics of steam power and, for all its ultimate failure, a pivotal episode in the construction of the idea of a national interest in foreign fuel supplies.

Thompson was a familiar figure in Washington. A native of Delaware, he had been raised in Philadelphia by an uncle who introduced him to the world of

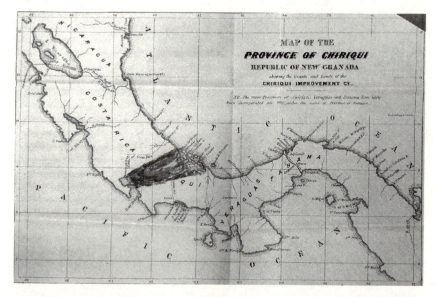

This hand-colored map from the mid-1860s shows the land grants claimed by Ambrose Thomson's Chiriqui Improvement Company. Thompson believed his grants offered coal, commodious harbors, and a shorter transit route across the isthmus compared with that of the Panama Railroad, marked as the dark line further east between Chagres and Panama. Critics assailed him as an unscrupulous schemer. "Map of the Province of Chiriqui, Republic of New Granada, showing the Grants and Lands of the Chiriqui Improvement Co.," box 43, Ambrose W. Thompson Papers, Manuscript Division, Library of Congress, Washington, DC.

commerce. When the uncle retired, Thompson joined a partner and opened a publishing and stationary firm, Hogan and Thompson, before turning his attention to steamships and the infrastructure of world trade. Though he ran coastal steamers out of Philadelphia through the 1850s and also patented his own design for a marine propeller, Thompson focused his energies on obtaining congressional funding for larger, more ambitious projects. Thompson was one of the many memorialists to Congress in the 1850s seeking subsidies to create a mail steam line, first proposing the establishment of one between Philadelphia, Norfolk, and Antwerp and another between San Francisco and China, both of which were quickly endorsed by the government of Pennsylvania. Two years later, he tried leveraging immigrant support to obtain federal funding for a mail steam line between four American ports and Ireland. When no bill passed, he turned his attention to opportunities in Latin America, where he focused on the Isthmus of Chiriquí.[4]

Today, Chiriquí is the southwesternmost province of Panama. Costa Rica borders it to the west, the Pacific Ocean to the south, and to the north and east, the indigenous Comarca of Ngöbe-Buglé and the provinces of Bocas del Toro and Veraguas. In the nineteenth century, however, these borders were fluid. During the second quarter of the century, the region was called Alanje, the westernmost of two cantons of Veraguas. Veraguas was itself a province of New Granada, the state that comprised what is today both Colombia and Panama. Alanje stretched across the isthmus, from Bahia Honda along the Pacific to Chiriquí Lagoon on the Caribbean. Its population was small—roughly twenty thousand in the 1850s, with a little more than four thousand in the capital, David—and its economy revolved around farming and ranching, supplying meat and tallow to regional markets in Panama City and the mining districts of Chocó in western Colombia. The indigenous population was considerable. Along Chiriquí Lagoon lived some fifty families of recently emancipated blacks from the West Indies, along with a motley assortment of foreign traders from England, France, Spain, Italy, Canada, the United States, and elsewhere in Latin America. The modest size and economy of Chiriquí helped make it a marginal region within the already marginal Panama. In a gesture toward granting greater autonomy, in 1849, the Congress of New Granada elevated Alanje to provincial status and in the process, replaced the name "Alanje" itself with the land's indigenous name: Chiriquí, the "Valley of the Moon."[5]

Two events of 1848 brought the Valley of the Moon to newfound prominence for international steam communication. The first event came from the United States. For Americans, interest in Chiriquí emerged as a consequence of the war with Mexico. With the annexation of California in 1848, Americans sought better and faster travel and communication between the Atlantic and Pacific. The discovery of gold in California later that year increased demand for transportation even more quickly than anyone had expected. With time at a premium, investors, schemers, politicians, and military officers fanned out across Central America, boosting chosen routes and disparaging others, investigating elevation, resources, and the diplomatic prospects of roads, railroads, and canals connecting the oceans. The most successful was William Henry Aspinwall's Panama Railroad Company, which Aspinwall founded in 1849 to connect his Pacific Mail Line of steamers with George Law's U.S. Mail Line in the Caribbean. This first North American transcontinental railroad would open in 1855 and become the premier way to travel between the East and West coasts of the United States until a domestic transcontinental track was completed in 1869. But Aspinwall faced intense competition from Cornelius Vanderbilt's Accessory Transit Company, which carried Americans across Nicaragua, and the Hargous Brothers' project to open transit across the isthmus of Tehuantepec in Mexico. Others, like Ambrose Thompson, looked to Chiriquí for still another route. Through the remainder of the nineteenth century, the debate over which route would dominate occupied a central place in commerce and diplomacy with Central America.[6]

The second event was an elaborate attempted land grab by neighboring Costa Rica. While Chiriquí was possessed and administered by New Granada, only weak ties bound many of its residents to Bogotá. Along with the rest of Panama, Chiriqueños periodically pushed for greater autonomy or even complete independence in the hopes that such self-government would make it more prosperous. Costa Rica hoped to exploit this disaffection and capture Chiriquí, its potential isthmian transit route, and its untapped resources. Around 1848, Felipe Molina, the Costa Rican envoy to Paris, produced dubious Spanish colonial documents asserting a Costa Rican title to the region. Aware of the weakness of these claims, Molina quietly schemed to issue a land grant to a French citizen, whom he instructed to bring infrastructure and new immigrants. According to the plan, once foreign capital (backed by a foreign army) reached Chiriquí, New Granada could be mollified with a financial settlement. In exchange, most of Chiriquí, with its valuable harbors on both sides of the isthmus, would remain part of Costa Rica.[7]

Whether such a plan could have worked, events in both Europe and Central America caused it to unravel. Molina's grantee was Gabriel Lafond, with whom he signed contracts in 1849 and 1850. After encouraging French maritime surveys, however, Lafond's efforts to raise capital suffered on account of the Crimean War. During the delay, officials in New Granada learned of the project and protested vigorously. In a prospective settlement in 1856, the two countries agreed to leave Chiriquí with New Granada, but rumors swirled around the region that the deal contained a secret provision for New Granada to sell the entirety of Panama to Costa Rica in exchange for Costa Rica assuming a portion of New

Granada's considerable debts to various European creditors (along with granting lucrative emoluments to New Granada's diplomatic representative, Pedro Herrán). Only a resurgent nationalist sentiment in New Granada following the anti-American "Watermelon Riot" in April foreclosed the possibility of the country's parting with the isthmus for at least the near future. This bloody riot, which had begun as an altercation between an intoxicated American traveler and a watermelon seller in Panama City, revealed the pent-up Panamanian resentments caused by increasing numbers of Americans crossing Panama en route to California. For a time, these resentments toward those who appeared to meddle inside Panama also meant Colombian leaders more jealously guarded the isthmus. As a result, Lafond's scheme collapsed for good. Still, regardless of which country eventually controlled the region, the leaders of both New Granada and Costa Rica knew that the development of Chiriquí would require capital and expertise from abroad. The only question was what the terms would be.[8]

Before its failure, Lafond's company helped publicize tantalizing discoveries of coal in the region. Writing of Chiriquí's northern Caribbean coast, a naval captain and senator from the nearby province of Veraguas observed that "small pieces [of coal] have been discovered in a river which falls into the Bay de l'Admiral, probably having been carried down by the current." Two French captains who explored the region on behalf of Lafond for six months in 1851 learned of newly discovered coal beds reported in Costa Rican journals. While they handled "some specimens of carbonized or anthracite wood," they refrained from confirming the existence of large deposits. Soon, however, news spread among Lafond and his associates of a coal discovery along the Pacific side of the isthmus, near the Costa Rican town of Terraba. The French admiral Odet Pellion, who reached Golfo Dulce in June 1852, investigated these discoveries, and while he lacked adequate time and resources to explore, he collected samples for his ship's surgeon to examine. Pellion, though hopeful, could only assure Lafond that existence of coal was certain. Its quality, he noted, would only be known after further exploration and experiment.[9]

Among those interested in developing Chiriquí were the people who called it home. Like other local and national governments across Latin America, the provincial one in Chiriquí itself sought to harness American, French, or British interest in new communication routes for its own purposes. A new route promised a flood of travelers, all of whom would need food, lodging, and supplies. Integration into global trade networks meant the prospect of attracting immigrants, and immigrants could further develop regional resources—from gold to gutta percha, gum elastic to tropical lumber, coconuts to coal. As a particularly underdeveloped, undervalued region, Chiriquí was among the most eager to encourage this sort of investment, especially as transit across nearby Panama, even before the completion of Aspinwall's railroad, was beginning to revitalize the long-underdeveloped province to the east.[10]

At least as early as 1841, Chiriquí began improving its infrastructure by initiating the reconstruction of an ancient, pre-Columbian Indian road across the isthmus. Six years later, the government in Bogotá indicated its interest by contracting to further reconstruct the road. After Chiriquí gained provincial status in 1849, the road reverted to local jurisdiction. The provincial legislature in David immediately began issuing road construction grants, first to a Pennsylvanian speculator named Theodore Moore, whose incivility and failure to actually build the road incurred the ire of both residents of Chiriquí and the government in Bogotá. The Panama Railroad, perceiving a threat to its transit monopoly, secured the annulment of Moore's grant in the supreme court of New Granada. In the meantime, Ambrose Thompson had purchased rights from Moore, and between 1851 and 1852, he secured three revisions to it from the legislature of Chiriquí.[11] What followed was a decade of tension between David and Bogotá over who would control isthmian transit, with the national government constantly pressured by lobbying from Aspinwall's Panama Railroad to prevent the emergence of competitors. The stakes remained high. The Panama Railroad believed its contract ensured a complete monopoly in lands under the control of New Granada. The government of Chiriquí, however, was not inclined to honor this agreement. "The Governor told me if I could place 4,000 foreigners armed at his disposal," reported one of Thompson's agents in David, "he would give all we asked in defiance of the Executive at Bogotá." No armed foreigners came to the governor's defense, but it was a possibility everyone across the country took seriously.[12]

After securing his initial road grants from Chiriquí, Thompson continued to collect others. Under the constitutional reorganization of New Granada of 1853, Thompson obtained yet another road grant from Chiriquí in 1854. That same year he wrangled title to a grant for lands along the Pacific coast intended for colonization. In 1855, he secured control of a series of coal mines around Chiriquí Lagoon. By mid-decade, he had placed all the grants in the portfolio of his newly chartered Chiriqui Improvement Company.[13]

At least as early as 1852, Thompson and his associates had concluded that coal would be one of Chiriquí's chief assets in what was then an unambiguously speculative venture. Thompson could secure harbor islands suspected of bearing coal for a mere $25 a year, islands that might soon be worth millions once Thompson opened them to regional markets with a new road. "If you wish to obtain all these immense advantages which in a year or so must certainly remunerate you a thousand fold," wrote an agent in David on the coal discoveries, "now is your time." By September 1853, Thompson was boosting Chiriquí coal to J. C. Dobbin, the secretary of the navy in Washington, and asserting the need for an American naval depot in Chiriquí Lagoon. "In the lands bordering on this Lagoon," he wrote, "large, and it is believed inexhaustible coal fields exist, both anthracite and bituminous," adding that similar advantages graced the Pacific side of the isthmus as well. Yet the ambiguous legal status of coal mines created yet another local-national tension between Chiriquí and Bogotá. Metal and precious stone mines belonged to the national government, quarries to local landowners—under which category did coal fall? To obviate the problem, Thompson also began promoting his coal operation to the national government in Bogotá by suggesting it would help displace the local Indian population with more desired new immigrants while also settling the boundary dispute with Costa Rica. The strategic value of coal in Chiriquí was not a simple geological or geopolitical fact but rather an argument that Thompson used to lobby the United States and New Granada to promote his speculative investment.[14]

Over the next several years, Thompson collected geological and engineering reports that endorsed his concessions. His correspondence and promotional documents were marked by the conspicuous inclusion of surveys by civil engineers, results of geological and chemical analysis, and testimony by professional scientists. A contracted civil engineer, James Cook, surveyed the prospective road for the company and reported on the province's "very abundant and rich" coal deposits. Led by Indian guides, Cook spent a month conducting geological examinations, offering sanguine analyses of Changuinola coal from the Pacific and Muerta coal from Bocas del Toro. In 1856, the company dispatched Newton Manross, a Yale graduate who had recently completed his doctoral studies at the University of Göttingen, to map the Chiriquí coal region. An Englishman, C. S. Richardson, examined Manross's specimens and announced that "practical" engineers like himself would be "unanimous in its favor." Armed also with the endorsements of the legislatures in both the province of Chiriquí and the state of Panama, Thompson turned his focus to lobbying Washington in mid-1857.[15]

As early as August, Thompson approached the new Buchanan administration's postmaster general, Aaron V. Brown, proposing to launch a mail steamer service through Chiriquí. Three months later, the navy sent the *Fulton* to examine Thompson's coal claims. The result was a favorable endorsement by the lieutenant in charge, John Almy. Even though he could only collect degraded surface specimens, Almy lauded the discovery, tying the development of a naval station there to a U.S. "ascendancy in the Central American States" and noting that with American engineering, "there will be found an abundance of coal of a superior quality, the value of which in that part of the world, under circumstances which may often arise, is beyond any calculation." The ship's engineer, James W. King, called the coal "of a superior order for ocean steam purposes." Yet the distance between dispassionate endorsement and biased promotion remained hard to measure; Almy was guided by one of Thompson's agents, and he based many of his conclusions not on the coal collected but an assumption (shared by Thompson) that higher quality coal lay deeper beneath the surface.[16]

Over the following year, Thompson continued his private lobbying efforts. In addition to promoting the value of transit rights across the isthmus, naval stations, and strategically located coal mines, he emphasized the international pressures bearing on the United States. The United States, New Granada, and Costa Rica were not alone in their interest in isthmian routes, Thompson reminded the government, for England was involved in Honduras and France in Nicaragua, and both Mexico and New Granada were soliciting Europe for capital and possibly offering control of transisthmian routes. Thompson waved a *New York Herald* editorial fretting that without government involvement, "we shall soon find ourselves driven from the American Isthmus, our communication with our Pacific empire cut off, and our prestige as a nation gone." It was a sentiment Thompson sought to cultivate in the administration. To President Buchanan, he emphasized that his Chiriquí route offered what "the United States *cannot secure at any other point in the Gulf of Mexico or Caribbean sea.*" This prize was vast stores of "inexhaustible beds of coal" to power American steamships; a single island, Thompson claimed, contained "a quantity of coal sufficient to supply the United States navy for centuries to come."[17]

Thompson needed an naval contract because he lacked the capital to fulfill the terms of any of his grants on his own. Despite his representations to Chiriquí and Bogotá, without hundreds of thousands of dollars from the United States, his projects would be worthless—he could build no road, establish no colonies, open no coal mines. This desperation for capital led Thompson to promise increasingly impossible terms to the various cabinet members to whom he pitched the project in the hopes that a clear commitment from the United States would persuade the government in Bogotá to consent later. What began as a conventional petition for a mail steam line quickly became an offer to sell coal and perpetual rights for a naval station. Soon after came promises of mining rights and the free, unlimited transit of government agents across the isthmus. He dangled the prospect of annexing New Granada, whose government, he claimed, itself desired it. When that country went on to reject key provisions of the Cass-Herrán treaty that created a process for settling claims created from the Watermelon Riot (including the seventh article, permitting the United States to establish a coaling depot in Panama), Thompson concluded that it only made his concession more vital to American policy. To Lewis Cass, the secretary of state, Thompson acknowledged that his project had begun "from motives of personal gain," but he insisted that now it had become clear it had national importance and that it "*must* soon be ripened into active policy"—all while warning of his imminent "pecuniary ruin" if the administration did not act soon. Yet despite verbal assurances, for two years, the administration took no definitive action.[18]

Thompson did what he could to speed things along. Like many seekers of government largess in the nineteenth century, he cultivated powerful "friends"— officials whom he supplied with cash or company shares to ensure their support when votes or endorsements were needed. In Washington, he courted the printer of the U.S. Senate, William A. Harris, a figure whose frequent access to lawmakers, like that of the Capitol's doorkeepers and postmasters, made him a regular targets of lobbyists.[19] One of Thompson's agents was Reverdy Johnson, the former attorney general, fresh from successfully defending John Sanford from his slave Dred Scott in the Supreme Court. Another was Richard W. Thompson (no relation to Ambrose), a former Indiana representative and lawyer to the Chiriqui Improvement Company, who joined Johnson in offering shares in the scheme to members of Buchanan's cabinet.[20] Francisco Párraga, Thompson's agent in Bogotá, informed Thompson that he "must prepare some friends" and asked for instructions on "how to dispose of money, should it be necessary." To Amalia Herrán, wife of New Granada's envoy extraordinary and minister plenipotentiary to the United States and daughter of Tomás Cipriano de Mosquera, a former (and future) president, he promised $50,000 of company stock, as well as prime land inside the two port cities Thompson anticipated founding on either side of the isthmus. "Mosquera" would rise on the shores of Chiriquí Lagoon and "Herrán" on Golfito Bay. Two weeks later, Amalia's husband, Pedro Herrán, offered a full-throated official endorsement to the United States of the authenticity and legality of Thompson's grants.[21]

These efforts paid off. After months of lobbying, Herrán's endorsement as the official representative of New Granada overcame the skepticism of Buchanan's attorney general, Jeremiah Black, who had questioned the legal status of the company as well as Thompson's extravagant terms, terms that appeared to violate the sovereignty of New Granada. In the process, the project's purpose shifted from securing a steam mail contract with the post office into a project with the navy for provisioning coal in the Caribbean. On May 21, 1859, without fanfare, Thompson signed a contract with the secretary of the navy, Isaac Toucey.[22]

Unfortunately for Thompson, Herrán's support proved short lived. At the end of December, the Buchanan administration released the annual reports of its cabinet departments, and from Secretary Toucey's report the press learned about the coal contract from the previous May. It was only then that Herrán realized the scope of the agreement. Its terms promised an unimpeded right of way across the isthmus for U.S. agents (that is, soldiers), land for naval stations, harbor use by government vessels, and unlimited access to coal at only ten cents per ton, paid to the Chiriqui Improvement Company. From the perspective of

New Granada, Thompson's grants had begun as an effort to bring needed infrastructure and capital to Chiriquí; they now appeared to have become a cession of sovereignty to the United States. An indignant Herrán dispatched the news to Bogotá, urging his government to reject the contract.[23]

While Thompson lobbied Washington, in New Granada, his agent Francisco Párraga had been involved in a simultaneous lobbying effort of his own. During the roughly two months it took for Herrán's letter to reach Bogotá, Párraga struggled to renegotiate the terms of Thompson's railroad grant to clarify its legal status. His work had started smoothly, with President Mariano Ospina Rodríguez and his cabinet fully endorsing it, even over the opposition of the Panama Railroad. Clear interest from the United States helped secure the administration's support, as did coordinated lobbying by George W. Jones, the new American minister to New Granada. Together, Párraga and Jones flattered Ospina's government by suggesting that approving the railroad grant would induce the imminent inflow of capital from Europe and the United States while securing the development of new harbors, coal mines, and a route from the Atlantic to the Pacific. All of these improvements were to benefit the country. But support from the executive did not translate into similar support from the legislature, whose approval was needed to authorize the president to contract directly with Thompson. In New Granada's senate, composed overwhelmingly by administration supporters, opposition came from friends of the Panama Railroad, which sought to preserve the company's monopoly over isthmian transit (a point underscored by the arrival of a representative of the company in Bogotá with $500,000, fine liquor, and a French cook to secure friends of his own). Despite an impassioned speech by the secretary of interior and war, Manuel Antonio Sanclemente (himself a future president), a senator from Panama, Rafael Nuñez, argued that the grant would simultaneously violate the transit monopoly promised to the Panama Railroad while also increasing the region's vulnerability to filibusters or even the cession of control of the isthmus to the United States. Nuñez also revealed that four years earlier, the country's supreme court had annulled the original grant from Chiriquí, a detail that somehow took everyone by surprise. Support further eroded with widespread rumors about Thompson—that he was a bankrupt speculator, that he and Párraga were really secret agents of the Panama Railroad engaged in a convoluted bid to further empower the company, that Párraga was an American spy seeking to wrest control of Panama for the United States. Combined with the arguments of another senator who had first negotiated the Panama Railroad contract in 1850 and now regretted it, support for the project collapsed. To the shock of Párraga, Jones, and the Ospina government itself, at the end of February the senate of New Granada rejected the proposed revisions to Thompson's grants unanimously.[24]

In the new year, the situation deteriorated even further. On February 28, 1860, Herrán's letter about Secretary Toucey's annual report from the Navy Department reached New Granada and members of the government learned in disbelief of Thompson's contract with the U.S. government. What had been a moment of political triumph in Washington turned into catastrophe in Bogotá. Despite the efforts of Párraga and Jones to renegotiate the legally suspect terms of Thompson's original grants, it became impossible to persuade President Ospina or his cabinet that Thompson had not illegally sold his privileges to the United States. Párraga reported that the contract "has been the *death blow* to all our projects." For nearly a year, he himself had lobbied without realizing that the contract Thompson and Toucey signed in May specified anything more than selling Chiriquí coal to the navy. "The President, the Secretaries, and *all* the members of Congress were highly indignant at the receipt of the news," wrote Párraga in dismay. "The contract has been discussed and canvassed every where, by the press, by the Executive Power, by the members of Congress in private meetings, and has been universally condemned, as an imprudent assumption of your Government, and of yourself." Instead of continuing to pressure the New Granada's senate to approve negotiating a new grant, the secretary of state requested that the supreme court annul the existing one.[25]

After the disastrous end of Párraga's negotiations in Bogotá, Thompson decided to pursue a fallback arrangement with Costa Rica. There, at least, he appeared to be in a position to salvage his project. His agent there, Thomas Francis Meagher, lobbied the new, revolutionary government in early 1860, winning the support of leading merchants and the new president, José María Montealegre Fernandez. Like the deposed government, the new one also saw the scheme as a way to claim territory in Chiriquí, as well as a means to attract foreign capital for local development. When the contract came before the legislature in San José on July 4, 1860, only two votes dissented in Costa Rica's house of delegates while its senate approved it unanimously.[26] Here too, however, success proved short lived. Without an appropriation from the U.S. government, Thompson was unable to supply the required $100,000 deposit to Costa Rica to execute the grant. When the new government declined to allow him more time, he appeared to have forfeited the last conceivable path for Latin American support.[27]

Despite the collapse of support from New Granada—to which the American minister, George Jones, alerted the administration—Buchanan's cabinet continued to pursue its coal and naval station in Chiriquí. Faced with the court decision in Bogotá, Toucey insisted to the Senate's Committee on Naval Affairs in June that the project retained his confidence and that he was actually "more strongly impressed with the importance of the measure and would earnestly urge that the necessary appropriations be made for its accomplishment." Members of

Congress were more cautious, when, less than a week later, the Senate proposed a $300,000 appropriation contingent on the president confirming the legality of Thompson's privileges and his rights to transfer them. A conference committee between the two houses quickly agreed instead to appropriate just $10,000 to fund an expedition to the region and have "some competent person or persons" investigate its coal, harbors, and prospective railroad route.[28]

In response, Buchanan selected Captain Frederick Engle of the navy to lead the survey, which left Norfolk on August 13 and arrived in Chiriquí ten days later. In addition to Ambrose Thompson's son (also named Ambrose), American assistants, local guides, and Indian laborers, Engle was accompanied by three technical experts: an engineer, first lieutenant James St. Clair Morton, who surveyed the railroad route; a hydrographer, Lieutenant William Nicholson Jeffers, who sounded the harbors; and a geologist, Dr. John Evans, who examined the region's coal deposits. Evans was an important selection; as the Smithsonian's official geologist for the Washington and Oregon territories, he had over a decade of experience in the Pacific Northwest as well as Iowa, Wisconsin, Minnesota, and Nebraska. His exploration around Chiriquí filled him with enthusiasm. "The coal is much better than I had anticipated," he recorded, collecting coal samples in a small canoe. Further, he noted that "the coal deposit is accessible to navigation and inexhaustible"; he also confirmed the presence of eight seams along the Changuinola River, "much of which is of excellent quality." Evans composed a glowing report, combining his field survey with a favorable laboratory analysis performed by Boston chemist Charles T. Jackson. Jackson praised the coal as "well suited for steam navigation," while Evans maintained that "this country offers a wide field for American enterprise, and is well worthy of the patronage of the government." The whole expedition left Evans so enamored with Chiriquí that he decided to collect his family and retire there. Unfortunately, he died of pneumonia just weeks after returning to the United States.[29]

The other technical surveys were also largely positive, but interested American newspapers reported rumors as facts. One account of the railroad survey had Lieutenant Morton discovering that available paths were too long and required cutting through too much elevation. Another report, however, from a commission member just returned to Norfolk, claimed that Morton *had* located a feasible path, that Chiriquí's harbors were "unequaled," and that coal was "discovered of superior quality and in inexhaustible quantities." The Panama Railroad, which had long denied Thompson's and Chiriquí's claims about roads and resources, expressed surprise at the findings. Meanwhile, Lieutenant Jeffers's hydrographic survey concluded that "no finer harbors than these can be found" and that since an earlier, published Royal Navy survey of 1838 had very precisely measured the hydrography of Chiriquí Lagoon, much of the hard work was already done. According to Thompson's hometown *Philadelphia Enquirer*, the report of the Chiriquí expedition was "exceedingly cheering," and it looked forward to the day when "American steam vessels will . . . soon be found in Chinese ports, and the trade with Asia may be diverted to San Francisco."[30]

As 1861 began, the Buchanan administration continued to back the project, but the encouraging scientific and engineering studies again met resistance in Congress. Thompson pleaded with the president that New Granada simply misunderstood the modest terms of his contract with the United States and that the actions of the government in Bogotá were really (and preposterously) "intended as a secret insult to the Government of the United States made by an attack upon the private vested rights of one of its citizens." The remedy, he explained, was to return fire with a diplomatic threat. Down Pennsylvania Avenue, the Senate was initially willing to appropriate Thompson his $300,000, approving the measure on January 17 by a wide thirty-eight to eight margin, with just a handful of skeptics dissenting, mostly Republicans. In the House, however, support fell apart.[31]

There, opponents painted Thompson's project as a corrupt scheme to defraud the government with grants that he lacked authority to transfer and that the local Chiriquí government alone could not make, all for coal and railroad routes that were not there. Maine Republican Freeman Morse, chairman of the Committee on Naval Affairs and an original supporter of the measure, tried amending the Senate appropriation to disavow any pretentions to sovereignty in either New Granada or Costa Rica, prevent the movement of troops or war materiel without the approval of either government, and require the administration to draft procedures for managing the contract so as to respect the public interest. Yet he found himself outmaneuvered on the House floor; opponents of the measure managed to seize most of the speaking time, during which they delivered ponderous speeches and prevented just about everyone from the Naval Affairs Committee that sponsored the appropriation from explaining why it deserved support. Yet for the most part, arguments for and against the appropriation served as proxies for views on the Panama Railroad and its current monopoly over isthmian transit in New Granada. The company—"a monopoly which has been the curse and bane of my State," according to one Californian representative—was believed to be lobbying vigorously against Chiriquí in Washington. Thaddeus Stevens of coal-rich Pennsylvania launched into a scathing critique of the project, witty and sarcastic, eliciting bouts of laughter from the chamber and making the appropriation toxic through ridicule. He articulated the usual criticism of Thompson's contracts but gave special attention to the coal, insisting that if it were really as abundant and valuable as its supporters claimed, American entrepreneurs would surely bring it to market and make it available to the government and everyone else at prices far lower than $300,000. In David, Bogotá, and San José, it was clear that the legality of the grants was just one political tool that could be used as needed to pursue the interests of party, province, or economic power; in the U.S. House, however, the legalistic arguments won the day, and the measure was defeated by a three-to-one margin.[32]

By February, the entire project seemed ruined. "I am in mourning for Chiriquí," wrote Párraga in February, "and have given up all hopes." Yet more bad news was still to come. In April, Thompson learned that his new (and already failed) contract with Costa Rica so infuriated Bogotá that the government there would soon seek the annulment of all his previous grants, not simply the railroad one. His failure over more than six years to bring development to the Valley of the Moon provided ample legal justifications for revoking his privileges to colonize the territory or mine its coal. On account of Thompson's serial secret negotiations, Párraga even speculated that "Costa Rica will be involved in troubles with New Granada."[33]

On March 4, 1861, a new administration came to office in Washington. "What is to be done next?" Párraga had asked just four days earlier. "What is the next move?" By April, the inability to form a lasting compromise around slavery had led eleven states to secede and the Civil War had begun. It was then that Ambrose Thompson began to see what his next move ought to be.[34]

A War of Industrial Power

Just eighteen days after arriving at Hatteras Inlet and beginning to blockade Pamlico Sound and the North Carolina coast, Union commander Stephen C. Rowan struggled to find coal. The Union navy had established the Atlantic blockade to suffocate the Confederacy by depriving its economy of foreign trade, especially with Britain. But blockade vessels, among them Rowan's USS *Pawnee*, depended on steam power, and supplies of coal were difficult to obtain when operating outside of northern ports. A failure to obtain coal would ruin the effort, a fact with which Rowan struggled daily. "I have already informed you that I wanted coal," Rowan wrote to the commander of the Atlantic Blockading Squadron in late September 1861. "I have now to state that unless I receive coal within the next ten days we shall not be able to move even the little tug *Fanny*." He requested an immediate shipment. Rowan would take any amount of coal he could get, of course, but he hoped the department would deliver regular cargoes and keep the ships of his blockade in action. Seeking to ensure that Rowan and hundreds of other naval commanders avoided the crippling paralysis that coal shortages would bring, Union officers and officials spent the first year and a half of the war cobbling together a system for fueling the first industrialized conflict in American history. Unfortunately for northern war efforts, however, the Union navy had entered the war without a mechanism for distributing fuel to its growing number of steamships.[35]

The navy's coal agents, established only in 1851, had been discontinued in 1859 after a political scandal that appeared to touch the executive mansion. As part of the same investigation that exposed the political perils of engine design, John Sherman's committee pursuing corruption in the Navy Department revealed that navy secretary Isaac Toucey had appointed an anthracite agent who knew nothing about coal but a great deal about sinecures. The man was Charles Hunter, a physician from Reading, Pennsylvania, who had campaigned for Buchanan and looked to the administration for a share of the spoils. Guided by a Pennsylvania representative who boosted him for the position, Dr. Hunter agreed to divide his commissions with two other local Democrats. Though the post of anthracite agent had been designed to leverage expert knowledge of the coal industry, Secretary Toucey deferred to President Buchanan, a native son of Pennsylvania who evidently understood the commission fees were to be shared, in appointing Hunter. The navy never complained about the quality of coal it received under Hunter's agency, but the entire requisition process was performed not by Hunter but by a single Philadelphian coal dealer, Tyler, Stone, and Co., with falsified papers and at rates Sherman's investigation revealed were higher than the market justified. The department was further embarrassed when Sherman revealed that a member of the firm was Toucey's nephew by marriage. For his part, Hunter had never left Reading to inspect any coal in Philadelphia. As newspapers broadcast the story, not only Hunter but the coal agency itself was tarnished; referring to "the gift of the Anthracite Coal Agency," the *New York Tribune* called it (with some exaggeration) "one of the richest within the power of the President." After the embarrassing exposure of incompetence and corruption, Toucey abolished the agencies altogether and began advertising bids for fuel to dealers directly. If Tyler, Stone, and Co. had no difficulty in supplying the navy with quality coal, why bother with a middleman?[36]

The return to fueling the navy with contracts to the lowest bidder proved adequate during peacetime but became an obstacle once the war began. What had been modest bids for 15,000 tons of anthracite in 1859 became requests for thirty to eighty thousand and finally regular orders for 100,000 tons by 1863. When the war began, no effective system yet existed to distribute fuel in such amounts from its principal markets to the vessels that so urgently needed it. As early as August 1861, Union naval commanders alerted Washington that if the navy expected to maintain a steam fleet in the Caribbean, it needed to send anthracite to strategically located ports like Kingston in British Jamaica or Danish St. Thomas. This need was especially acute for vessels pursuing Confederate raiders like the *Sumter* and *Jeff Davis*. As both Confederate and Union ships cruised the Caribbean on unpredictable schedules, they were forced to rely on whatever kinds of coal might be available in local markets, which were usually bituminous. These coals, "the smoke from which can be seen even in the darkness of night," according to Admiral David Dixon Porter, proved a liability when ships sought the element of surprise. What was more, these soft coals were typically expensive and performed poorly in many American boilers designed for anthracite.[37]

The scope of these challenges was only a matter of speculation when Ambrose Thompson first met the new president, Abraham Lincoln, on April 10, 1861. In their meeting, Thompson does not appear to have mentioned Chiriquí, his failed contract with Buchanan's Navy Department, or colonization. He simply wanted the president's ear. Instead of discussing his speculative schemes, he offered opinions on how Lincoln's administration could peacefully address the crisis of secession.[38]

War in the United States gave Thompson a glimmer of hope for salvaging his Chiriquí scheme, as did war in New Granada, which had broken out in 1860. After his company's humiliating collapse in political support in Bogotá in early 1860, it had appeared that regaining the confidence of New Granada's government would be impossible. Thompson's agent, Francisco Párraga, left the capital and was reduced to working as a clerk in the small New York office of a London life insurance agency. But then the political situation changed dramatically. In May 1860, Tomás Cipriano de Mosquera launched a Liberal revolution against the Conservative Ospina government. By July 1861, his forces would seize Bogotá. Mosquera had long spoken favorably of the Chiriquí project, but Lincoln's administration was not yet prepared to recognize his revolutionary government. Meanwhile, between February 21 and March 31, the Panamanian provinces of Veraguas and Chiriquí had declared the independence of Panama from New Granada, with residents at David claiming (on behalf of the whole isthmus) that since joining New Granada forty years before, the capital had done "nothing but impoverish and divide it into political parties." Taken together, Thompson had reason to hope for still another opportunity.[39]

Over the following months, news of Chiriquí eventually made its way back to the Navy Department. Gideon Welles, the incoming secretary, may have learned of Chiriquí from New York's Charles B. Sedgwick, the new chairman of the House's Committee on Naval Affairs. In 1860, Sedgwick had penned a minority report to his committee's evaluation of the Thompson-Toucey contract, enumerating a list of objections, both legal and geological, and had emerged in the House as one of Thompson's harshest critics. Barely a year later, in August 1861, that criticism had turned into praise. "I know of no other coal deposits on the Atlantic Coast south of the Potomac," Sedgwick wrote to Welles, "& I consider it of the last importance that a supply of coal depots should be secured from for the use of our Navy in the Gulf & on the Pacific." His opinion had shifted, he claimed, after additional grants from New Granada eliminated questions about Thompson's titles, while the government's geological expedition in 1860 convinced him of the character and availability of the province's coals. He made no mention of New Granada's supreme court voiding Thompson's road grant.[40]

Thompson had likely coordinated Sedgwick's appeal to Welles. Only a day after the representative dispatched his letter, Thompson sent his own note to the navy, proposing new terms to lease Chiriquí to the government. Thompson promised Welles that his company could supply as much coal as the navy desired in Chiriquí Lagoon for half the cost paid at any point over the preceding decade. With a railroad across the isthmus in the works, Thompson guaranteed coal delivery on the Pacific at similar prices. A company memorandum detailed how opening Chiriquí mines for the Union would fuel the blockade and save over $1.3 million a year in coal and transportation costs, not to mention provide a permanent naval station. When Thompson composed a draft indenture with the Navy Department, he only detailed the terms for supplying coal. With this first formal broaching of the subject, Thompson presented the advantages of Chiriquí just as he had to the Buchanan administration, namely, as a matter of fueling American power on either side of the isthmus, only now he also adverted to the imperatives of wartime demand.[41]

The matter quickly reached Lincoln, who referred it to his brother-in-law and confidant, Ninian Edwards. Edwards read approvingly the reports of the 1860 Chiriquí Surveying Expedition, noting the government affiliation of its leaders and the respectability of Charles Jackson, the Boston geologist and chemist who had examined the collected coal samples. Published statements from the government of New Granada assured him of the legality of Thompson's grant. Most importantly, Edwards described the "vast saving" to be expected from Chiriquí coal. The coal itself would be cheaper than what was then available in the Caribbean while the navy would no longer need expensive supply ships floating off the ports of Aspinwall and Panama. Edwards believed the government stood to save half a million dollars a year with depots on shore alone. Furthermore, Chiriquí offered a new source of valuable timber for naval shipbuilding. The vast harbors of the Chiriquí Lagoon promised unprecedented strategic advantages in the Caribbean as well, where it "might save whole squadrons." And during wartime, by then no mere hypothetical event, Edwards looked to the prospects of fueling the Union navy—with savings, he suggested, in the neighborhood of nearly $2 million a year. The only problem for Edwards was that Thompson had proposed a series of separate contracts, one for each element of the deal—one for supplying coal, another for leasing the land for a naval station—thus saddling the government with leases it would not need if, unexpectedly, the coal later proved inferior. Since the "main object" of the plan was fueling the navy,

Edwards suggested to his brother-in-law that a single contract be drafted that would ensure that receiving quality coal remained the key stipulation, without which the government guaranteed nothing. Almost as an afterthought, Edwards noted that the lands might also be employed for colonization.[42]

Edwards's mention of colonization brought up a subject that Lincoln himself had long taken an interest in but never taken an active role in furthering along. Unlike his attorney general, Edward Bates, Lincoln had never been a vice president of the American Colonization Society. Nor had he been an outspoken supporter of colonization like Republican party elder Francis P. Blair or either of his two sons, Frank, a representative from Missouri, and Montgomery, the postmaster general. Yet the idea was central to Lincoln's thinking on race and politics. During the 1850s, Lincoln had viewed colonization as the ideal solution to the nation's racial problems but had been unable to imagine a practical manner of carrying it out. "If all earthly power were given me," he noted in an 1854 speech in Peoria, "I should not know what to do as to the existing institution." Immediate emancipation followed by a mass exodus to Liberia seemed the obvious solution, he explained. "But a moment's reflection would convince me that whatever of high hope (as I think there is) there may be in this, in the long run, its sudden execution is impossible." Lincoln thought too practically to be enticed by a utopian vision. Practical details mattered. How would the colonists make a living in their new home? Who could provide sufficient ships to move them or capital to pay for the project? With such questions unanswered, the colonization vision remained just that—only a vision—until Lincoln was in a political position to try to make it a reality.[43]

Some time between August and October 1861, Lincoln began to see Thompson's coal contract as an opportunity to attempt a colonization project. The idea had already been floating around the cabinet. As early as June, his postmaster general, Montgomery Blair, had contacted the Mexican chargé d'affaires about colonizing American blacks in the Yucatán and Tehuantepec, but the Chiriquí project appeared to offer far more advantages. Following Edwards's endorsement, the president referred the matter to Welles in the Navy Department, instructing Welles to support the plan and to pay Thompson's company $50,000 to begin mining and colonization. An 1819 law already required the government to return Africans captured in the illegal slave trade to Africa, and the president reasoned that sending them to Central America would incur far less expense and require only minimal legislation from Congress. The Chiriquí mining contract would permit a pioneering settlement of free blacks from the United States to "be the introduction to this," in Lincoln's words. Once there, colonists would open new sources of coal for the Navy, War, and Post Office departments. But Welles was cool on the project from the beginning, later recalling that his inquiry into the Chiriquí plan convinced him "that it was a speculating, if not a swindling scheme" about which he warned the still-enthusiastic president. He dissembled to Lincoln that his department was restricted by law to purchase coal on yearly contracts from the lowest bidder. Only after learning of this supposed restriction did Lincoln reluctantly rescind his order.[44]

Lincoln then turned to his secretary of the interior, Caleb Smith. After nearly two weeks of study, Smith returned with a very different opinion from Welles. He ridiculed Welles's claim that an 1843 law requiring competitive bidding for "all provisions and clothing, hemp, and other materials of every name and nature" should be interpreted to include coal. Fuel was too important, Smith argued, to be subject to such a binding constraint and classified with mundane supplies. "It can hardly be supposed," he asserted, "that Congress designed to prevent the Navy Department from purchasing such supplies of fuel as might be needed at remote Stations thousands of miles from our own coast except upon contracts made after advertising." In addition, the navy's Appropriation Act of 1850 explicitly gave the secretary special authority over coal, granting him the "power to discriminate and purchase, in such manner as he may deem proper, that kind of fuel which is best adapted to the purpose for which it is to be used." Finally, and most concretely, the navy's Appropriation Act of 1845 explicitly excused "ordnance, gunpowder, medicines, or the supplies which it may be necessary to purchase out of the United States" from the requirements of competitive bidding. Lest the president mistake the aim of his legal argument, Smith explained he was "strongly impressed" with the Thompson contract and the coal and harbors Chiriquí had to offer the United States. Smith also professed his concern that the navy might forsake a rare opportunity to secure the coal before another nation claimed it for itself.[45]

For Lincoln confidant and party elder Francis Blair, the stakes were even higher. To him, Central America was destined to become the India of the United States. In India, according to Blair, a private company supported by a rising geopolitical power consolidated its rule over a fractured polity. In consequence, Britain acquired a vast empire, political stability on the subcontinent, countervailing power against European rivals, and, of course, commercial opportunities. Thompson's offer of Chiriquí gave the administration a home for emigrating blacks, a reliable path across the isthmus, and a buffer against the incursion of European influence on the continent. He insisted that "Chiriqui may be made the pivot on which to rest our lever to sway Central America and secure for the free states on this continent the control which is deemed necessary for the preservation of our Republican Institutions." The U.S. minister to Guatemala had alerted him of that nation's willingness to support a colony of freed blacks, if only to counter the British settlement of Belize. An additional land grant from Honduras, along the Honduras-Guatemala border, would offer yet another transisthmian railroad route and commercial opportunities. To confirm the value of Thompson's grant, Blair suggested that the president appoint Henry T. Blow, a man with experience in business, railroads, and steamships who was then serving as Lincoln's minister to Venezuela; he was "a practical miner," Blair said, adding that he had "made his fortune in that line." A positive report from Blow about the earlier Chiriquí surveys would add measurably to the government's case for both coal and colonization. "I can show Mr. Blow the coal," Thompson wrote, "the magnificent harbors, the splendid cotton lands, fertile beyond conception"—yet despite support from Blair, the proposed expedition never took place.[46]

Events in Washington caused the Chiriquí proposal to lose political ground. On April 16, 1862, Congress emancipated the slaves of Washington, DC. Slave owners in the district who declared allegiance to the Union were eligible for up to $300 for each slave freed. As part of the act, Congress also appropriated $100,000 to allow the president to begin colonizing not only the newly free but also all "such free persons of African descent now residing" in the district. Congress mentioned Haiti and Liberia specifically but added "such other country beyond the limits of the United States as the President may determine," allotting him up to $100 per emigrant. Yet the law caused a problem for Thompson: the administration now had authority to fund colonization along with its existing authority to purchase coal overseas. Thompson's project was first and foremost a speculation—what was to force the government to colonize Chiriquí as opposed to competing destinations? What was to stop the government from simply negotiating directly with New Granada, Costa Rica, or any other Central American nation? Thompson needed a coal contract; what did Lincoln need from him? Thompson's advisor Richard W. Thompson noted the constellation of forces that would mobilize against his project—the Panama Railroad, of course, but also proponents of colonizing Haiti or Liberia, abolitionists, and New Granada itself—and advised his client simply to try to persuade Lincoln to contract for coal and a naval station as war necessities while avoiding mention of colonization altogether. But even he remained skeptical. "There is not much prospect of this," he wrote, "though it is barely *possible.*"[47]

The remainder of April and May were occupied by negotiations between Thompson and Caleb Smith on the terms of a possible contract. Ten days after Congress appropriated funds for colonization, Smith formally inquired whether the Chiriqui Improvement Company would be willing to accept free black "colonization and settlement."[48] Thompson, desperate, expressed his pleasure to do so, and Smith turned to the task of detailing to the president why Chiriquí was the superior choice for a colony. In its favor, he noted that it was both close to the United States and also beyond the reach of the Clayton-Bulwer treaty of 1850, which prohibited the United States from occupying territory in Central

America. Yet the secretary was also interested in the potential of gaining some form of sovereignty over the colony, precisely the prospect that terrified Latin American governments. Here, even Thompson demurred, noting that while New Granada would accept emigration and bestow rights onto foreigners, it would never grant such rights to governments. Putting a positive twist on it, Thompson noted that the United States should not desire such rights anyway, for then Britain and France should be expected to assert similar sovereignty claims elsewhere in Central America. Still, he added, it made sense for the Interior Department to establish "a Colonial branch" tasked with formulating rules for colonists while using his company as the intermediary. Smith affirmed the arrangement to Lincoln, noting that de jure sovereignty was hardly the only means to power. "The settlement in that province of a colony of colored Americans, whose sympathies would naturally be with this country," Smith explained, "would ultimately establish then such an influence as would most probably secure to us the absolute control of the country." All it took was approving the coal contract.[49]

Supporters of the project lobbied the president over the summer. For his part, Thompson drafted a contract for the Navy and Interior departments that he hoped would appeal to the administration. Its first part detailed how the company would provide the government with coal, promising rates a dollar less than offered at nearby Aspinwall and discounted until the mines had delivered 50,000 tons to repay an initial advance of $300,000. It further detailed terms for building a coaling wharf, leasing mining rights to the United States, supplying timber, and providing land for coal depots and naval stations. The second part specified sites on both coasts for colonization and detailed terms for transporting and provisioning emigrants in successive parties of five hundred. A later revision shielded the government from spending any money on mining before an official survey could be undertaken and the coal tested for its capacity to raise steam. Thompson received support from the president's inner circle in the form of John P. Usher, then serving as Smith's assistant secretary of the interior. In late July, Usher reported that Lincoln still had not resolved his mind on the issue. He encouraged the president by pointing out both that the country would support the plan and that the effective House resistance the plan encountered during the Buchanan administration had dissipated, Charles Sedgwick as well as Thaddeus Stevens having become friends of the project. According to Usher, who may have had a financial interest in the project himself, the pecuniary success of Ambrose W. Thompson and the political and strategic interests of the nation at war aligned. Colonization and mining reinforced one another. "The advantage you gain in this," Usher wrote, "is the employment of the blacks and the obtaining the coal, when the government must want it in large quantities."[50]

Usher's mention of government wants pointed to the challenges the Union continued to face in supplying coal, especially for the blockade along the Atlantic and Caribbean coasts. The navy had more ships in the Caribbean than ever before, but coal remained expensive whether obtained from foreign merchants or via shipments from Pennsylvania. Additionally, with most Caribbean islands controlled by European powers and the southern coastline in rebellion, Union ships had few harbors to rely upon for delivery and resupply. A particular problem was British Nassau, the most important foreign port for supplying Confederate blockade-runners. In December 1861, the American consul there, Samuel Whiting, tried persuading colonial officials to permit the transfer of coal from one leaking coal ship to a naval screw steamer. As he noted in his report to Washington, "The request was courteously refused." When the screw steamer's commander protested that declaring coal contraband was equivalent to detaining the ship and thus unfairly aiding the rebels, the island's acting colonial secretary replied that "the real question here is not whether coal is or is not contraband of war, but whether the United States armed vessels are to make this a coaling depot, for the better facilitating their belligerent operations against vessels of the Confederate States." Union officials believed that by denying the Union's requests, the island was keeping its thumb on the scales.[51] In Jamaica, Lieutenant David Dixon Porter had encountered similar obstacles. "There is an indisposition on the part of the Government to furnish us with coal," he wrote Gideon Welles, "and there is none for sale except at most exorbitant prices." In Danish St. Thomas, a small Union coal supply was far from the docks and inaccessible to ships. As for the dock owner, an English merchant who kept some 2,000 tons of his own, the commander of the USS *Iroquois* grumbled that he "probably has an eye to a small profit, say 500 or 600 per cent."[52]

As naval coal demand increased, Lincoln's government further worried about its supply from the anthracite districts of Pennsylvania. The foremost concern was labor. For over two decades, the region had experienced ethnic, religious, racial, and especially class strife. In Luzerne, Carbon, and Schuylkill counties, and particularly in Schuylkill's Cass Township, the population of predominantly Irish, Catholic, Democratic miners already resented coal operators' growing power. In anthracite country, as elsewhere, the new industrial capitalism clashed with an older, producer ideology, as miners sought—and fought for—their share of income, dignity, and control. The war exacerbated this class conflict. When Lincoln ordered states to conscript soldiers during the summer of 1862, anthracite miners stood poised to bear the brunt of the hazards of war. Enumeration errors that placed ineligible men on the rolls (as many as a third of drafted miners in Schuylkill County alone) further added to the sense of injustice. Violence was limited, though, as most draft resisters simply fled or refused to give their names; it was women (believing themselves less likely targets for retribution) who were implicated in most reports of throwing stones at government agents and other acts of resistance. Still, Republican newspapers described miners as threats to the war effort. Historian Grace Palladino has observed that as government policy was based predominantly on Republican informants, "these perceptions and expectations of draft resistance often proved more potent than the resistance itself." By late October, the state administrator of the draft only averted what he feared would become massive riots by following Lincoln's instructions to leave the miners alone and provide falsified affidavits to Harrisburg claiming that the township's draft quotas had indeed been met.[53]

Yet if threats to the Union's coal supply existed, there were also those who were adamant against solutions that involved Chiriquí. In 1862, Joseph Henry was perhaps the most widely known and distinguished scientist in America. Sixteen years earlier, he had left a distinguished career at Princeton, where he had carried out a series of innovative experiments in electromagnetism, to lead the new Smithsonian Institution, a position he still held during the war. In May, Secretary of State William Henry Seward contacted him for his scientific opinion on Chiriquí coal. To his friend John Peter Lesley, the state geologist of Pennsylvania, Henry admitted that he was "some what suspicious" of Lincoln's plan to settle freed slaves to mine Chiriquí coal for the navy, but he agreed to look into the matter to "be true to my self and the government." What he found, however, only confirmed his suspicions. Henry explained that geologically, Chiriquí coal could never be compared with the familiar steaming coals of Britain and the United States because the geological age of isthmian strata was simply too young to produce a high quality fuel. Henry asked Lesley to produce a chemical analysis and to provide "any other reliable information" that might aid the government, which from Henry's pen may have meant information on Ambrose Thompson himself.[54]

If Henry's analysis gave Seward reason for caution, the message had not yet reached the president. Henry was baffled by Lincoln's widely publicized speech to the black delegation of August 14. "I was much surprised to find that he believed in the humbug coal mines of the Isthmus," Henry wrote to Alexander Dallas Bache, a fellow Washington scientist and head of the U.S. Coast Survey. Henry believed he had already conclusively rejected the findings of the earlier government reports, and moreover, claimed that Thompson, whom he mockingly referred to as "St. Ambrose," had dangled before him "a direct offer . . . of a share in the speculation" if Henry produced "a favorable report" to the government. It was all evidence that money, not sound geology, was driving the scheme.[55]

By early September, Henry had received Lesley's analysis. Writing to Frederick Seward, the assistant secretary of state, Henry bluntly explained the results. Questioning the skill and morals of the isthmus's previous surveyors,

Henry expressed hope that "the government will not make any contract in regard to the purchase of the Chiriquí district until it has been thoroughly examined by persons of known capacity and integrity." He further noted that Lesley, whom Henry identified only as "a gentleman who has been extensively engaged in geological surveys and has published a work of much merit on coal," had delivered an unequivocal chemical assessment. "There can be little danger of going wrong in an opinion upon a Tertiary coal," Lesley observed. This coal, readily distinguished from the more desirable anthracite and bituminous mines of his native Pennsylvania, was above all younger—more recently formed—than most Pennsylvanian coal, sometimes by hundreds of millions of years. Coal that young had usually not been exposed to the underground pressure and chemistry required to remove impurities like sulfur and concentrate the mineral's combustible carbon. The coal also traveled poorly. "A boxfull [sic] sent to the Academy of Sciences, Philadelphia," Lesley noted, "has slacked down to a boxfull of coal dirt." Worse, the coal was also prone to spontaneous combustion, a disastrous liability for a coal transported by wooden ships. With the authority of "the experience of the world in the use of coal," Lesley suggested that the defects of Chiriquí coal, like all other soft, young coals, would prevent their economic use as a steaming fuel, as semibituminous and anthracite coals, even when exported from the United States or Great Britain, would always price the lower quality coal out of the market. He called Chiriquí coal "as nearly worthless as any fuel can be" and asserted that he believed "the property will always be of little or no value to its owners." With chemistry and geology on his side, Henry considered the case of Chiriquí coal closed.[56]

Yet at the same moment, the project found a new champion. Twelve days after Lincoln's August 14 speech to the black delegation at the White House, Samuel Pomeroy, an abolitionist senator from Kansas, published a public appeal titled "To the Free Colored People of the United States." Pomeroy had been appointed by Lincoln to carry out the Central American colonization project. In the appeal, approved by the president, Pomeroy represented himself "as one awake to the momentous revolution in American history" and solicited volunteers to begin a voyage to become "free and independent beyond the reach of the power that has oppressed you." Pomeroy sought a contingent of five hundred colonists to steam from New York to Chiriquí in early October, promising government-funded food and domestic animals, along with equipment for farming and mining. The *Hartford Daily Courant* called Pomeroy, who had previously led migrant settlements to Kansas and Colorado, "the Moses to lead the blacks out of this house of bondage."[57] When Pomeroy consulted Welles about the plan on September 10, the navy secretary related his suspicions about Thompson and his skepticism about the legality of his grants. Should the administration persist in its commitment to colonizing the isthmus, as Welles recalled the conversation in his diary, negotiations should be undertaken with the government of New Granada or other Central American states directly, "not through scheming jobbers" like Thompson. Pomeroy, though, much to Welles's consternation, continued promoting the plan.[58]

In addition to Welles, however, there was a more influential opposition to Pomeroy's Lincoln-supported expedition. Governments across Latin America were ambivalent about immigration—nearly all welcomed it in principle, but only on their own terms. Between June 1861 and the following summer, the governments of Mexico, Guatemala, Costa Rica, and Honduras had all expressed their interest in establishing colonies of free blacks, but they remained wary, suspecting that Americans really intended to make a land grab (and some Americans, like Francis Blair, certainly did, even if he was in the minority). In every instance, the Latin American states welcomed new labor while condemning a potential loss of sovereignty. Either through his ministers abroad or directly himself, Seward assured each government that the essence of colonization was voluntary—free blacks would choose if and where to go, and foreign countries would choose if and how to accept new arrivals. Yet the closer Lincoln came to approving a colonization venture, the more individual Latin American governments disavowed their interest. First Mexico walked away, then Guatemala and El Salvador, and finally Costa Rica, Nicaragua, and Honduras.[59] Newspapers reported that Central American diplomats likened the colony idea to the disastrous filibustering expeditions of William Walker in the 1850s.[60]

New Granada was something of a special case, for it was then undergoing a revolution of its own. Its minister in Washington was still Pedro Herrán, the official who had turned against Thompson in 1860 and remained vigilant against his constant scheming. Herrán had told Lincoln in mid-June that both New Granada and Costa Rica would object to any colonization scheme carried out by Thompson's company. But the government from which Herrán was accredited no longer existed. His father-in-law, Tomás Mosquera, whom he opposed, had led a revolution against the country's Conservative administration (changing the name from New Granada to the United States of Colombia in the process). Unprepared to recognize Mosquera's new government, the United States simply continued to treat Herrán as the country's official representative. It is possible that an awareness of Herrán's dubious position allowed Lincoln to press on with plans to colonize Chiriquí in the face of his ostensible diplomatic opposition. Regardless, the situation allowed Francisco Párraga, Thompson's former agent in New Granada then living in New York, to act out a farce in support of the plan. In the fall of 1861, Párraga had tried to finesse Herrán's support by quietly financing his return to Washington, but once Párraga saw that Herrán remained staunch in his opposition to the plan, Párraga began to claim that he, Párraga, was the real, "yet unrecognized" representative of Mosquera's new government. His claim to speak for Mosquera led Párraga into some embarrassing exchanges, however. When he learned of Pomeroy's solicitation of black emigrants, he blasted the proposal, launching into a vituperative denunciation of the racial dimensions of the plan. "We have no hatred or prejudice against the negro race already existing amongst us," he told Pomeroy, but "there is in Chiriqui a great dislike for the colored race" that would lead Colombia "to ruin and desolation for the sake of the negroes." When informed that Thompson himself had negotiated the deal, Párraga demurred and endorsed it, at least publicly, although he still thought adding black colonization to the scheme was a bad idea. Noting that the government and people of Colombia would oppose the immigrants, he told Thompson that "God grant that bloodshed may not be the result."[61]

Despite the controversy surrounding the measure, on September 12, 1862, Thompson signed a contract with Caleb Smith and the Interior Department. This final version placed black colonization at the center and detailed Samuel Pomeroy's responsibilities for ferrying emigrants and establishing them on farm plots in Chiriquí. This time, however, reflecting the resistance of Welles in the Navy Department, Thompson reserved coal lands to himself and the subject of a naval station never came up. If the president could confirm the viability of the colony, and if the ambiguously worded "existing government" evinced no opposition, the administration would release $50,000 to Thompson to begin opening the coal mines. Despite reports that this project was for black colonists to mine coal for the government, the contract established mining as an undefined second stage, contingent on the success of the first.[62]

Among the prospective subjects of colonization, some supported it, others were indifferent, and still others were outright hostile. Above all, there was a debate. Most prominent leaders of the black community denounced it, from Frederick Douglass to the leaders of Washington's influential Social, Civil, and Statistical Association. Thousands of others were willing to try emigrating. But once again, political developments changed the circumstances on the ground. On September 22, Lincoln issued his Preliminary Emancipation Proclamation, promising to abolish slavery in the rebel states on January 1, 1863. Lincoln mentioned continuing to pursue colonization in the proclamation, but for many, the assumptions that had produced the politics of colonization earlier in the century had begun to crumble. The window during which black Americans had shown a greater willingness to consider colonization began to close. On Thanksgiving Day, Pomeroy, along with Harriet Beecher Stowe and other abolitionists, attended a banquet for freed slaves in Washington. The senator spoke glowingly of Chiriquí colonization, displaying samples of the region's resources before the gathered crowd. As he concluded his presentation, a preacher and self-emancipated slave from Virginia rose and attacked the project—and Pomeroy himself—calling colonization in Central America a trap for an even worse form of slavery. The senator could offer no response.[63]

He would also never set sail. Pomeroy's original departure date from New York, October 1, came and went. Why he and his erstwhile colonists remained in the United States is not entirely clear. In early October, Usher informed Pomeroy that Lincoln had at least temporarily halted the project; a week later, Seward forwarded Caleb Smith a recent message he had distributed to his American ministers in Europe, announcing what appeared to be a new call for colonization proposals. The key clause stated that negotiations for colonies were to happen only between accredited representatives of each country involved. No longer would the administration work through intermediaries like Ambrose Thompson. An officer in Thompson's company speculated that Seward was "interested probably in some other piece of property and is only waiting for you to offer him a larger interest in this for him to advocate Chiriquí." Had Seward been offered an interest in the first place? It would not have been beyond Thompson to have made such an offer, but if he had, there is no evidence that Seward ever accepted. Either way, Seward was well aware of the intense opposition to Thompson's project that had emerged across Central America. His letter to Smith appears to be a subtle instruction to abandon the Chiriquí scheme. Just as importantly, it became increasingly clear in the State Department that there was a danger the United States would be unwittingly drawn into the boundary dispute between Colombia and Costa Rica.[64]

In the months that followed, Thompson continued to press for his signed contract to be put into effect and for colonists to sail for Chiriquí, but none would ever go. Pomeroy managed to secure Lincoln's approval to advance $14,000 to Thompson, though what happened to the money no one ever figured out. Lincoln would pursue colonization elsewhere, such as, Île à Vache off the southwestern coast of Haiti, where 453 former slaves settled in 1863 before some hundred of them died and the remainder returned to the United States. But even if diplomatic opposition had doomed the Chiriquí project, what of Thompson's original premise about the value of Chiriquí coal and naval stations to the war effort? Under Gideon Welles's leadership, the navy was answering that question in a way no one had expected before the war.[65]

Fueling the War

From the moment he learned of it, the secretary of the navy had not hidden his disapproval of Thompson's Chiriquí project, and he resented when other cabinet secretaries meddled in the affairs of his department. This resistance is why Thompson's final contract was exclusively with the Interior Department and why it barely mentioned the coal or naval stations that had originally constituted the ostensible prime value of Chiriquí. But Welles and the navy hardly ignored the obstacles to securing coal and in fact worked hard to ensure a vast distribution network, particularly for Pennsylvania anthracite. At the same time, for want of finances, lack of labor, and dearth of fuel itself, the Confederacy proved unable to organize its fuel network on remotely the same scale. The Union success in managing its fueling during the war helped shape the terms of debate over foreign supplies of coal and the value of distant coaling stations for over three decades after it drew to a close.

The navy began to systematize fueling in Philadelphia in the fall of 1862. As the outlet for hard coal, the city became the epicenter for fueling the Union navy as well as the army. After more than a year of haphazard fuel distribution, in the fall of 1862, the navy's new Bureau of Equipment and Recruiting opened a special office on Philadelphia's Walnut Street, near the city's bustling docks, to coordinate the shipment of coal to naval vessels in the Atlantic and Caribbean. This office, under the supervision of Henry A. Adams, a naval captain and native Pennsylvanian, contracted directly with private ship owners to carry fuel to coastal points and the West Indies. To entice them into this risky service, the government insured these ships at a cost of nearly $50,000. Still, though they sailed in dangerous waters, only 3 of 168 coal carriers fell to either privateers or southern war vessels during the first year the government provided the coverage. In this manner, the navy allocated 130,000 tons of fuel a year to its ships, not to mention tens of thousands more for heating and iron work at navy yards. The operation was so successful that even in the crucial port of New Orleans during the winter of 1863, the department could supply David Farragut's Western Gulf Squadron with coal for only $12.63 per ton, with freight accounting for some two-thirds of the cost.[66]

Yet domestic supply was not without its difficulties. Both coal production and adequate muster rolls were threatened when a new Union conscription law went into effect in 1863. As early as January, Pennsylvania anthracite miners struck for higher wages and greater safety in Schuylkill, Luzerne, and Carbon counties in a campaign that would continue for months. Provost marshals—federal agents who enforced the draft—increasingly called for troops to put down the strikers, and by August, the army's Department of the Susquehanna created the Lehigh District specifically to enforce lawfulness with bayonets. Still, the strikes continued. In Hazelton in November, miners briefly shut down coal production; only the arrival of federal troops restored it. Afterward, coal mine operators pleaded with the army for more soldiers to give them space to "discharge the bad characters and employ new men," a stratagem designed as much to help the operators with their labor management issue as the government with its draft enforcement problem. The consequences of inaction appeared catastrophic. "If commenced," wrote Union major-general Darius Couch to Washington, "the troops must not be withdrawn until the work is thoroughly done, otherwise two-thirds of the anthracite region would stop sending coal to market." With approval from the War Department, soldiers arrested roughly a hundred miners and subjected them to military trials, all with the enthusiastic approval of their former employers. Similar suppression of miners happened elsewhere in the region. Simultaneously, coal demand, production, and prices all rose. The federal troops would remain until the war ended.[67]

There was also the prospect of the anthracite region's military vulnerability. When Robert E. Lee invaded Pennsylvania in June 1863, he sought a crushing victory that would cripple the Union war effort while cultivating a northern desire for a peace settlement. Union generals, however, could only speculate on the location of Lee's army and the purposes behind its movements. In mid-June, Darius Couch reported the movement of three columns of southern troops: one heading towards Chambersburg, one towards Gettysburg, and ominously, "the other in the direction of the coal mines."[68] By the end of the month, Lee had troops outside of York preparing to destroy the railroad link between Harrisburg and Baltimore and had others near Carlisle, ready to march on the state capital itself. If all went as planned, Lee expected his maneuvers would attract a fatigued Union army, which he could crush once and for all. Of course, all did not go as planned, and on July 1, as a division under A. P. Hill pursued a desperately needed supply of shoes at Gettysburg, it unexpectedly encountered two Union brigades. The three-day battle that would turn back the Confederates had begun.[69]

Yet as the battle continued, leaders in Richmond remained unaware of it, and the talk of city was instead what Lee would do once he captured Harrisburg (rumors of which circulated in the Confederate capital as the battle itself raged in Pennsylvania). On July 2, the second day of fighting, the influential and critical *Richmond Whig* speculated that if Lee had succeeded in taking the northern capital, his next target would be the nearby coalfields. The paper anticipated Lee cutting vital railroad lines and wrecking mining machinery before turning to an industrial scorched-earth policy. "He might set fire to the pits," the paper explained, "withdraw the forces sent out on this special duty, and leave the heart of Pennsylvania on fire, never to be quenched until a river is turned into the pits, or the vast supply of coal is reduced to ashes." Such action would not only strangle the Union navy but vast numbers of factories and railroad lines as well. The dependence of the war effort on fuel was clear: "Northern industry will thus be paralyzed at a single blow." Unsurprisingly, northern papers viewed the desire to "turn the Keystone State into a Gehenna" as nothing less than barbaric.[70]

Despite both labor unrest and military vulnerability, into 1863, the navy's Bureau of Equipment continued to refine its coal operation. In addition to fuel, it distributed iron buckets and warehouse trucks. It ensured the availability of firemen and coal heavers. It developed streamlined procedures for ordering,

inspecting, and distributing coal, working with coal dealers, ship commanders, and various navy yards to coordinate the vast operation. The navy's operation was even large enough to continue supplying fuel from reserve stocks, if
at a slower pace than usual, when striking miners once again slowed the delivery of coal in April 1864.[71]

After Gettysburg, the Union's anthracite supply remained secure, though not
without the presence of the army's Lehigh District keeping its boot on the neck
of restive miners. Confederate access to coal, meanwhile, continued to decline.
Just as the Union navy created its nerve center in Philadelphia, Confederate navy
secretary Stephen Mallory handed authority over coal to the Office of Orders
and Detail in September 1862. A year later, unlike its counterpart up north, the
office's chief confessed to Mallory that "the general supply of coal for the Navy
for the past year has been inadequate to its wants, both production and transportation being deficient." The Union occupation of Chattanooga in August
1863 isolated the productive coal mines of Tennessee that had been supplying
Georgia, South Carolina, and much of the Confederate navy. The Office of Orders and Detail dispatched what it could from Richmond and North Carolina,
but it was hardly enough to make up for the loss out west. Mines in North Carolina and Alabama increased production only slowly, though the efforts of a
new Confederate mining bureau helped substantially.[72]

The declining southern economy increasingly squeezed Confederate mining
operations, however. One coal operator in Alabama confessed in early 1862 that
without a government contract to supply the Confederate navy, he would likely
have to shutter his mine. Privation in Mobile simply made the civilian coal market too weak to make mining a profitable enterprise. A year later, having received the contract, the operator found that labor shortages reduced his production to a mere third of its capacity. Pleading with the superintendent of coal
contracts for an advance large enough to purchase a slave gang, the operator
complained that "there are scarcely any professed coal miners in the country
here and no man will hire his negroes to work under ground."[73] When a Confederate naval engineer conducted a favorable coal examination in 1863 and
reported that "if a sufficient quantity of the coal can be had great economy
would be the result," the telling word was "if."[74]

The most revealing evidence of the Confederate navy's struggles to secure
coal is that the majority of its fuel invoices were not for coal at all but for cord
upon cord of wood. As early as 1861, the Confederate navy began substituting
wood for coal in its steamers, and by 1863, wood consumption was especially
high in the fuel-starved ports of Charleston and Savannah. Confederate invoices
reveal that most orders were for fewer than 25 cords at a time, though nearly
thirty document purchases of at least 100 cords apiece. Most orders that noted
a variety specified pine, in line with what one would expect given the ecology

of southern forests, but the orders occasionally requested oak or ash if they were available. Altogether, the Confederacy purchased over 14,000 cords of wood for its navy, roughly the energy equivalent of 14,000 tons of coal and a mere fraction of what the Union navy could obtain from a single bid in Pennsylvania. Only 7 of 1,370 extant Confederate invoices indicate an order for more than $10,000, and at least 2 of those were for coal somehow supplied from Pittsburgh (possibly wholesale orders already in the South when the war began). Taken together, these records suggest that the Confederacy spent a little over $600,000 on fuel between March 1861 and December 1864 (and this with a steadily depreciating currency). In contrast, over the course of the war the Union navy allocated nearly $18 million toward coal for steaming alone, with much of that figure covering transportation and storage. While the Union navy was about five times the size of its Confederate counterpart, its ability to muster fuel resources was disproportionately larger still.[75]

As for Ambrose Thompson, he pleaded with Secretary Seward in March 1863 to advance his plan, asserting that few blacks would join the Union army and that those who had already registered for emigration were suffering "deep disappointment." Still, he admitted, few blacks would ever colonize Chiriquí either. "I do not believe that any very great numbers would emigrate if the way is opened," he confessed, "but enough would go to relieve the question, and there would act upon the remainder, as a safety valve," setting an example of liberty for the majority who would inevitably remain in the United States. He then returned to enumerating the commercial and geopolitical advantages of the scheme that had motivated him for a decade. It was not hard to lighten the focus on colonization, as black emigration had never been the point of the project, merely a timely addition that cultivated the support of Lincoln and a handful of his advisors. In response, Seward ignored him.[76]

In 1864, Thompson announced to the government his intention to offer his Chiriquí land to the government of Great Britain. He wrote John P. Usher, by then secretary of the interior, asking for copies of the government's own surveys of Chiriquí so that he could present them in England. Perhaps he needed the reports, perhaps he hoped to invite a counteroffer from the Lincoln administration. He received neither. Usher acknowledged Thompson's new plan to travel to England but noted that the contract with the president was still in effect whenever the president decided to invoke it. "I entertain a high opinion of the International value of this property and of the benefits to be derived thereupon by extended commercial relations between maritime nations," he wrote. Yet while the land remained important, with the emigration of freed blacks no longer a realistic proposal, the time for coal and colonization had passed.[77]

The war did highlight the challenges to fueling in distant waters during wartime. By 1865, Americans had a very clear picture of the material demands of industrial warfare. If Thompson had only offered a speculative scheme for his own enrichment, might other distant coaling stations be of value nevertheless? Could the anthracite mines of Pennsylvania be counted on into the distant future? How frequently would American naval and commercial vessels require refueling and where would the fuel come from? If coal would not, in fact, support colonization, would it support national defense instead? In the years after the war, these questions of fueling American vessels—and the wisdom of obtaining distant coaling stations—would emerge as an important debate in American foreign relations.

CHAPTER FIVE

THE DEBATE
OVER COALING STATIONS

Then said William Henry Seward,
As he cast his eye to leeward,
"Quite important to our commerce
Is this island of St. Thomas."

Bret Harte, "St. Thomas (a Geographical Survey, 1868)"

In May of 1891, Admiral Bancroft Gherardi and Frederick Douglass, the U.S. minister to Haiti, were completing what would become a failed negotiation to lease or purchase the Môle St. Nicholas, a bay tucked into the extreme northwestern peninsula of Haiti, as an American coaling station. As an undeveloped harbor, it offered a tantalizing location for a Caribbean naval base. Before learning that discussions had broken down, the *New York Times* reflected on the importance of establishing foreign coaling stations like this one. Calling it a "generally recognized necessity," the *Times* distinguished the pursuit of naval bases from an earlier period of continental conquest. "This policy is quite distinct from a general mania for annexing territory," the paper noted, "although it might in some cases pave the way to the latter, and has often been opposed on that ground." According to the *Times*, the hunt for coaling stations "has its origin in a state of things quite outside the experience of the founders of the Republic." What had changed was technology. "The introduction of steam as the motive power of ships created the need of foreign coaling stations, which, acquired in times of peace, could be relied upon also in war." The recent shift to ships that relied solely on steam, having no sails at all, appeared to the *Times* to increase the necessity of these stations still further.[1]

These 1891 observations by the *Times* raise two important sets of questions. First, why did the paper suggest the search for coaling stations was new at the beginning of the 1890s? Ambrose Thompson had lobbied for such a station at Chiriquí as early as 1857. The Civil War had not even ended when William Henry Seward began an ultimately failed diplomatic quest to secure the Danish West Indies island of St. Thomas as a coaling station. Before the decade was over, Americans had debated coaling stations in Santo Domingo and the Pacific atoll of Midway. The 1870s brought continued interest in Santo Domingo's

Samaná Bay, Hawaii, Samoa, and along the Central American isthmus. These debates around coaling stations continued into the 1880s around stations in the Caribbean, Pacific, and even far-flung Africa. Where had the *New York Times* been for twenty-five years? Did the paper observe something new and different from what had come before?

Second, was the necessity of foreign coaling stations indeed a "generally recognized necessity"? Was a need for them indeed created by the technology of steam? The United States had been building oceangoing steamships since the 1840s—why did the subject of coaling stations become so prominent only decades later? And if these stations were so essential, why had the United States been so unsuccessful for so long in securing them? Or were coaling stations less a perceived necessity than the *Times* implied?

Part of the confusion revolved around language, for coaling stations meant different things to different people. One model was provided by British stations in places like Gibraltar, Malta, Aden, and Bermuda. These coaling stations were strategically located, fortified, available in peacetime and defensible during war. If they were not all fully colonies, they at least guaranteed extraterritorial legal rights. Another model saw the pursuit of a station as prelude to a more general territorial annexation, as happened eventually in Hawaii. But coaling stations could also mean a more commercial arrangement, as when the American government leased coal storage in foreign ports or negotiated access to coal on behalf of American shipping lines. The United States pursued depots of this kind in Yokohama, Japan; Pago Pago, Samoa; and Pichilingue, Mexico. Naval vessels could access these coaling stations during peacetime but probably not during war—not necessarily because of conflict with the host nation itself but because of new developments in international law during the 1870s that placed strict restrictions on neutral nations' fueling of belligerent vessels with coal.

Did Americans of the Gilded Age believe the pursuit of coaling stations was a national priority? One dictated by the demands of new technology? Certainly some did. But delving into what Americans were talking about when they were talking about coaling stations reveals a more muddied picture. The ostensible need for coaling stations was not obvious in the late nineteenth century. The idea had to be constructed, prepared, explained, and justified. Over time, the preponderance of reasons given by Americans who sought coaling stations changed. The boosting of American commerce became territorial expansion; expansion became a preoccupation with American vulnerability. Over three decades, the need for coaling stations was not an uncontested fact but an argument. As Americans argued, they helped construct the very idea of an American national interest in coaling stations in the first place and finally in coal itself.

If some Americans believed that steamships demanded coaling stations, there was one group who often believed otherwise: the people who designed, built, and commanded the steamships themselves. The most striking thing about the debate over coaling stations in the late nineteenth century is not how often successive administrations pursued them, but how frequently engineers, mathematicians, and naval officers tried addressing geographical limitations with technical innovation instead. New kinds of engines, new ships, new methods of navigation, new naval strategies—all these approaches addressed the challenges of providing commerce and security through coal in ways that avoided the need for new territory.

By the late 1890s, Americans faced the irony that just as they began a colonial project justified in part by the need for coaling stations, a confluence of engineering innovations, changes in strategic thought, and developments in the practice of navigation had made distant bases less pressing of an issue than any time since the end of the Civil War. To understand this irony we must first understand how the subject of coaling stations looked to two powerful rivals in the cabinets of Lincoln and Johnson: William Henry Seward, the secretary of state, and Gideon Welles, the secretary of the navy.

Entrepreneurial Diplomacy, 1865–90

Secretary of State William Henry Seward embarked on a tour of Caribbean islands in January 1866. He had personal reasons for seeking a vacation—both he and his son Frederick, his assistant secretary of state, had narrowly escaped an assassination attempt the night Lincoln was shot the previous April. Moreover, the lurching executive transition from Lincoln to Andrew Johnson had been exhausting, politically and personally. Seward, however, had other motives in mind as well. His itinerary included the Danish island of St. Thomas, along with Santo Domingo, Haiti, and finally Cuba. As the visits signaled, Seward believed the United States was destined to expand into the Caribbean and beyond into Alaska, Canada, Greenland, and elsewhere in the Americas. "Events had compelled the United States to become a great maritime power," the Danish minister recalled Seward saying a year earlier. And for a great maritime power, Seward concluded that "a harbor and depot in the West Indies had become a necessity." It is possible, too, that Seward believed that Austria desired to seize the islands to settle accounts following its recent war with Denmark. The prospect of losing St. Thomas to another European power, along with the Union navy's difficulties coaling during the war, suggested to Seward the value of a Caribbean naval station. His desire for a tangible legacy certainly contributed as well.[2]

Commercial geography also drew Seward's attention to St. Thomas. The island lay at the intersection of major sea routes: from England to Central America, Spain to Cuba and Mexico, and the United States to Brazil. It was among the most convenient ports for reaching the Lesser Antilles. Frederick Seward observed that "St. Thomas has come to be a place where steam lines converge" and recalled the local maxim that the island was "the place which is on the way to every other place." Trade was only one advantage of St. Thomas, however. The Sewards described unique strategic advantages for the nation that controlled it. Graced with a commodious harbor and a narrow and easily defended entrance, the island possessed what they believed to be an ideal location for a coaling station. "It would have been of great value to the United States," the younger Seward observed twenty-five years later, "had they owned it during the civil war."[3]

While seeking the cession of St. Thomas, the elder Seward simultaneously pursued the lease or cession of Samaná Bay in Santo Domingo. As with the Danish islands, Seward was partly motivated by a fear of European intervention. He had received notice that the vast bay in the northeastern corner of the island might be ceded to France, and Britain had recently approached the United States to form an agreement to keep the peninsula altogether neutral. The fragile government of Santo Domingo itself lobbied Seward, desperately seeking a $2 million loan in cash and armaments to help secure the independence of the country. Between the end of the Civil War and leaving office in 1869, Seward divided his attention between these two prospective acquisitions.[4]

Through his 1866 tour and subsequent diplomacy, Seward tried cultivating a sense of inevitability and necessity around the annexation of a Caribbean island for a coaling station. In many respects, he was largely successful. Over the coming years, comments like "the necessity of our possessing a naval station somewhere in the West Indies has long been apparent" soon became ubiquitous. But was such a necessity as apparent as Seward believed? The department most concerned with national defense was not Seward's but that of Gideon Welles. Welles did not object to acquiring a Caribbean coaling station, and department policy approved of it. Welles himself saw the value of a coaling station on either Martinique or Guadeloupe. Yet it was a matter of utility not necessity, let alone destiny. Welles certainly disapproved of Seward's theatrics. "I am amused and yet half-disgusted with Seward's nonsense," he wrote before Seward sailed to St. Thomas, certain that his colleague's public maneuvers were likely to increase the cost of the islands while decreasing his chances of obtaining them.[5]

For Welles, the need for Caribbean stations was far less urgent than they appeared to Seward. When Seward first raised the Santo Domingo matter in Johnson's cabinet, Welles was the only skeptic. When Seward and Secretary of War Stanton pressed the issue, Welles replied with reports that Samaná Bay was in fact rife with disease, not to mention out of the way of the New York-Aspinwall route to the Pacific, lacking in population, without local commerce, and more costly than the nation's fragile postwar finances could justify. Welles's policy for securing coaling stations had been to proceed "prudently, carefully, and at little cost," and the prospect of spending $2 million for a single, out of the way harbor made little sense to him. To Welles, Seward hinted at what he called "political reasons" to favor the plan, mentioning the enthusiasm of leading congressional Republicans Thaddeus Stevens in the House and William Fessenden and James Grimes in the Senate. Welles suspected that the purpose of Seward's negotiations in St. Thomas and Santo Domingo was to gain favor with the radicals, who had long since broken with President Johnson, perhaps by making a magnanimous gesture toward "the negro element there and here," perhaps simply by finding any subject on which the executive and Congress could agree. As for the harbor's strategic value, however, Welles pulled no punches. "There is no object," he concluded, "naval or commercial, in getting Samaná." In case of war with Santo Domingo or any other European power in the Caribbean, Welles believed it both easier and cheaper simply to seize a West Indian island for coaling if needed. Steamships needed coal, Welles knew better than anyone, but how to fuel them was hardly geographically determined.[6]

Whatever Seward's private motivations, both negotiations ultimately fell through. In the fall of 1867 a series of natural disasters engulfed St. Thomas and dashed prospects for annexation. On October 29, a massive hurricane pummeled the island. Barely two weeks later, as representatives of the United States and Denmark met on nearby St. Croix, the island suffered the additional devastation of a series of eighty-nine earthquakes in a twelve-hour period. The aftershocks lasted weeks. The largest triggered destructive tsunamis, one of which wrecked the 2,000 ton USS *Monongahela*, anchored at St. Croix to support the American delegation. Four of the crew died. It was a catastrophe that could not be countered even by the island's residents signaling their approval of annexation in a plebiscite 1,244 to 22. The treaty died in the Senate. In Santo Domingo, Seward continued negotiating for Samaná until the end of January 1868, but the two parties could not agree on the final terms of a lease, as the island's fragile government feared transferring sovereignty would spell revolution. Seward broke off discussions with the Dominican minister and the two countries signed a commercial treaty instead.[7]

The following two decades witnessed additional attempts to expand American territory overseas, formally or informally, mostly ending in failure. Some plans revived earlier schemes. Others involved places new to the American geographic imagination. Many were driven by speculators with personal stakes in the projects, businessmen who conducted a kind of entrepreneurial diplomacy that framed private advantages in terms of public interests, especially the idea that securing coaling stations was vital to the government. Seward's St. Thomas and Samaná Bay negotiations established this popular perception of the unquestionable value of foreign coaling stations. However, the many attempts to obtain them later often had little to do with technological necessity; instead such efforts appealed to security arguments to increase public and congressional support for various speculative projects.

Nowhere was this more true than in Santo Domingo, where Seward's efforts to secure Samaná Bay quickly grew into a larger project to annex the entire country, a scheme that became one of the defining foreign policy failures of the Grant administration. Grant's old comrade in arms and advisor, David Dixon Porter, had first visited Santo Domingo as a thirty-three year old lieutenant in 1846, two years after the country had broken away from neighboring Haiti. Twenty-one years later, he returned with Frederick Seward. Porter had long ago concluded that the United States ought to annex Santo Domingo altogether. On his first visit, he had observed unworked mines of copper and coal, forests of mahogany and lignum vitae, and fields of sugar, coffee, and breadfruit. Porter imagined boundless opportunities for American investment. After his second tour, he continued to support the more general annexation with a naval coaling station as an added bonus. Porter had perfect awareness of Samaná's weaknesses as a station, however. Thus, instead of attempting to justify annexation by reference to the imperatives of new technology, he instead appealed to the inadequacies of tropical geography to American health. After the Civil War, the American squadron in the West Indies sometimes coaled in harbors at Key West and Havana, but regular outbreaks of malaria and yellow fever there prevented their regular use. In contrast, according to Porter, Samaná Bay was "perfectly healthy," as "the tide ebbs and flows regularly every day thereby carrying off all impurities and removing those causes which create sickness in the places heretofore alluded to." It was a matter of life and death. "We have to send our ships away from there or else everybody would die." It was hard to come up with a more dire-sounding argument.[8]

In the reports of General Orville Babcock and General Rufus Ingalls, dispatched by President Grant to survey the island in 1869, a coaling station at Samaná Bay likewise figures in almost as an afterthought to the larger annexation project. Like Porter, the generals emphasized abundant resources and reassured Americans back home that most citizens there were white, religiously tolerant, peaceful, and above all, greatly desirous of annexation to the United States. As for Samaná Bay, Ingalls emphasized its value "as a depot for the commerce" at least as much as he emphasized its significance as a naval station.[9] Hamilton Fish, Grant's secretary of state (and personally ambivalent about, if not opposed to, annexation), admitted privately that negotiations regarding Samaná Bay were merely intended to encourage residents to support annexation and provide collateral for a $150,000 American loan to the government of Buenaventura Báez in Santo Domingo.[10]

As for those Americans invested in Santo Domingo and aggressively lobbying Grant and Congress for annexation, it was clear that they were simply interested in absorbing new territory for personal speculation, as the generation before them had done in the western continent. Samuel L. M. Barlow, a New York lawyer invested in William Cazneau and Joseph Fabens's Santo Domingo Company, characterized what he called "the real value of the island" as its potential for cultivating sugar and other tropical fruits, as well as harvesting timber. Gold, if found, would encourage the migration of white Americans. To possible congressional worries about race, Barlow responded that "it is a White-man's Island" and "practically uninhabited." Barlow, like Seward, viewed the acquisition of Santo Domingo as just one more step in the providential annexation of the other great and prosperous islands of the Caribbean. It was what he called "a National point of view." As for its value as a naval station, Barlow said it was a long-standing "military necessity," adding "whatever that may be."[11]

Historians have explained the Senate's failure to ratify the Santo Domingo treaty as the result of a combination of factors. Annexation, and tropical expansion more generally, had little popular support, especially at a time of western development and unsteady government finances. Many in Congress were nervous about establishing a precedent for island acquisitions. Racial concerns also played a key role—either a fear of making new Americans out of supposedly inferior stock or of ultimately undermining the independent, largely black republic of neighboring Haiti. Others were skeptical about the role of corrupt speculators (like Cazneau and Fabens) and the legitimacy of the plebiscite almost unanimously expressing Dominican approval of annexation. On top of everything, the personal relationship between President Grant and the powerful chairman of the Senate Foreign Relations Committee, Charles Sumner, was deteriorating. For both the Congress and the press, these issues were indeed central.[12]

However, objections to one of the central premises of the strategic value of the island have been largely ignored. Annexationists' proclamations about the strategic necessity of island were challenged by those who saw the claims for the pretexts they really were. As Justin Morrill noted in the Senate, the only reason the Union navy had difficulty operating in the Caribbean during the Civil War was that all the American states that bordered the Gulf of Mexico and South Atlantic coasts had become the enemy—hardly a future prospect anyone cared to plan for. As for protecting American commerce, he noted as others had that there was no American commerce anywhere near Samaná. The bay was far from existing trade routes to Panama or Vera Cruz and so remote from existing communications that the recent commission investigating the island—accompanied by nine newspaper reporters—spent thirty-three days incommunicado. In June, the Senate finally voted on the treaty and gave Grant a humiliating defeat—twenty-eight to twenty-eight. As a treaty, it had needed a two-thirds majority to pass.[13]

In the case of Samaná, the *idea* of needing a coaling station loomed larger than any material realities. But the project revealed how linking speculation to national priorities and technical necessity might influence diplomacy. Seeking coaling stations became a tool in the foreign policy toolbox, especially for entrepreneurial diplomats who wished to influence Washington. As the project for Caribbean coaling stations temporarily faded, these speculators turned their attention to the Pacific. William H. Webb, recently retired as one of New York's largest shipbuilders, conceived of a plan in 1871 for an American steamship line to Australia. As the line would pass through Polynesia, Webb settled on seeking a coaling depot on Samoa. Through an agent, Webb enlisted Richard A. Meade, a U.S. navy commander aboard the *Narragansett*, to secure the Samoan harbor of Pago Pago for the United States. Meade obliged, and though the Senate rejected Meade's unilateral attempt at annexation, Webb still managed to arrange the appointment of a special agent, ostensibly to report on the island to the American government but really to represent the interests of investors in the line, the Central Polynesian Land and Commercial Company (CPLCC). The agent, Albert Steinberger, quickly schemed to win the confidence of Samoans against a competing German land appropriation at the same time that he attempted to bring the island into an official relationship with the United States that would protect and promote the CPLCC. By 1875, Steinberger had become premier of the Samoan government. Though he was removed by British marines in 1876, two years later, the government he had created was induced by the CPLCC to sign a treaty with the United States, turning the United States into Samoa's interlocutor with the outside world and establishing American rights to Pago Pago as a naval and coaling station. In the meantime, the private interests of the CPLCC had become framed as public ones. Speaking of Samoa, the secretary of the navy, George Robeson declared that "as an available station for coaling and supplies for our national and commercial marine in that part of the world, it is far the best to be found within a sweep of many thousand miles." Robeson called the harbor of Pago Pago "not only far the best and safest, but absolutely the only land-protected harbor among the islands of the South Pacific."[14]

Yet how much the United States really needed the island as a national coaling station is suggested by how little it was subsequently used. In 1880, the navy deposited some 2,000 tons of coal there, the quantity of which was justified as both saving money and facilitating naval cruising in the South Pacific. But for nearly a decade the coal simply sat on a rented lot, and neither the Navy nor State departments secured land for an actual station—no wharf, no lighters, no structure for an official to superintend the supply. But there was the matter of national pride. When tensions with Germany around Samoa flared up in 1888, Secretary of State Thomas Bayard observed that preserving Samoan neutrality and respective treaty rights around the coaling station was essential. "It is of special importance to the United States," he explained, "for in no other part of Polynesia is a right of this nature possessed by them."[15]

But preserving a right was hardly the same as needing to exercise it. Did Americans need coal in Samoa? In 1889, Winfield Scott Schley, the commodore who headed the Bureau of Equipment, reportedly claimed that until the end of 1888, "not a particle of it had been touched." But in fact, the coal had actually run out years before and no one had bothered to replenish it. As Americans and Germans began to contemplate war over the islands, in the Senate, John Sherman orchestrated a vote for $500,000 to be appropriated "for the maintenance of American rights in Samoa" and another $100,000 for a naval station, though Schley was forced to deny he had any knowledge of reports that the government had already sent 10,000 tons of coal to the island. Rumors swirled that navy secretary William C. Whitney had dispatched Philadelphia merchant ships—or were they from Australia?—to supply the navy. As in St. Thomas in 1867, however, nature intervened in foreign relations. When a hurricane wrecked six German and American vessels on March 15, no coal deliveries had yet arrived. "It is called a coaling station, but this seems to have proved a misnomer," dryly observed the *New York Herald*. If it was any consolation, the German navy had little coal there, either.[16]

The whole conflict between Germany and the United States over the islands was reduced to a profession of technological necessity. But the argument, captured by Benjamin Harrison's inaugural address on March 4, 1889, when he declared that "the necessities of our Navy require convenient coaling stations and dock and harbor privileges," was not based on any requirements the U.S. Navy then had. Its ships rarely visited Samoa. There was little commerce and little pent up demand for trade with Australia. It was costly for the government to supply Samoa with coal and patrol the neighborhood regularly. What was worse, war would make the remote depot vulnerable to enemy assault and an easy target for capture. "A coaling-station at Pango-Pango is of little use in time of peace and is a great expense," concluded one lieutenant in 1889.[17]

Unlike in Samoa, there was no company pushing for a steam line to Hawaii. Instead, the prospects of establishing a government coaling station there were intimately tied up with efforts by a small group of expatriate Americans to annex the islands. For decades, the islands had been sites of sugar cultivation and American Protestant missionary activity. And while there had long been talk of eventual annexation, the subject of Hawaiian coaling stations only arose in the early 1870s as a way to entice Americans back home into noticing the islands at all. Henry A. Peirce, the longtime Hawaiian resident and influential American minister in Honolulu, claimed that Hawaii was "valuable, perhaps necessary, to the United States for a naval depot and coaling station," and he described the

Engraving of the American naval coaling station in Pago-Pago, Samoa, in 1889. Though the supposed vital interest of a mid-Pacific coaling station nearly led the United States to war with Germany, during the late nineteenth century, the American navy barely made use of the tiny and rudimentary outpost. Readers of *Frank Leslie's Popular Monthly*, where this engraving by Ernest Wilkinson appeared, likely would not have missed the ironic juxtaposition of this image—with its single pile of coal, a small flag, and a few thatch huts—alongside an article discussing the great power conflict over Samoa. "Samoa, and the Troubles There," *Frank Leslie's Popular Monthly* 27, no. 4 (1889): 489.

vast commerce anticipated between the United States and Asia. But Peirce knew most Americans were not listening. When Secretary of State Hamilton Fish raised the subject of Hawaii at a cabinet meeting in 1870, he observed that "no one responds & the subject is dropped." On the islands, where native Hawaiian opposition to annexation was strong, American annexationists were cognizant of the fact that the United States had failed to acquire St. Thomas and Santo Domingo and of the general American lack of interest. Annexation, wrote Henry Peirce, "they deem impossible of attainment.[18]

Two men on the mainland had other ideas, however. In 1872, John Schofield and Barton Alexander were worried about a future "war with a maritime nation." The two soldiers, both of whom were generals and Civil War veterans, were stationed together in San Francisco, where they discussed how the country might go about prosecuting the next war, a war that could require operations in the Pacific. Alexander, the army's senior engineer on the Pacific coast, believed that the United States lacked knowledge about harbors across the island Pacific, knowledge that could prove decisive in a conflict, particularly with Britain. The problem was especially acute in Hawaii, and Alexander suggested it would be wise to negotiate with the kingdom before war broke out and better still to determine in advance which harbors were most valuable for what he called "temporary possession." In March, disavowing any interest in promoting annexation, he volunteered to examine the islands for the War Department. Schofield selflessly offered to accompany him.[19]

The navy does not appear to have been involved in the planning of the mission and the subject did not particularly interest either the secretary of war, William Belknap, or President Grant, but both eventually approved it and issued formal instructions in June. Given the sensitivity of prospective war planning, Schofield and Alexander were advised to travel under the cover that the trip was "a pleasure excursion." In late December, Grant gave a final endorsement of Schofield's proposed voyage.[20]

With Hawaii's economy so dependent on sugar exports, Americans there were desperate for a reciprocity treaty with the United States. Short of outright annexation, which was then not widely popular outside of the community of American expatriates, all the Hawaiian government could conceivably offer the United States in exchange was a naval station on the Pearl River. It was a trade favored by the American elite in the government of the islands, as it stopped short of annexation but accomplished the broadly desired reciprocity treaty.[21]

When Schofield and Alexander returned to California, they reported that while Honolulu offered an excellent commercial harbor, only the nearby anchorage along the Pearl River could accommodate naval vessels and be adequately defended in combat. It was capacious, possessed deep water for ships and fresh water for their engines, and bordered enough space along the bank for defensive artillery and supplies like coal. Only a tangled network of dead coral prevented the largest ships from entering. Considerable labor would be required to remove it, but the generals advised further study. Yet given existing American indifference to the islands, Schofield had to work to build up the significance of such a station. "The value of such a harbor to the commerce of the world and especially to that of the United States is too manifest to require discussion," reported Schofield to William T. Sherman (who had not previously thought the islands particularly valuable). "It is the key to the Central Pacific Ocean, it is the gem of these islands, valueless to them because they cannot use it, but more valuable to the United States than all else the islands have to give." With Alexander, Schofield further hinted that everything the army and navy might desire "would probably be freely given by the Government of these islands as a *quid pro quo* for a reciprocity treaty."[22]

Striking that quid pro quo proved more difficult than Schofield had imagined. It was not until 1875 that the Senate approved a hard-fought reciprocity treaty with Hawaii, and it took yet another year before Congress passed the required legislation to put it into effect. This treaty did not cede Pearl Harbor, as this proposal proved still too controversial—among native Hawaiians because they were concerned about losing national territory and among some American expatriates because they worried it could put the ultimate goal—complete annexation—further out of reach. The Senate did add a provision, however, prohibiting Hawaii from granting land to any other power. Making Hawaii off-limits to others, especially Britain, was what American strategic thinkers considered most essential. If American sugar planters sought reciprocity (and annexation) as a path to lucrative commerce, people like John Schofield in the army and David Dixon Porter in the navy saw reciprocity as primarily a way to manage an American vulnerability—so long as Hawaii remained independent. When Porter, the highest ranking admiral of the navy, justified commercial reciprocity in 1875, he presented American strategic interests there as essentially defensive. The kingdom was fast loosing population and what population it had was aging (nearly half, he claimed, were over forty) and about a thousand— 2 percent of the population—resided in a leper colony. Its destiny, according to Porter, was to submit to the political orbit of a stronger power. The contenders were obvious. Hawaii fell between British Columbia and the new British colony of Fiji, and Porter asserted that if Britain acquired Hawaii as well, "it would complete a chain of naval stations which would practically close the Pacific ocean to the American navy and commerce." Germany, too, he worried, "is seeking outposts for naval depots and stations." The problem was the colonial policies of European governments. When he explained that "the present conditions of naval service is such that fleets cannot keep the sea a long while and outlying posts are necessary to effective service," he meant that the American navy would be unable to defend American interests against another power that was itself based in Hawaii. Doing so would be possible only at "great expense on the part of this Government" Fear of future constraints, not technological imperatives, guided Porter's analysis.[23]

Preventing Britain or another power from establishing a coaling station in Hawaii mattered far more than developing an American one there. Even after reciprocity had produced a more than twentyfold increase in sugar exports during the last quarter of the nineteenth century, the American government had done little to create a coaling station along the Pearl River. White American elites on the island tried their best to draw the United States in, however. A coup against the Hawaiian king Kalakaua in 1887 brought land-owning Americans on the island to power. The new rulers quickly agreed to an addendum to the 1875 reciprocity treaty that included an article giving the United States "the exclusive right to enter the harbor of Pearl River, in the island of Oahu, and to establish and maintain there a coaling and repair station for the use of vessels of the United States," yet still no construction took place. Henry Peirce's successor, John L. Stevens, a veteran diplomat who had been appointed minister to Hawaii in 1889, seized on this languishing provision to argue that this coaling station was in the interest of the United States. Stevens hoped to use such a station to advance the prospects of annexation. Conveniently, he had an attentive ear near the White House, as his former newspaper partner was James G. Blaine, at the time serving as President Harrison's secretary of state. Blaine and Stevens alike favored Hawaiian annexation. Watching Harrison press for American coaling rights in Samoa, Stevens raised the subject with his former partner. "If it is well to have a coaling station at the Samoan group," he wrote, "how much more important it is to have one at Honolulu." For Stevens, the "well-known truths as to the necessity of good stopping places and coaling stations" made them essential for commerce and defense alike. But his argument for these stations in Hawaii was essentially economic: dramatic fluctuations in the price of coal on the islands injured both commercial vessels and naval ones. His perception of the value of a Hawaiian coaling station was very different from Porter's just fifteen years earlier. Stevens proposed a long-term harbor lease to supply American ships at the lowest possible cost. It was a matter of economy— ideas "which business men would carry out in their private affairs." It was a step in the direction of annexation, but it would aid the country even if Hawaii long remained independent. And whatever the United States ultimately chose, Stevens simply wished it would choose *something.* "Napoleon's axiomatical remark that 'an army marches on its belly' has an equally forcible application to commerce as to war," he wrote in 1891. "Whether the agencies of transport are caravans, railroads, steamers or electrical forces, there must be feeding places, coaling stations, and storehouses." Providing for these new agencies of transport would further help keep the islands within an American commercial and political orbit.[24]

By the end of the century, Congress, at least, had been persuaded of the value of Hawaii to the American navy. Other American efforts to argue for the necessity of coaling stations were less successful. When Ambrose Thompson returned to Washington in the 1870s, he again peddled his Chiriquí grant for its supposedly valuable coal and strategic location for a coaling station on the isthmus. Thompson presented his scheme to Ulysses Grant in 1874 and again to Rutherford Hayes in 1877, but he did little more than lobby until February 1880, when Ferdinand De Lesseps arrived in New York to raise money for his Panama Canal project. The prospect of foreign capitalists or foreign governments having control over a prospective isthmian canal sent fear through Washington. In response, two months later, the House Committee on Naval Affairs reported a joint resolution demanding the navy secure naval and coaling stations on either side of the American isthmus. Within a year, Hayes and Congress had agreed to appropriate $200,000 for the stations. Their location was unnamed in the legislation but was widely understood to mean along the Chiriquí coasts— Thompson's lobbying had again convinced members of Congress that his leases remained valid. Again, however, the scheme foundered when the incoming Garfield administration distanced itself from the project after the old legal and diplomatic objections again came to light. Sharper naval officers pointed out that in claiming that the project would save $250,000 a year on isthmian coal Thompson seemed to be overlooking the fact that the department spent $300,000 to supply the entire navy and only $15,000 a year on the isthmus.[25]

The project, however, refused to die. After Thompson's own death in 1882, his Chiriquí grants wound up with a group of investors led by Sylvanus C. Boynton and former California senator William McKendree Gwin. Like Thompson, the new speculators sought to turn government investment into a lucrative private payoff. Boynton, Gwin, and their associates, who included disgraced Kansas ex-senator Samuel Pomeroy and former Nevada senator William Sharon, revived the idea of building a railroad across Chiriquí to counter the French-funded Panama Canal project and again offered the U.S. government the inducement of naval stations and local coal at deeply discounted rates. William Chandler, the new navy secretary, appeared to support the scheme, and a former American minister to Colombia, Ernest J. Dichman, agreed to lobby Bogotá and San Juan. The promoters argued that they needed American naval and coaling stations to build their isthmian road, as only a material display of American naval strength could bring the security necessary to safely begin construction. "We only await the presence of the American flag," noted Sharon to President Chester A. Arthur, "without the protection of which we could not build the road." Of course, the investors also needed government capital to fund their project.[26]

Unfortunately for the former senators, they proved even less persuasive than Ambrose Thompson had been. "What a sad government we have, for so great a nation," sighed Gwin's attorney when Arthur's administration seemed to allow France and England to tighten their grip on transit across Panama. "I am sick when thinking of it." Arthur had evinced an early interest in the project but grew concerned that stock of what had been named the Isthmus Pacific Railroad might still fall to English capitalists. Such vagaries were an inherent limitation of conducting foreign policy via private investors. Gwin replied with an offer to allow the United States to purchase the railway itself, but the suggestion made little difference. Ultimately, Arthur and his secretary of state, Frederick Frelinghuysen, abandoned the plan, reasoning that above all, it jeopardized a more promising treaty with Nicaragua to build an isthmian canal there instead.

By then, however, it was clear that the supposed importance of Chiriquí as a coaling station was disconnected with either plausible expansion opportunities through diplomacy or a coherent conception of American strategic needs. It was instead simply an excuse for a flailing profit-making scheme.[27]

Yet if Arthur rejected the Chiriquí coaling station scheme for the last time, his administration hardly remained uninterested in pursuing these stations elsewhere, including elsewhere in Central America. During his time as Arthur's navy secretary, William Chandler sketched out the most expansive global map of prospective American coaling stations the country had yet seen. In his report of 1883, Chandler asked Congress to establish coaling stations "at some or all of the following points; Samana Bay, or some port in Hayti; Curaçao, in the Caribbean Sea; Santa Catharina, in Brazil; the Straits of Magellan; La Union, in Salvador, or Amapala, in Honduras; Tullear Bay, in Madagascar; Monrovia, in Liberia; the Islands of Fernando Po; and Port Hamilton, in the Nan-how Islands of Corea. . . . Similar stations should in addition be maintained, one at the best point on the Atlantic side of the Isthmus of Panama and another at the islands of Flamenco, Perico, Calabra, and Ilenoa on the Pacific side, now owned by American corporations." He reiterated this request in his 1884 report. The length of the list suggests a voracious imperial appetite, but a closer look reveals the familiar aspirations of increasing American trade by establishing steamship infrastructure and lowering coal expenses for naval and merchant vessels.[28]

It was, in fact, simply a more detailed list of prospective depots that Chandler's recent predecessor as navy secretary, Richard W. Thompson had described as valuable "to promote commercial intercourse where it is already established, or to invite it where it is not." According to Thompson (Ambrose Thompson's former lawyer), the significance of these stations was economic: to help American vessels avoid the burdens imposed by "extortionate monopolists" who were selling coal abroad and supposedly stifling American commerce. Thompson had been influenced by Commodore Robert Shufeldt, the chief of the navy's Bureau of Equipment and Recruiting, which supplied coal to the fleet. Observing both the general decline in the navy and the economic turmoil of the mid-1870s, Shufeldt believed the answer to these difficulties rested in increasing American trade—"the re-creation of our commerce through the absolute necessity of procuring a market for our surplus products." With a new subsidized fleet of swift mail steamers and a network of foreign coaling stations to fuel them, the commodore imagined both rebuilding the reserve force of the navy and increasing American commerce to manage the chaos caused by industrialization. To further this vision, in 1878, Shufeldt began a two-year cruise aboard the *Ticonderoga*, part of which took him around Africa, India, and east Asia (with special visits to Liberia, Fernando Pó, Madagascar, Korea, and elsewhere). The prospective stations requested by William Chandler in 1883 and 1884 were the result of

Shufeldt's recommendations, based on what he learned during this mission. Taken together, the Curaçao-Santa Catharina-Straights of Magellan route would have provided way stations down the eastern South American coast, around Cape Horn to the Pacific. The African and Korean depots would have served to open American trade in potentially vast markets. The isthmian stations, debated since the 1850s, would have given the United States greater leverage over a prospective canal. At Port Hamilton, a cluster of islands southeast of the Korean peninsula, Chandler explicitly called on Congress to establish a mail steam line connecting Korea with the Pacific Mail Steamship Company's harbor in Japan. This wish list of stations thus represented a theory that government funding for the infrastructure of trade would inevitably increase trade itself.[29]

That Chandler's prospective coaling stations would primarily serve economic rather than security purposes suggests why similar stations had already been secured with little controversy. The American depot at Yokohama, for example, had been established on behalf of American steamship lines after the Civil War. In 1865, Congress began again to grant subsidies for mail steamers, allocating $500,000 a year for a transpacific mail, freight, and passenger service. The legislation left the choice of contractor up to the postmaster general, yet no one was surprised when the Pacific Mail—by far the western ocean's dominant transportation line—won the ten-year contract. When the American chargé d'affaires in Yokohama learned of bill's passage, he joined his British counterpart and the French minister to secure coal depots in the growing commercial port for commercial steam lines of each country. The governor of Kanagawa prefecture, which embraced modern Tokyo Bay, quickly consented. The action was, in part, a way to ensure equal access to the trade of Yokohama, as the British Peninsular and Oriental Steamship Company had already established its east Asian line there and the French Messageries Imperiales was expected to do so soon as well. The move supported American commerce by securing coastal land that the expatriates expected would quickly rise in value. The U.S. government first leased half, then in 1871, all the land to the Pacific Mail, expecting that the arrangement would additionally ensure naval vessels a reliable and less expensive coal supply when cruising far eastern waters.[30]

The Diplomacy of Limits

As Americans sought coaling stations to build the infrastructure of commerce, they also dealt with the separate but not unrelated problem of fueling the American navy during wartime. While commercial stations could be pursued in the neighborhood of other nations or the stations of other maritime powers, industrial warfare presented certain problems around coal that were not faced during peacetime. The claim that modern warfare made coaling stations necessary elided a subtle distinction. Steamships required coal, of course, but the barriers to refueling during wartime were as much legal and political obstacles as technological or geographic ones. The first obstacle stemmed from how international legal conventions incorporated fossil fuel into the law of contraband.

Since at least back to Grotius in the seventeenth century, scholars of international law had debated the scope of what constituted contraband, the articles a neutral power was prohibited from trading with a belligerent. Some items, like munitions, were unambiguously prohibited. Beyond that, however, opinions diverged. Some jurists, like Emerich de Vattel, included timber and naval stores; others, like his older contemporary Cornelius van Bynkershoek, disagreed, noting that ultimately, any material could be fashioned into an implement of war. Chief Justice Salmon Chase, ruling on a contraband case in 1866, acknowledged that it was a subject that "has much perplexed text writers and jurists."[31]

Steam power created new problems for defining contraband as warfare grew more dependent on coal. "Here is an article," wrote Secretary of State Lewis Cass during the Second Italian War of Independence in 1859, "not exclusively nor even principally used in war, but which enters into general consumption in the arts of peace, to which indeed it is now vitally necessary." Merchant ships needed coal, Cass observed, and as some nations possessed mines while others did not, the coal trade ought to expand, not contract, even during wartime. Cass expressed the position of a neutral nation with vast coal resources that sought to export coal to other nations at war. Yet international law had not kept up with technological change. "The attempt to enable belligerent nations to prevent all trade in this most valuable accessory to mechanical power has no just claim for support in the law of nations," he argued, protesting that contraband ought to mean only actual "arms and munitions of war."[32]

The experience of the Civil War tempered American enthusiasm for that narrow definition, however. Union cruisers had little trouble entering foreign ports in the Caribbean but found that British restrictions on coaling changed over time and were applied differently in different places. Lincoln's government might have wished for greater latitude in purchasing coal, but it also would have welcomed greater restrictions on Confederate vessels. The war, however, revealed how a neutral state trading in coal could lead to undesirable entanglements with belligerents. Later, during the Franco-Prussian War in 1870, the United States effectively adopted the British position, placing strict constraints on fueling warring ships. As far as the law of war was concerned, the question remained in what ways choosing or refusing to coal a belligerent ship violated neutrality. Was coal entirely contraband? Or should it be furnished to all belligerents indiscriminately? If only under certain circumstances, who should decide and by what standard? Could a neutral state be held accountable for acts of war committed by the recipient of that nation's coal?[33]

These questions around the legal status of supplying coal during wartime oc-cupied a significant amount of time at the first modern international arbitra-tion tribunal ever held, in Geneva in 1872, to settle the claims leveled by the United States against Britain for its conduct during the American Civil War. It was well known that British shipyards built steamers for the Confederacy, most notably the *Alabama*, and in the United States, Americans spoke generally of the need to resolve the *"Alabama* claims" against Britain. But American represen-tatives at the Geneva arbitration pressed larger claims that raised questions as to what neutrality meant in an era of widespread steam power. Much of their case against Britain revolved around how its colonies had facilitated blockade-runners and Confederate cruisers. The American representative, Charles Fran-cis Adams, best captured the challenge of articulating international law on this subject. The legal place of coal in warfare was new, he observed, and yet "has become one of the first importance, now that the motive power of all vessels is so greatly enhanced by it." This was true for all nations, but according to Ad-ams, Britain's massive coal industry, global distribution system, and network of fortified coaling stations gave the country a unique place in modern industrial warfare in those conflicts in which it remained neutral. In times of war, Brit-ain could simply decide that neutrality required withholding coal from all bel-ligerents. Adams believed that this position would have perverse consequences. Not only would it be seen "as selfish, illiberal, and unkind" but "it would in-evitably lead to the acquisition and establishment of similar positions for them-selves by other maritime powers, to be guarded with equal exclusiveness, and entailing upon them enormous and continual expenses to provide against rare emergencies." That is, Adams understood that most countries could not and believed they should not aspire to emulate Britain's network of naval outposts because doing so would result in an arms race and worldwide land grab, mili-tarizing the globe with jealously guarded fortresses and costly colonies. The ex-ceptional times of war should not excessively constrain the normal times of peace. Instead, he argued, Britain, like any neutral, should act responsibly when fueling belligerent ships, while ensuring that its aid would not directly assist in acts of war. These judgments, of course, could never be perfect, but for Adams, Britain ought to be safe by international law when it could demonstrate it acted "in response to a demand presented in good faith."[34]

Ultimately, the tribunal agreed that during wartime, coal could be neither entirely contraband nor perfectly unconstrained but must be seen as something in between. Whatever the terminology, no country, including the United States, wished to relinquish its own discretion in those circumstances in which it might find itself the neutral party. But the participants in the arbitration also agreed that neutrals could not indiscriminately open a port to belligerents and make it "a base of naval operations" either. Instead, the tribunal voted that "supplies should be connected with special circumstances of time, of persons, or of place, which may combine to give them such character." In other words, the coaling question could not be answered in any general way. This decision was not unanimous; the Brazilian representative, Marcus Antônio de Araújo, Viscount d'Itajubá, signed on to the majority decision but admitted he believed that every neutral should chose how it handled coal for itself. Sir Alexander Cockburn, representing Britain, agreed with Viscount d'Itajubá, so long as the neutral treated all belligerents equally. The majority, however, accepted that circumstance must determine whether any neutral nation illegally allowed warships to use their ports as bases of operations. These views guided the law of war into the twentieth century.[35]

Gradually, Americans adopted the English construction of coal as the most important article classified as "conditionally contraband." But while Charles Francis Adams saw no reason a coal-fired navy would require its own fortified coaling stations, others took a different view. It all depended on the kind of war a nation wished to be able to fight. Confederate cruisers like the *Alabama* and *Florida* had devastated American commerce. If subject to the kind of constraints the Grant administration had imposed during the Franco-Prussian War—only providing enough coal to belligerent vessels to reach the nearest home port and refusing to refuel a ship within three months unless it touched a home port first—it was not clear whether commerce destroying as a naval strategy would be possible. In 1885, an exasperated Benjamin Butler wrung his hands at the constraints the United States had submitted to, what he called a "fraud upon the American nation" and "the greatest loss of all!" If future neutral nations were held liable for damage committed by belligerent steamers they refueled, how would the United States ever fight a naval war again? Whatever the solution, the origins of the problem lay in international law.[36]

The second political obstacle surrounding coaling diplomacy had even deeper roots in American political culture. Ever since Washington's 1796 Farewell Address admonishing the United States to make its "true policy to steer clear of permanent alliances with any portion of the foreign world," Americans had resisted peacetime promises of future military assistance. Washington, of course, did not advocate a studious isolation from world affairs. Far from it. He urged instead a dispassionate engagement with other nations solely according to American "duty and interest." Both excessive fondness or undue hatred, be they expressed through trade or treaty, led only to conflict, jealousy, and unforeseeable obligations. Washington believed both tendencies made the nation a slave. Washington also considered foreign partners unreliable. "There can be no greater error than to expect," he explained, "or calculate upon real favors from nation to nation." These ideas hardly prevented Americans from advocating war in the nineteenth century, but they did act as a profound constraint on how Americans pursued coaling stations.[37]

The avoidance of what came to be known as "entangling alliances" (the phrase came not from Washington but Jefferson's first inaugural) did not prevent the acquisition of foreign coaling stations but instead prevented the country from entering into the kinds of agreements that would ensure the availability of coal in distant ports during wartime. Yokohama was one of the uncontroversial stations to which Americans knew they would probably lose access during war. Another was Pichilingue, Mexico. Some time before 1867, the American consul in La Paz, the small town at the southern end of Baja California, negotiated for a naval coaling depot on the nearby island of Pichilingue. Beginning in 1871, a collier arrived yearly from Philadelphia to replenish the depot's supply. But the arrangement revealed the weak hand of the United States. When the chief of the navy's Bureau of Equipment sought to relocate the depot to Magdalena Bay on the Pacific side of Baja a decade later, the Mexican government refused. The Mexican depot was little more than rented land, entirely subject to Mexican authority, and there was no guarantee of access during wartime.[38]

In other cases, the terms for acquiring stations represented too great a stretch for American political culture. Even for an ardent commercial expansionist like James G. Blaine, there were limits to what he would do to build a global coaling infrastructure, an issue he faced during both his terms as secretary of state. In Lima, the American minister, Stephen Hurlbut, had negotiated the use of the Pacific harbor of Chimbote as a coaling station in 1881. Peru was then in the midst of the devastating War of the Pacific and desired American assistance. In exchange for $2 million in cash and stock, the agreement granted a naval station and ceded a flailing railroad line to carry coal to the harbor. Learning about the prospective agreement during his first, short tenure in the State Department, Blaine demurred. His first concern was timing. In Peru, the agreement was likely to give the appearance of American coercion at a time of need, while in neighboring Chile, it might suggest sympathy for her adversary. Later, Blaine acknowledged that the proposal was simply of little value to the United States. By its terms, Peru insisted on maintaining legal jurisdiction, the right to grant similar concessions to other states, and the authority to terminate the agreement unilaterally. "A naval and coaling station on the South Pacific coast," Blaine explained, "carefully chosen, with the aid of the professional knowledge of those specially qualified to determine its capacity to answer the wants of our national ships, and over which we might exercise proper and necessary jurisdiction, with a secure tenure, would be of undoubted value, and this government, at a fitting time, may be willing to negotiate upon fair terms for such a privilege." This deal, however, effectively only granted the United States the right to coal at Chimbote—which the United States could do already. There the negotiation abruptly ended until 1889, when Blaine returned to head the State Department.

Learning from the his new Peruvian minister that Lima again sought to grant the United States a coaling station at Chimbote, Blaine, at President Harrison's urging, asked the minister to negotiate informally. This time, the terms again represented an impossible burden for Congress—a American guarantee to protect Peruvian borders and funds to meet a coming $10 million payment to Chile to help dispose of the lingering dispute over the provinces of Tacna and Arica. American political culture at the time was simply not prepared to accept such terms.[39]

Entrepreneurial diplomacy also resulted in proposals that were disconnected from strategic needs as perceived in Washington, let alone political possibilities. In 1891, Whitelaw Reid, the former *New York Tribune* editor serving Harrison as minister to France, quietly revealed a scheme even bolder than the one proposed for Chimbote. Portugal, then embroiled political turmoil, a financial crisis, and a colonial struggle with Britain in southern Africa, sought to entice the involvement of the United States with offers of coaling stations in its far-flung empire. Portugal's minister of finance, Cyril Mariano de Carvalho, offered depots in both Mozambique and Angola, along with a third in the Azores. To Reid's surprise, the president immediately rejected the proposal, suggesting instead a modest program of trade reciprocity. Blaine called the idea "entirely inadmissible." Reid was familiar with talk of the needs of the New Navy of steel and steam, but he wrongly assumed that any prospective additional coaling depots were inherently valuable independent of naval strategy or the patterns of American commerce. He similarly underestimated the aversion to tying American security to that of other nations in peacetime. As much as Harrison's administration of foreign affairs appeared to mark a break from tradition, when it came to the problems around coal and security, it drew a very familiar line, resisting international arrangements that might bind the United States in the future.[40]

Whatever the constraints, most Americans concerned with the new limits and possibilities of industrial energy after the Civil War did not look to facilitating security by obtaining foreign territory. Instead, they turned to mathematics and engineering, pursuing technical solutions to challenges of coaling. These efforts undercut the sense of inevitability that attended the idea of acquiring coaling stations and reveal how technological innovation often had a greater appeal than territorial expansion.

Engineering Experimentation, 1865–1900

"It is not generally recognized" observed George Washington Littlehales in 1899, "that science, employing the mathematician and the engineer alike in the problem of shortening the duration of ocean transit, has accomplished as much by causing ships to travel fewer miles as by causing them to travel faster."[41]

Littlehales, an engineer in the U.S. Hydrographic Office, was commenting on one of the most far reaching transformations in ocean navigation since the introduction of steam power itself. Traveling fewer miles by sea had become possible not only because steam power allowed ships to travel independently of the wind but because new mathematical techniques developed in the nineteenth century allowed navigators to calculate new routes more directly than ever before. These new techniques facilitated navigation along a great circle.

A great circle is a mariner's fiction. It describes the imaginary path tracing the shortest distance along the surface of the earth. Imagine sticking two pins into an ordinary globe, one at San Francisco and the other at Yokohama. Stretch a thread tightly between the pins. The path traced by the thread describes an arc of a great circle. Mathematicians dubbed it an orthodromic curve. This path between San Francisco and Yokohama does not run through the central Pacific as a map with an ordinary Mercator projection suggests, but much farther north, near the Aleutian Islands. If this thread extended beyond both pins and met again on the other side of the globe, the complete curve would represent a circumference of the planet, the largest circle one could measure on the globe: a great circle. The equator traces a unique great circle, everywhere equidistant from the poles. Meridian lines are great circles as well, stretching from one pole to the other and back again, crossing the equator at right angles.[42]

Navigators had understood for centuries that routes following a great circle minimized distances. Littlehales himself speculated that "knowledge . . . of the great circle must have been coeval with the knowledge of the spherical form of the earth." The earliest English authors on navigation were certainly aware of them. The Elizabethan John Davis called great circles "the chiefest" of all possible routes. In the seventeenth century, Henry Phillippes called them "the most exact way." Yet knowledge of great circles in theory did not translate into their use in navigational practice. Many standard navigational texts as late as the early nineteenth century barely discussed great circle sailing, if they mentioned it at all. Characteristic of this neglect was Nathaniel Bowditch's *New American Practical Navigator,* "the seaman's bible," well into its eighteenth edition in 1848 before its editors even added a section on great circle sailing. Before steam power first augmented and then came to dominate ocean propulsion, great circle routes could be traversed only when they coincided with favorable winds— which was rare.[43]

Sailing ships depended on wind power and wind circulated in great cells. Catching the wind thus required different routes for each leg of a round trip voyage, one leg frequently being much longer in distance and transit time than the other. Sailing ships between China and the United States experienced particular challenges. Beginning in China, navigators were advised to follow the great circle, catching the mighty, warm Japan stream northeast toward the Aleutian

Archipelago before veering southward along the North American coast. The journey from the United States to China was more circuitous. Ships left from San Francisco or Puget Sound bearing south toward Mexico. There, they picked up the northeasterly trade winds between 15° and 20° north, a band of latitude that includes most of the Yucatán Peninsula. Navigators took care to avoid the quiescent "horse latitudes" a bit further north. Crossing the Pacific westward, they sailed south of the Hawaiian Islands (or they might stop there for trade or supplies) and then north of the Marshalls. Seasonal weather determined what came next. Southwest monsoons between May and October forced ships to travel between the Caroline Islands and the Marianas, where ships could find respite at the archipelago's southernmost island of Guam. Otherwise, between October and April, northeast monsoons drove vessels north of the Marianas, past the volcanic peaks of the Farallones de Pajaros before they finally entered the Philippine and East China Seas.[44]

The introduction of steam power did not immediately obviate the popularity of this central route across the Pacific. Authors of navigational texts blamed the continued avoidance of shorter great circle routes not on ignorance of the concept but instead on the difficulty of performing the navigational calculations it required. More conventional routes, however lengthy or roundabout in practice, required only simple determinations of the ship's course. They were easy to chart and straightforward to travel. Great circles, in contrast, required ponderous, repetitive calculations. Navigators had to plot frequent course adjustments every one or two hundred miles. "It has been found impossible to introduce the general use of great circle sailing" lamented one mathematician, citing the difficulty of "fresh calculations or constructions by no means simple" when a ship inevitably deviated from an originally plotted course. John Towson, the developer of one technique that simplified great circle sailing, explained that course adjustments needed "so often to be repeated as to preclude its being generally adopted." In the Hydrographic Office, Littlehales complained of "tedious operations" and "the want of concise methods for rendering these benefits readily available." If the technology of steam engines liberated ocean transit from the wind, it left more fundamental mathematical problems of navigation unsolved.[45]

Yet on long voyages far from domestic coal markets, fuel was expensive, and as both commercial and naval steamships grew in size and strength, minimizing a ship's steaming distance offered a way to stretch resources beyond existing limits of engine design and available coaling depots. When Congress approved a new steam mail subsidy for a line between San Francisco and Shanghai in 1865, legislators required the contractor to touch at Honolulu on both the outgoing and incoming voyages. The contractor, the Pacific Mail Steamship Company, quickly protested the Hawaiian detour, arguing it would add days to the voyage compared with the northern great circle route—almost four heading west and six returning home—and unnecessarily demand distances so vast that any ship built for the route would be burdened with what the postmaster general called "heavy expenses" of coal. An immediate concurrence came from the Senate Committee on Post Offices and Post Roads, which oversaw foreign mail contracts. All endorsed the great circle route instead. Charles H. Davis, the rear admiral who superintended the Naval Observatory, remarked that the United States could certainly construct gigantic vessels with enormous engines that plied sea routes ignoring natural advantages, but then the country "must also be prepared to leave this field of enterprise, at no distant day, to those who will obey the laws governing the navigation of the great seas." The great circle route was not perfect—it passed though less familiar, colder, foggier waters and adverse weather could strain a ship's engine, forcing it to exceed the time needed on the longer, more southern route. According to a Coast Survey assistant in Alaska, only "the discovery of deposits of good coal among the Aleutian Islands, or within a reasonable distance of the harbor nearest the great circle route" could establish the superiority of the great circle. For others, the solution was to maintain technological hybridity, following a great circle "provided the steamers carry sufficient sail to enable them to take advantage of favorable wind, and sav[ing] their coal for use when the wind comes out ahead." After a debate, weighing the burdens on the Pacific Mail and the value of cultivating economic and political ties to Hawaii, Congress voted to allow the company to steam the great circle route.[46]

When the Pacific Mail did seek a depot more centrally located in the Pacific, it settled at first on a pair of islands that came to be called Midway. In 1859, the captain of a sealing bark, N. C. Brooks, had sighted the islands while sailing out of Hawaii in the north Pacific. Not appearing on his maps, Brooks planted a flagstaff, named the islands for himself, and after discovering extensive guano deposits, claimed it for the United States under the 1857 Guano Islands Act. Brooks noted that the islands' location, about a third the distance between Honolulu and Japan, offered the only opportunity for a coaling station between the United States and China besides Hawaii itself (and skipping Honolulu altogether in favor of Brooks' islands would have, in fact, saved five hundred miles of steaming). After the Civil War, the Pacific Mail, now holding a transpacific contract, persuaded the navy to order a survey. On August 28, 1867, Captain William Reynolds, accompanied by an agent of the Pacific Mail, fired twenty-one guns, raised the flag, and formally took possession—not as a guano island but a prospective coaling station. The Pacific Mail's expectation of developing the island was dashed, however, when the naval survey revealed that extensive shoals blocked access to the harbor. As the costs of dredging far exceeded what Congress was willing to appropriate given the politics of Reconstruction and the massive war debt, Americans quickly lost interest in the island and the company continued to steam the more northern route.[47]

To help navigators aboard both merchant and naval vessels, mathematicians throughout the second half of the nineteenth century worked to devise increasingly simplified methods for calculating great circle routes. By 1900, this flurry of mathematical activity by German, French, British, and American investigators had resulted in more than two-dozen techniques to improve steam navigation. The British astronomer royal, George Biddell Airy, for example, devised a method for superimposing an approximate great circle track on an ordinary Mercator chart using a table he prepared and simple geometry. While teaching at the Naval Academy in Annapolis, the mathematician William Chauvenet constructed a great circle protractor in 1854 that allowed the user to plot routes by rotating a pair of disks, one representing the globe and the other the path of any course desired (the research drained his finances so severely he was rescued only by the Hydrographic Office buying the device). Taken together, one set of methods for calculating great circles entailed examining specially prepared tables and charts. Another cultivated the geometry of the gnomonic projection, a cartographic technique that produced distorted maps in order to make great circles appear as straight lines. This approach was favored by the innovative Gustave Herrle in the navy's Hydrographic Office. By 1889, George Washington Littlehales could collect various methods in a single volume.[48]

The navy's Bureau of Navigation also championed the development of great circle navigation by producing special nautical charts. After years of interest in the project, it completed its first chart in 1865. Nevertheless, since "seamen are not generally the first to perceive and seize advantages," the department's chief hydrographer remained skeptical as to whether the techniques it offered would be adopted quickly. Over time, though, interest grew. By 1885, the navy was designing a series of five great circle charts, one each for the northern and southern portions of the Atlantic and Pacific, along with one for the Indian Ocean, as well as a kind of improved protractor, designed by Commander Charles D. Sigsbee, for graphically calculating great circle routes. A few years later, in 1894, Sigsbee had risen to become the Bureau of Navigation's hydrographer himself, noting the "great demand" for his office's great circle supplement to its North Atlantic pilot chart. With these developments, navigators increasingly traveled along great circles.[49]

These technical innovations in navigation were accompanied by developments in ship design and steam engineering that likewise allowed ships to travel further on the same quantity of coal. The subject was pervasive throughout the post–Civil War era. When Gideon Welles stepped down in 1869 after eight years heading the Navy Department, the new president, Ulysses Grant, named Adolph Borie the new secretary. But Borie, a successful Philadelphia merchant with little experience in public life, was little more than a cipher. Grant had in fact wished to appoint his old friend, the navy's vice admiral David Dixon Porter, but demurred to respect the expectation of civilian control of the department. Instead, he gave Borie the post and Porter the power. By Grant's instructions, matters that ordinarily went straight to the secretary passed across Porter's desk first. What followed was an extraordinary three-month period during which Porter issued (over Borie's signature) forty-five general orders. Many were trivial. Number 92 ended the shellacking of decks. Number 93 directed spars be painted black instead of yellow. Number 123 prescribed "gold-embroidered shoulder-loops" for junior officers. Other orders, however, signaled significant changes in policy. Number 128, issued on June 11, announced a near-complete return to sail. Number 131, distributed a week later, informed the fleet that given the burden of supplying coal, all naval vessels already capable of sailing "should not use their steam, except under the most urgent circumstances." Ships dependent on steam engines would soon be retrofitted. Any use of coal was to be thoroughly explained to the department. Though Porter justified the measures as economizing coal and training young sailors the art of seafaring, they have been widely condemned as backward by historians. Porter's actions, especially with respect to steam, have been called "reactionary measures" caused by "his nostalgia for the old seamanship under sail," and Porter himself was labeled "infamously foremost" of naval officers seeking to abandon steam.[50]

Historians typically view what Robert Albion calls the "Navy's Dark Ages" as nearing a close in 1882, when Congress agreed to funding to build three new cruisers, the *Atlanta, Boston*, and *Chicago*—the beginnings of the New Navy of steel and steam. But change came slowly. Porter's restrictions on coal consumption were not immediately reversed. Instead, they were largely reiterated late into the century. Even as newer vessels grew in size and power, the department continued to demand restraint. When the navy's Bureau of Equipment and Recruiting faced drastic cuts in its congressional appropriation in 1887, the department's acting secretary issued a blanket order to conserve fuel. "As coal is the largest item of expense to the Bureau," it read, "Commanding Officers of squadrons and of vessels acting singly will exercise the greatest economy of its use." Most steam-powered naval vessels in the 1880s maintained masts and rigging for sail, including the new cruisers, and the order instructed ships to rely on wind whenever possible. According to the circular, resort to coal "will be limited to occasions when dispatch is absolutely required, or to emergencies, but steam will never be used under the ordinary circumstances of cruising." When officers did fire their engines, they were to alert Washington immediately. The restrictions were only slightly loosened four years later, when a more detailed order again requested officers "to practice the utmost economy in the use of coal." The secretary, Benjamin Tracy, reaffirmed that vessels fitted for sails were to use them whenever possible while reporting any unusual circumstances to the department. Only when entering and exiting ports were ships to use steam. When they did, Tracy added that commanding officers across the navy should undertake experiments to maximize fuel economy. He requested lengthy reports that included fuel consumption across a range of speeds, details on the nature of the coal consumed, accounts of the physical condition of the ship, as well as "such remarks in reference to the most economical rate of steaming of the vessel as the experience of the Commanding Officer may suggest." Reminders to conserve coal continued into the early twentieth century.[51]

Porter's 1869 order was extreme but it was grounded in his war experience. During the 1850s, Porter had been an enthusiastic supporter of steam and between 1851 and 1853, he had commanded the Law Line's mail steamer *Georgia* between New York and Panama. Yet the war moderated his views. In 1861, when Porter was a lieutenant in the Caribbean, he encountered the kind of obstacles securing fuel that imperiled the Union effort. "There is an indisposition on the part of the Government to furnish us with coal," he wrote Gideon Welles from Jamaica, "and there is none for sale except at most exorbitant prices." The only solution, Porter suggested, was to gain access to a coaling yard from a Jamaican resident and maintain an American depot there. "Without some arrangement of this kind our steamers can not cruise in these waters." A year later, British neutrality regulations with respect to coaling would make Porter's suggestion impossible. As he rose in rank, he would spend the rest of the war ensuring his ships had coal, and the work was often a struggle. As he told William T. Sherman in 1862, "every bushel is worth its weight in gold."[52]

Ordering ships to largely abandon steam in 1869 hardly signaled a rejection of progress. Even then, sail was far from an antiquated technology. A decade after Porter's order, an exposition in the *International Review* (edited by a young Henry Cabot Lodge) explained that ideal vessels for the navy "should be designed especially for speed, and fitted with sufficient sail power to cruise under sail and to work well, because they can thereby economize in fuel, and can always carry enough coal in the bunkers for exigencies when speed is required." The debate over the use of sail continued into the 1890s, when Rear Admiral Stephen B. Luce, one of the leading intellectuals advocating naval buildup, continued to advocate for attaching sails to cruisers. And he was hardly alone. Until the 1890s, naval intellectuals infrequently considered securing distant coaling stations. Back in the 1860s, the exclusive use of steam power was not yet an inevitable future; for many ships, it was an undesirable present.[53]

Improving that undesirable present occupied naval engineers from the Civil War through the turn of the century. As chief of the navy's Bureau of Steam Engineering during the war, for example, Benjamin Isherwood had continued his research into improving fuel economy. His bureau examined nearly all varieties of coal from the eastern United States. It explored the value of engines designed to work steam expansively. It investigated the possibility of replacing coal with petroleum, a new fuel in commercial production in the United States only since 1859, before abandoning the idea after discovering that it released uncontrollable and explosive volatile gases. This feature presented a liability so grave that, according to Isherwood, "it is manifestly useless to experiment upon the best form of apparatus for burning it."[54]

After the war, Isherwood continued championing technological ways to conserve valuable coal. In 1881, he described a peculiar new invention by John Gamgee, a member of a scientifically prolific English family who was then visiting the United States to consult on a variety of public health matters. Gamgee had proposed what he called the zeromotor, a device that produced mechanical motion not by the combustion of conventional fuels but instead by boiling pressurized ammonia using the heat latent in ordinary room temperature water. As Isherwood noted in his report, the consequences of such a machine were enormous, especially to the U.S. Navy. An engine that ran without coal "would produce an industrial and consequently social and political revolution equal to what was effected by the introduction of the steam engine." With a zeromotor, no longer would the United States operate with the handicap of lacking coaling stations, and no longer would the British navy maintain a strategic superiority over the U.S. fleet. "If coal . . . can be dispensed with," Isherwood noted, "we are at once placed on an equality in this respect, and our cruisers enabled to penetrate the remotest seas as easily as those belonging to countries having possessions there." His description was vague but enthusiastic, and he asked that the navy offer Gamgee the use of the Washington Navy Yard for his research.[55]

Most observers greeted early reports with cautious interest. Isherwood's excitement was such that Gamgee was invited to demonstrate the device before President Garfield and two of his cabinet secretaries. Yet as American engineers learned more, they quickly grew critical. Next came puzzlement that Isherwood, one of the nation's most prominent engineers, could have endorsed the idea. The editors at *Scientific American* were especially scathing, going so far as to title their publication of Isherwood's own positive report "The Gamgee Perpetual Motion." The mathematician Simon Newcomb of the Naval Observatory explained Gamgee's error in terms of a violation of the second law of thermodynamics, a mathematical language unavailable to American inventors before the Civil War. As Newcomb explained, once the gasified ammonia raised a piston, there was no reservoir of cold to liquefy it again. Any mechanical means of compressing it would take the same power Gamgee hoped to realize from expanding it again. In consequence, Newcomb dismissed the zeromotor, explaining "it may be pronounced a chimera with as much safety and certainty as we call perpetual motion machines by that name."[56]

Isherwood maintained that the zeromotor was "far from chimerical." There were, however, more conventional ways to engineer solutions to America's lack of coaling stations. Without these stations, the long-running debate over abandoning sail or retaining it often hinged on pursuing inventions to facilitate coaling at sea. Into the later part of the century, coaling was a time and labor-intensive process. On St. Thomas, often cited as possessing model coaling facilities, advanced steam technology still depended on traditional sources of labor: "The ships were coaled by women," recalled one retired rear admiral, "who formed a procession from the coal-pike, each on carrying a basket on her head. In this way a ship was rapidly coaled." Island resident and booster Charles Edwin Taylor encouraged his readers to observe the ships coaling at midnight, when one might "watch the dusky figures of hundreds of women, each with a basket of coal on her head, swarming up the steamer's side busy as bees, and running back again with them empty, to be refilled." Taylor lauded the women's industriousness and praised their singing "in a quaint minor key." On St. Thomas, the newest form of energy remained bound entirely to the oldest.[57]

Instead of focusing on how coaling could be accomplished in geographically fixed ports like St. Thomas, some engineers pursued an alternative of coaling at sea. "The *desideratum*, I submit," argued Sigsbee in 1890, "is not sail power, but a method of fueling in a seaway." "WHY SPEND MONEY FOR COALING STATIONS?" blared a *New York Herald* headline in 1891. Unlike Britain, the United States lacked colonies it needed to defend, and even if it were to possess impenetrable depots around the world, an enemy's blockade would immediately make refueling there impossible. The *Herald* promoted instead having a fleet of swift steam colliers that could transfer fuel in open ocean. The North Atlantic flagship cruiser *San Francisco* and the steamer *Kearsarge* tested one such device in 1893. A cable connected the vessels while sailors launched sacks of coal weighing almost two hundred pounds apiece from one ship to the other. Under the carefully designed circumstances of the experiment—a calm sea and a short distance between the vessels—the delivery proceeded slowly, but it offered hope for a more robust technological solution in the future. "Any one who will devise a method of rapidly and safely coaling our cruisers at sea will add to the navy's efficiency and, no doubt," noted *Scientific American*, "will receive an abundant reward in dollars from the government."[58]

The *San Francisco* and *Kearsarge* experiment was part of a larger process of engineering experimentation. Between 1880 and just after the turn of the century, American and British engineers devised a range of inventions to facilitate coaling at sea. The devices were typically systems for connecting two ships together by ropes or wires and transferring bundles of coal from one to the other. The most significant innovation came in 1893 from Spencer Miller, a young American civil engineer. Miller's design connected the stern of a warship to the bow of a collier by a cable held taut by a motorized winch on the collier. An endless rope shuttled bundles of coal attached to the cable. It took five years, but the American navy eventually began examining the device, discovering that even during a violent storm, some 375 tons could be transferred over twenty-four hours. The advantages of coaling at sea were manifest to Miller. "It is not my intention to enter into any controversy with the advocates of coaling stations," he wrote Admiral George Dewey in 1902, "but the facts that have been established should certainly be taken into account in considering the question of coal supply for the U.S. Navy." Miller joined those who saw in coaling stations "a source of weakness to the Government owning it, rather than a source of strength." Not to mention that maintaining a fleet of colliers capable of coaling at sea was vastly less expensive than acquiring and maintaining distant stations. And as Miller had hoped, within a few years, all American naval colliers carried the device for rapid coaling.[59]

Even more importantly, the lack of coaling stations abroad before 1898 forced Americans to think differently about naval strategy and ship design. Some, like

View from the quarterdeck of the USS *Massachusetts* during one of several tests of Spencer Miller's apparatus for coaling at sea in 1900. Here, bags of coal run down a zip line from the attached collier *Marcellus*. In the late nineteenth and early twentieth centuries, coaling at sea offered Americans a way to maintain naval steamers around the world without an extensive network of coaling stations. Spencer Miller, "The Coaling of the U.S.S. Massachusetts at Sea," *Transactions of the Society of Naval Architects and Marine Engineers* 8 (1900), folder 2, box 12, GDP.

the officers of the Naval Advisory Board of 1884, argued that the resolution to the question of sails aboard cruisers was abandoning canvass, increasing coal capacity, and acquiring coaling stations, at least in Hawaii and the Caribbean. But even the Naval Policy Board, which released its own detailed proposal for naval construction in 1890, only mentioned the value of a single station in the mid-Pacific. Instead of a wide pursuit of stations, the bulk of this report emphasized building ships of greater coal endurance. However, it also noted that the shift to battleship fleets served fundamentally defensive purposes, and thus these battleships would only need to operate close to continental shores (and thus close to domestic coal supplies). The officers of the board, led by Commodore W. P. McCann, believed that the United States could achieve security despite building a fleet weaker than the strongest navy in the world, so long as it remained capable of destroying foreign coaling stations in the Caribbean that prospective enemies would need for their operations near the American coast. With this view, the officers of the board joined most of the naval establishment, and into the 1890s, the board placed great emphasis on the technological solution of designing new vessels with vastly larger steaming radii in case the navy needed to steam far from home ports.[60]

When Congress began funding new naval construction in the early 1880s, plans drawn up by both private contractors and the navy itself reflected American limitations and turned them into technical assets. First, naval engineers placed a premium on coal endurance for cruisers, the fast vessels designed to hunt enemy merchant ships. The *New York*, for example, was built with an unprecedented coal capacity of 1,500 tons, which would carry the ship thirteen thousand miles—or more than seven trips across the Atlantic—before it would have to stop to refuel. Observing the engine room's eight boilers during the ship's steam trial in 1893, New York journalist Franklin Matthews noted "that the coal was licked off the shovels by the draft as the firemen threw it in." Such an appetite for fuel allowed the ship to exceed twenty-three knots, as fast as any naval vessel in the world. Normal cruising would only allow ten knots, but the ship was expected to be able to run down nearly any vessel during wartime.[61]

Even more impressive were protected cruisers nos. 12 and 13, later christened the *Columbia* and *Minneapolis*. These sister ships carried 2,000 tons of coal and were built to be the fastest ships afloat. These ships were outfitted with the newly invented triple expansion engines, unthinkable even after the New Navy building program was inaugurated in the mid-1880s, and boosted a 20 percent improvement in fuel economy. Writing of the *Columbia*, navy secretary Benjamin Tracy observed in 1890 how new technology circumvented older geographical constraints. "She needs neither colliers nor coaling stations," he wrote, "for she carries both between her decks." A year later, he added that these ships were "a peculiarly important addition to a navy destitute of coaling stations abroad."

Promotional photograph of the USS *Brooklyn*, an armored cruiser, advertising Pocahontas coal, the trade name for a semibituminous "smokeless" coal from around the Virginia-West Virginia border, some time between 1895 and 1897. Since the 1840s, coal producers had jostled for navy contracts, both for the profitable orders they entailed as well as for the chance to promote their distinctive products around the world. Cruisers like the *Brooklyn* were built with unprecedented coal endurances. *Warships of the United States Navy Which Made Their Trial Trips with Pocahontas Coal* (Philadelphia: J. Murray Jordan, 1897), 23, courtesy William L. Clements Library, University of Michigan.

Engineers designed these ships so that they could steam for over one hundred days at a cruising speed of ten knots before having to stop, allowing them to circumnavigate the globe without once dropping anchor. In actual operation, the ship's coal endurance and steaming radius were lower due to everything from the conditions of the ships to the practices of engine crews to the qualities of coal used, but American cruisers still far outperformed and outdistanced their British counterparts.[62]

As for the battleships, their purpose was to operate close to American shores, making it, as Secretary Tracy noted, "unnecessary to emphasize the feature of coal endurance" that was so important for the cruisers. The *Indiana*, *Massachusetts*, and *Oregon* came first, followed by the *Iowa* in 1897. Any of the ships could easily cross the Atlantic—they could steam at least five thousand miles without stopping—but as they were designed both architecturally and strategically to remain in American coastal waters, they had no need for large coal bunkers or distant coaling stations. Engineers likewise designed the next generation of battleships, contracted during the 1890s but not commissioned until after the turn of the century, with these constraints in mind. Throughout the 1880s, advocates of the naval building program had argued for constructing these battleships instead of commerce-destroying cruisers. In their view, cruisers would be ineffectual not only because potential maritime rivals were building powerful battleships to protect their commerce but because the range of the early cruisers was limited without distant and defendable coaling stations. In their main journal of debate and ideas, the *Proceedings of the United States Naval Institute*, discussions of coaling stations in the 1880s are almost entirely absent; most naval reformers advocating building battleships, not acquiring a large network of coaling bases.[63]

As a result, prior to 1898, there was a growing divergence between what many writers believed the navy needed and what the navy was actually doing. Thanks to developments in ship and engine design and choices about naval strategy under steam, American naval strategists increasingly saw coaling stations as unnecessary, while the public frequently read that such stations were necessary for any significant naval power. It was a state of affairs that would have been difficult to imagine even a decade before. Yet at the same time, the decade of the 1890s brought a shift in the debate over coaling stations. Their ostensible purpose became more narrowly defined, and the numbers of them that advocates claimed the country needed shrunk. In a further shift in thinking, advocates demanded them less from a desire to support distant American commerce and more from a sense of deep and inescapable vulnerability to foreign attack.

Vulnerable Giant, 1890–98

For many Americans in the late nineteenth century, seeking foreign coaling stations made little sense when the technical choices of rival nations appeared to bring increased security to the United States. In 1887, *Scientific American* scoffed at the danger posed by new, foreign steamers. Exercises by the Royal Navy's most powerful ships had resulted in "several collisions and many breakdowns"—not to mention the rapid exhaustion of coal supplies. The magazine, one of the most diligent chroniclers of developments in naval technology, concluded that steam created such weaknesses that Americans had little to fear from foreign attack. "Few of these large ships could carry anything like enough coal to bring her across," the magazine noted, "and those so capable would be compelled to coal at some station here before ready for aggression or, barring the supply, be unable to get home again." Blockading Britain's few coaling stations in the western Atlantic would quickly disable the fleet. Adding insult, the magazine taunted the Royal Naval by observing that its *Terror* and *Imperieuse* ought be rechristened the *False Alarm* and *Impotent*. Theodore Ayrault Dodge agreed. A

Union veteran and military historian, Dodge noted in 1891 that despite the massive size of many European navies, few of their ships held sufficient coal to reach the United States without soon needing to refuel. "The coal question is the most difficult one," he wrote. "So much of the flotative power of the big vessels is consumed by machinery, armor, guns, turrets, and ammunition, that there is not much left for coal." If Caribbean coaling stations could allow an attacking European power to compensate for these deficiencies, Dodge suggested building more ships to patrol domestic waters, enhancing coastal defenses, and creating a more regular reserve militia.[64]

This comfort, to the extent that Americans shared it, proved short lived. Into the 1890s, the weight of argument for acquiring foreign coaling stations shifted from seeking commercial or territorial expansion to thwarting potential attacks from abroad. In the 1860s Seward had spoken of the United States emerging as "a great maritime power," and for decades, the pursuit of coaling stations was designed to open markets and serve American commerce. It was an argument for flexing American economic muscle. In contrast, by the 1890s, arguments favoring coaling stations overwhelmingly emphasized the problem described by one Signal Corps captain in a prize-winning essay as being that "the United States is by nature and by neglect one of the most vulnerable nations of the world." These defensive arguments were not entirely new, and older aspirations for overseas territorial growth did not disappear altogether, but in the last decade of the nineteenth century, strategic thought with respect to coal shifted from the advantages it offered to the weaknesses it revealed.[65]

No figure associated with naval expansion has attracted more attention from historians than Alfred Thayer Mahan, and no figure has been more associated with justifications for acquiring distant colonies and foreign coaling stations. Yet these claims are largely misreadings of Mahan's writings before 1898. As his most thorough biographer has noted, he is an odd figure to label "imperialist." Mahan abstained from the New Navy debates of the 1880s, consistently supported free trade against government support, and both opposed American colonialism and doubted his country would ever acquire colonies anyway. Through the 1890s, his consistent focus was on securing, legally, an isthmian canal and defending the nation from a range of imagined prospective naval threats. When he did advocate acquiring coaling stations (focused entirely around Hawaii and an isthmian canal), it was not because he thought steam technology somehow demanded it but rather because he sought to prevent future adversaries from securing them instead. Like most of his contemporaries, he was unable to countenance foreign alliances and distrusted the United States' ability to keep places like Hawaii from enemy hands during wartime.[66]

Mahan had personal experience with the challenges of steam power. While commanding the *Wachusett* along the Pacific coast of Central America in 1885, he had struggled to coal his ship. The long, fourteen-hundred-mile stretch between Acapulco and Panama lacked any reliable infrastructure for fueling, and Mahan meekly suggested to William C. Whitney, the navy secretary, that the problem "may be worthy the Departments attention." Less than a decade later, however, new naval vessels were capable of steaming five or ten times the distance of the old *Wachusett*, and Mahan turned from thinking about facilitating American ships to denying coaling stations to other naval powers.[67]

Part of the confusion over Mahan's thinking comes from his own writing. In his famous first book, the 1890 *Influence of Sea Power Upon History*, Mahan observed the requisites for becoming a great, global power. "It is vain to look for energetic naval operations distant from coaling stations," he wrote. "It is equally vain to acquire distant coaling stations without maintaining a powerful navy; they will but fall into the hands of the enemy." But did he mean for the United States to become such a power as Britain? His observation was not, in fact, an argument for an American pursuit of coaling stations on the British model. Far from it. It was instead a critique of the American policy of basing national defense on commerce-destroying cruisers, for it was *they* he believed could not operate effectively far from secure bases during wartime. Mahan called this policy "the vainest of all delusions." If he believed colonies allowed the kind of globe-dominating sea power he so admired, he did not expect the United States ever to achieve that kind of sea power. Simply put, "such colonies the United States has not and is not likely to have." Mahan was neither lamenting nor goading but simply stating his view of the place of the United States in the world. Strategizing within the constraints of American political culture, Mahan advocated a program of battleship construction to protect the Atlantic coast, from which colonial European coaling stations sat just a few hundred miles away. To protect the Pacific, his solution was to forestall any naval power from securing a coaling station closer than three thousand miles to San Francisco.[68]

Mahan remained resolute that distant coaling stations were both expensive and nearly impossible to protect during wartime. It was an acute awareness of these vulnerabilities that led him in 1890 to draft a war plan against Britain that centered around disrupting the Royal Navy's ability to fuel in American waters. "The controlling element in modern naval strategy is fuel—coal," he wrote. In the Atlantic, he imagined cutting Britain off from the extensive coalfields of Nova Scotia, forcing Britain to steam instead out of Bermuda. Mahan supposed that in order to supply the island and fleet, vulnerable colliers would have to transport tens of thousands of tons of coal across the open ocean. In the Pacific, American ships could coalesce around Puget Sound to prevent Britain from coaling at Vancouver with coal from the nearby Nanaimo mines. With these ships based in the secluded San Juan Islands and nearby Port Orchard, wrote Mahan, "we so menace her coal and communications as to paralyze her." In

Mahan's view, Britain's dependence on coaling stations was not a global strength, but a potential weakness.[69]

Throughout the decade, Hawaii remained one of Mahan's central interests. He never argued for commercial advantage or the need for a mid-Pacific way station. Rather than explain what the United States could do with a Hawaiian coaling station, he argued instead that the United States would be vulnerable should the islands be seized (which was an inevitability, in his mind) by another naval power, a danger he believed would be "a national misfortune amounting to a national humiliation." In this sense he was rediscovering David Dixon Porter's arguments from two decades before, but Mahan was not only anxious about industrial Britain or Germany but the "Yellow Peril" as well. Mahan first articulated his views in early 1893 by imagining the threat posed to the United States by a resurgent China. In a brief note to the *New York Times*, Mahan cautioned that for too long, Americans had understood Hawaii mistakenly in the context of European rivals alone. "China, however, may burst her barriers eastward as well as westward," he argued, "toward the Pacific as well as toward the European Continent." According to Mahan, Hawaii lay directly in this eastward path. And the danger went beyond giving greater mobility to Chinese naval vessels close to North American shores. Mahan offered his countrymen a choice: Hawaii as "an outpost of European civilization" or under the thrall of "the comparative barbarism of China." Of course, annexation would require a larger navy to keep the islands American, and Mahan could only wonder if America was up to the task.[70]

Two months later, Mahan expressed concern over Britain rather than China. In this more familiar argument, Mahan explained that Britain desired Hawaii to better connect her colonies in British Columbia with Australia and New Zealand. Americans must never forget, he warned, "the immense disadvantage to us by any maritime enemy having a coaling-station well within twenty-five hundred miles, as this is, of every point of our coast-line from Puget Sound to Mexico." In the Caribbean, a multitude of islands claimed by a variety of sovereigns lessened the importance of any single one of them. In the Pacific, Hawaii was an isolated island group, as yet unclaimed by a naval power, surrounded by the vast ocean. "Shut out from the Sandwich Islands as a coal base, an enemy is thrown back for supplies of fuel to distances of thirty-five hundred or four thousand miles,—or between seven thousand and eight thousand, going and coming,—an impediment to sustained maritime operations well-nigh prohibitive." This sense of vulnerability persisted; by early 1898, Mahan was worrying in similar terms about Japan.[71]

As Europe renewed its scramble for colonies in Asia and Africa during the 1880s, many Americans came to share a version of Mahan's anxiety, perceiving steam power as suffocating the United States. They feared that as Britain,

Germany, and Russia grabbed islands, especially in the Pacific, the vast distances that had for so long protected the country would rapidly shrink.[72] They feared that Britain in particular was closing in and began to think that Hawaii was the only place where America could still stake a territorial claim. How could the United States remain safe with the islands claimed by another power? Or vulnerable to capture during wartime? An American claim would preempt that worry. "No fleet would dare cross the Pacific, and leave this powerful naval fortress in their rear," wrote one Californian engineer, "and none could threaten our coast without fear of having its coal supplies or transports attacked from it." John Schofield, who had first surveyed Pearl Harbor in 1872, was certain that without American control, "the enemy" would "occupy it" and attack the West Coast and future canal. There were many political and commercial reasons why the Senate's Committee on Foreign Relations endorsed annexation in March 1898, but when it came to security, the committee's final report likewise focused on forestalling rivals. The committee argued that "the navies of to-day are all steamers with limited coal-carrying capacity" and that "with foreign countries barred out of Hawaii, the Pacific coast and its commerce is almost absolutely safe from naval attack." A Hawaiian coaling station also offered a kind of economy. If Hawaii were annexed "we should require fewer war-ships in the Pacific and fewer fortifications on our Western and Alaskan coasts," wrote John R. Procter, a former state geologist of Kentucky and commissioner of civil service.[73]

When George Melville, the influential chief of the navy's Bureau of Steam Engineering, looked to the scramble for island colonies in the Pacific, his concern was likewise not that America was not partaking in the feast but that European acquisitions brought future adversaries closer to American shores. During the 1880s, Germany had annexed the Marshall Islands, Spain the Carolines, and France the Marquesas and Tahiti. Britain, observed Melville, "declared protectorates over island after island and group after group—the Gilbert, Ellice, Phoenix, and many others." With transcontinental railroads, Melville conceded that a Pacific invasion was unlikely, but he argued that the long and largely unguarded coastline offered many bays and inlets susceptible to capture. If anything protected the country, it was the vast span of ocean, difficult to cross for most vessels of prospective enemies. If a rival nation were to seize Hawaii, however, the vast span became much more manageable. Without Hawaii, ships posed little threat because of the paucity of coal they would carry after steaming across the ocean. Melville considered an attempted attack on the United States without control over Hawaii first "midsummer madness."[74]

With Hawaii, however, these ships became a great threat. As Mahan had also argued, the significance of Hawaii was not its potential for projecting American strength but its ability to defend against its weakness. "If, then, there were no Hawaii," Melville explained, "if it could be blotted wholly from the map—the Pacific coast would be at this time entirely safe from transoceanic attack. Since these islands are, however, a permanent feature of the sea-scape, this security can be had only by their transfer to the United States and such guarding thereafter as will prevent their use, in war, by any foe." It was due to this vulnerability that Melville advocated Hawaiian annexation. "This right is not tangible in law," he admitted, "nor recognizable by treaty, but it is yet inherent through the possession of an imperial territory which bounds, almost wholly, these northern waters, which looks to them for commercial outlet, and which, from them, is susceptible to attack in war." Just before Congress voted to annex the islands in July 1898, Melville gloated as a Canadian officer acknowledged that the islands "should be our coaling station and cable pier in the Pacific" but that instead "their value has been grasped and appreciated by others."[75]

Opponents of annexation challenged the claims of figures like Mahan and Melville. In both houses of Congress, anti-annexationists pointed out the weakness of the strategic argument for a Hawaiian coaling station, noting that the islands were out of the way of transpacific commerce to Asia and that Unalaksa in the American Aleutian islands already provided a convenient depot along the shorter, more traveled northern route.[76] Fred Dubois, a Republican politician who toured the islands in 1897 in the midst of a brief hiatus from the Senate, concluded them undesirable for any country, the United States or otherwise. "They are too far away from any other country to be of service as a coaling station in time of war," he explained, "and they are worthless for any other purpose, in comparison with the cost and danger of maintaining an alien government there." The costs of annexation were not entirely theoretical. Already, the 1887 convention between the United States and Hawaii that granted exclusive rights for a naval station at Pearl Harbor had not resulted in the construction of a coaling depot, while the commercial reciprocity provisions of the earlier 1875 treaty resulted in a loss to the Treasury of over $20 million from duty-free Hawaiian sugar.[77]

Toward the end of the century, then, the strategic argument for coaling stations revolved mainly around securing Hawaii from foreign foes and protecting a future American isthmian canal. When senators asked more generally "if we can not have our own coaling stations why expend millions of dollars per annum to build and equip a navy?" and advocated modeling the acquisitions of these stations on Britain's global network, their appeals to the needs of modern steamships rang hollow. They were not expressing the view of the young naval officers designing the New Navy nor the strategic thought of Mahan and those who followed him. Instead, they were making a political argument about their desire to project growing American power in the world.[78]

Most Americans, though, were at pains to distinguish their foreign policy from those pursued by Europe. "They are not colonies," Mahan said of the new empire in 1898. "They should more properly be termed dependencies." It was a comforting thought. Mahan, representing a naval strategy board, had recommended immediately after the war that the country secure coaling stations in the Caribbean and Pacific. These suggestions exceeded any recommendation he had ever previously made—he not only called for one on Hawaii and for one each on the periphery of the Caribbean and another to guard the future canal but also for ones on Guam, Manila or a nearby harbor, and Chusan in China. It remains hard to square his pre-1898 warnings about the difficulty of defending distant bases with these recommendations. But as the historian W. D. McCrackan had concluded just five years earlier, limiting overseas acquisitions to strategically located coaling stations would be difficult. If there was a determinism at work, McCracken feared it was not of technology but power. After coaling stations, he wrote, "annexation is the next step, and an era of conquest must inevitably follow in its wake. . . . And we shall have a train of mean little wars to our credit. The United States will figure as the bully of the western hemisphere."[79]

Yet the idea, whether by conviction or clumsy shorthand, that steam power itself necessitated coaling stations remained a justification for empire building both at the time and in the histories of the period that followed. The idea followed the chronology of new naval construction that was truly dependent on coal. It rationalized the unprecedented acquisition of island territories following the Spanish-American War in 1898 as a product of technological inevitability instead of political choice. But the widespread consensus that coal necessitated coaling stations for strategic (as opposed to economically speculative purposes) only came *after* 1898, not before. Prior to the war, Americans preferred technical and mathematical approaches to the constraints of coal. When, in the 1890s, minority voices like Mahan did argue for coaling stations, the stations they demanded were few in number and justified on the basis of a perception of American weakness, not strength.

After the war, the United States staked its sovereignty over a network of islands. In the Caribbean, it secured Puerto Rico, along with a foothold in Cuba at Guantanamo Bay. In the Pacific, it secured Hawaii, Guam, Samoa, and the entirety of the Philippines. In the years that followed, American naval planners discovered that building a network of coaling stations created far more problems around industrial energy than solutions.

CHAPTER SIX

INVENTING LOGISTICS

There can be no naval fighting unless the logistic possibilities are realized. Men have been known to starve and yet fight and again advance to battle, but history has been silent upon the act of a warship, without coal and oil, or without ammunition, doing such heroic acts of devotion to duty.

Yates Stirling, *Fundamentals of Naval Service*

Despite decades of debate over coaling, when the United States declared war on Spain in 1898, the navy found itself largely unprepared to fuel combat. In February, assistant navy secretary Teddy Roosevelt cabled Commodore George Dewey in Hong Kong instructing him to "keep full of coal" and remain alert for imminent battle. Though Roosevelt's supporters later identified this telegram as evidence of his foresight and enterprise, he was, in fact, only restating long-standing department policy. It was a directive quickly repeated by Roosevelt's chief, navy secretary John D. Long, to Dewey and fellow commanders in Barbados, Honolulu, Lisbon, and Key West.[1]

Keeping full of coal proved a hideously complex task, however. Dewey alerted Washington that there was a "great scarcity of coal within the limits of the station," requesting the immediate dispatch of more from San Francisco. In Hong Kong, he secured some fuel on his own before pleading to Long that he needed still more because "other governments have bought all good coal." Long permitted Dewey to purchase 5,000 tons and, if needed, place orders directly from England. With this authority, Dewey attempted coaling arrangements with Japan, whose representatives responded by noting international law prevented providing belligerents with coal during wartime. He tried as well in China, even at the risk of what he called "international complications."[2] One analysis of Dewey's campaign concluded that he might have proceeded by "abusing the unbidden hospitality of a neutral power" by secretly refueling "from colliers in smooth water either at sea or under the lee of some remote corner of the world," but admittedly, he had no colliers, either. When Dewey's efforts to secure coal failed, Long permitted him to purchase two coal-laden supply ships, laying out £32,000 for the steamer *Nanshan* and another £18,000 for the *Zafiro*. Dewey further endorsed the commander of the American gunboat *Monocacy*, anchored in Shanghai, who found merchants willing to sell some 2,000 tons of local coal and risk the chance of diplomatic protest in the future.[3]

Had Dewey failed to defeat the Spanish garrison at Manila so quickly on May 1, the precariousness of his fuel supply would likely have proved disastrous. At first, after destroying the Spanish fleet and seizing the Cavite naval station, Dewey cabled Long that he could "supply the squadron coal and provisions for a long period." But even with his hastily gathered coal supply and the thousands of tons captured from the Spanish garrison, by the end of May, it was clear that fuel was harder to acquire than Dewey had hoped. The navy's Bureau of Equipment could not ship him coal until the end of June; even then it proved to be an inferior grade from Australia. Dewey believed this supply could last until a delivery from American mines, but this shipment did not even leave for Manila until August 10 and would not arrive for a month and a half. Even when coal finally did arrive, there was no guarantee it would be in any shape for actual use. The December shipment from the coal dealers Castner, Curran, and Bullitt arrived both burnt and waterlogged. Dewey rejected the delivery and the dealers were forced to sue their insurers.[4]

Everyone acknowledged that coal was essential to warfare, but war exposed how ill prepared Americans were to manage its supply. "In fact," recalled one young veteran of 1898, "it was a saying, as I remember it, at that time, that coal was king. That seemed to be the worry above everything of commanding officers. They took coal whenever they could." Back home in the United States, however, newspaper reports uniformly trumpeted the foresight and judicious planning of the Navy Department and Dewey's ample supplies. As no fuel crisis actually occurred, there was little publicly available evidence of how precariously Dewey kept his ships running during the American occupation.[5]

When Dewey triumphantly returned to the United States a national hero, joyous Americans greeted him with parades, banquets, and calls to run for president. The results of the war had been unexpectedly spectacular. As part of the Treaty of Paris ending the conflict, Spain ceded Puerto Rico, Guam, and the entirety of the Philippines to the United States. With American attention suddenly on the Pacific, Congress passed a resolution annexing Hawaii (a treaty to the same effect with the ruling American cabal having previously failed to garner enough support in the Senate). By 1899, Americans claimed an overseas empire and Dewey basked, if a little uncomfortably, in his nation's glory. Jubilant editorialists revived manifest destiny to give the acquisitions an air of inevitability. It was an inevitability even attributed to Dewey himself. When the international high society palm reader, William John Warner, popularly known as Cheiro, examined Dewey's right hand, he announced the prominence of "the Sign of Empire."[6]

But the new empire brought new bureaucratic challenges. How could Americans maintain their possessions? How could they protect them from rival powers? How could they ensure supplies of the lifeblood of the navy—coal and, within a few short years, oil? Over the coming years, there would be many attempts to answer these questions. Dewey, drawing on his experiences in east Asia, argued that the navy needed a reorganization, one that would allow the department to undertake more systematic studies of strategy, security, and supplies. When railroads grew from tiny stretches to vast operations, Dewey noted that "there must be a group of men who have, as their principal work, to *think* for the railroad, to observe rival lines, to consider the local laws of towns and states which their tracks traverse, and above all, to *watch the future* and prepare their system to draw all possible advantage from events." Thinking for the railroad was the job of the board of directors. For the navy and the nation, it was the job of what Prussians had created when they formally established a general staff in 1814 and used so effectively after the formation of the German empire in 1871. Beginning in 1900, Dewey served as president of the navy's new General Board, an advisory panel of senior officers that studied department problems and issued recommendations. But he imagined something more, a more powerful body that would subsume the existing activities of the General Board while adding still new functions—from recruiting and training sailors to synthesizing foreign intelligence to developing a library of war plans. Efficiency, mobilization, readiness, and method became the new organizational language. "With our development as a world power," Dewey wrote, "it has become a vital necessity that we should be prepared for all emergencies." A critical element of this preparation was fueling the American empire.[7]

Historians who have analyzed how questions of supply, provisions, and resources have shaped warfare have typically done so by examining past campaigns—their organization, flows of materiel, and bureaucratic management. It is a subject we recognize as "logistics." But logistics itself has a history. The concept did not exist before the nineteenth century, and even then, it was not until the turn of the twentieth that logistics began to receive extensive analysis by war planners, at least in the United States. The invention of logistics arose from shifts in how Americans perceived resources, bureaucracies, and the virtues of planning. Logistics introduced a science of resource flows and mechanical processes into thinking about war. It also facilitated a new approach to how American war planners connected security to fuel—not only coal but gradually oil as well.[8]

The study of logistics and the centrality of fuel was so intimately connected to America's insular acquisitions in the Pacific that some officers began to invert the causality of empire building. While expansionists of the late nineteenth century had once argued that the need for coaling stations justified building an island empire, logistics theorists like the Naval War College's Carl T. Vogelgesang argued that possessing an island empire justified building coaling and naval stations. Vogelgesang noted that "providence has so guided our destiny in the Pacific that we find ourselves the sole possessors of stepping stones that lead across that ocean." These stepping stones—Hawaii, Guam, Samoa, and Kiska in Alaska—provided an opportunity to engage in logistics preparations in peacetime that could protect the nation in times of war. All that was needed, claimed Vogelgesang, was to fortify these islands with naval bases and coaling stations. "By properly grasping and faithfully solving the logistic problems of the Pacific," he concluded, "we will properly link up our outposts in the Pacific with the home country by fortifying, garrisoning, and storing those positions." Logistics, in short, would "supply that present day expression of moral force that alone can ensure and guarantee peace." Thus the American insular empire slipped from being a consequence of naval buildup in the past to a cause of naval buildup in the future. But before that could happen, Americans had to rethink their conceptualization of war.[9]

From the Art of War to the Science of Logistics

For nineteenth-century American students of war, strategy reigned supreme. Despite the importance of supplies and provisioning to warfare, the subject of what European military theorists had begun calling logistics was slow to attract interest in the United States. For uniformed and armchair planners alike, strategy announced a nation's vision of what it believed possible and desirable in warfare; logistics merely sought to make that vision a reality. Stephen Luce, the founder of the Naval War College, explained in the Naval Institute's influential *Proceedings* that when it came to building and fueling new steamers, "the underlying principle is that logistics should conform to strategy, not strategy to logistics." As for Alfred Mahan, he remained attached to the idea that logistics belonged under the purview of junior officers. "While as vital to military success as daily food is to daily work," Mahan wrote in 1912, "yet, like food, it is not the work." Luce and Mahan were hardly out of step with their peers; through the quarter century after 1890, American naval officers rarely studied logistics. Articles in the *Proceedings*—the most important intellectual exchange for American naval thought—barely used the term. Between 1890 and 1912, only eleven articles even mentioned it, and six simply referenced studies at the Naval War College in Newport. It was not until 1913 that a single article appeared with logistics as its self-designated subject—and that was for an article originally delivered at the War College as a lecture.[10]

The Naval War College was, in fact, central to the development of American logistics. The college had been founded in 1884 on Coasters Harbor Island, on the western shore of Newport, Rhode Island. It was the creation of then

Commodore Luce, who envisioned an advanced program to study the science of war. A small class of eight met in September 1885, followed by three years of irregular sessions. Until 1900, however, the college's future remained uncertain. At times, it was combined with the adjacent torpedo station; others times it remained separate, and during many years it went without students at all. After the turn of the century, however, its program became more formal and regular, with a little over twenty young officers meeting for what came to be called the Summer Conference, or short course, which some two-thirds completed every year. In 1911, a handful of students began a new year-long curriculum, or the long course. Yet even then, the navy had yet to place the program on a solid foundation. Only some of the early long-course students completed it. Officers attending the shorter Summer Conference could be deployed at any time. After the election of Woodrow Wilson, however, the new navy secretary, the North Carolinian editor Josephus Daniels, gave the college a more permanent, formal status. Daniels first directed a contingent of officers of the Atlantic Fleet to enroll. Then in January 1914, Daniels, advised by the college's new president, Admiral Austin Knight, detailed a complete faculty of seven—all of whom were themselves War College graduates—and directed fifteen officers to begin the year-long curriculum every six months. Officers not needed for other duties attended the short course. For those at sea, the college began a correspondence program, which by 1916 enrolled over five hundred officers around the world. With these changes, the college finally became the institution Stephen Luce had imagined thirty years earlier, and Daniels anticipated that "the time will not be far distant when it will be practicable to deny responsible commands to officers who have not taken a course there."[11]

Even during its early years, however, the War College was the most important American institution thinking about the mechanics of warfare. Its officers worked especially on the importance of fuel supplies with a focus unmatched elsewhere in the government, including naval leadership. When Teddy Roosevelt, as assistant secretary of the navy, proposed a war game to the college in 1897, he suggested simulating a conflict in the Pacific in which Japan placed "demands on Hawaiian Islands" and incurred a naval retribution from the United States. "What force will be necessary to uphold the intervention," Roosevelt inquired, "and how shall it be employed?" In his own musings, Roosevelt asserted "that the determining factor in any war with Japan would be the control of the sea," insisting that the United States would have to "smash the Japanese Navy." In response, War College president Caspar Goodrich tactfully explained the complex demands of naval warfare. "That you are right as to the desirability of smashing the Japanese fleet is a matter of course," Goodrich noted, "but with the qualification, which was doubtless in your mind, although unexpressed, that the fleet should enter upon the proposed theatre of operations." Reaching this theater was a tricky business, and Goodrich sent Roosevelt a memorandum on coal supplies to explain why. Defeating Japan would depend on maintaining a steady fuel supply. It meant coordinating naval vessels with commercial colliers, placing coal orders for railroad and steamship transportation, and establishing temporary bases. It meant tripling the fleet's Pacific coal consumption from 11,000 tons a year to over 30,000 and allocating 20,000 tons more for operations in east Asia. Studying the coaling problem "emphasizes the difficulties attending a crossing of the Pacific," Goodrich wrote, concluding with "regrets that facts seem to forbid a rapid, vigorous, aggressive war." Counting, measuring, preparing—these were the skills Goodrich cultivated at the War College.[12]

Yet even there, instructors evinced an ambivalence about the scope of the subject. Goodrich described the central difficulty of fueling modern warfare, but was he describing logistics? Naval thinkers of the late nineteenth century took for granted the importance of fueling modern warfare, yet "logistics" referred to the systematic study of the problem and its potential solutions, and that barely existed. In his ten closely argued pages to Roosevelt, Goodrich never used the word once. In his own War College lecture notes, probably from 1900 and 1901, Goodrich substituted an earlier reference to logistics as a foundational element of warfare with the phrase "convoy and supply." The word "logistics" perhaps seemed too laden with military connotations to have much applicability to war at sea. Whatever the reason, logistics was simply not a subject instructors like Goodrich enjoyed teaching or students enjoyed studying. It was tedious, unglamorous, and unlikely to inspire young officers. Goodrich himself almost apologized for being compelled to cover it. "It is not an interesting subject I grant," he acknowledged in a lecture, "nor does it demand complete treatment at our hands, yet it cannot be wholly ignored." To his students, he promised that he did "not contemplate wearying you with an elaborate volume on supplies, their nature and use." His goal was more modest, simply "to offer a few suggestions pertinent to the topic" before quickly returning to lessons in strategy and tactics. He cataloged the things a commander should have, from fresh soft water to provisions to shore storage facilities. Of coal, he emphasized that "fuel is the ever present want of a modern navy in comparison with which others are insignificant in bulk at least" and that other materiel "rarely occasion so much anxiety." But even during this brief diversion into what he called "the second great sub-division of the Art of War," Goodrich's focus remained on strategy, as in how disabling a foreign power's coal supply was the best way to protect the American coastline. Despite mentioning the sorts of things logistics must be called upon to supply, he never once explained precisely how it might supply them.[13]

The subject was so little understood that writers disagreed about the scope and even the etymology of the word. The influential Swiss strategist Antoine-Henri Jomini wrote of "logistique," explaining it derived from the office of the *major général des logis*, the officer responsible for directing, housing, and feeding troops on the march. "Logis" was linked etymologically to the English "lodge" and thus the literal lodging of troops. According to Jomini, the French rank corresponded to the German *Quartiermeister* and from there to the familiar English and American quartermaster. Others traced the word further back. In his *Military Encyclopedia*, former West Point instructor Edward S. Farrow derived the term from the Latin "logista," meaning "the Administrator or Intendant of the Roman armies." Under the purview of the *logista* fell "all details for moving and supplying armies," from ordnance to medicine, from provisions to pay. The Prussian general Rudolf von Caemmerer offered another derivation, this time from the Greek word for "calculation," noting that "calculations form an important part of the labours of a General Staff." Yet despite this ambiguity around the intellectual genealogy of the subject, these writers all agreed that logistics involved the material aspects of war. Still, they disagreed on what duties devolved to the logistician and how the subject related to the more fundamental project of strategic planning.[14]

By the 1910s, however, American naval thinkers were introducing changes to the study of logistics. They rejected its older subordination to strategy and instead elevated it to coequal status. Strategic planners who failed to collaborate with logistics officers would end up "making demands that logistics could not supply," wrote Rear Admiral Bradley Fiske in 1916, "or, through an underestimate of what logistics can supply," not "demanding as much as could be supplied"[15] Where Jomini had written of an "art," logistics offered a science—"facts and not fancies," according to Vogelgesang. "Here the demands of the Art are calculable and solvable by rule and method," he informed his students, "and cause and effect are only separated in many cases by a problem in simple arithmetic."[16] With this simple arithmetic, it was possible to fuse practical calculations with the planner's vision and make sense of the size, scale, and mechanization of modern war. When it came to outfitting "a great over-sea expedition," explained War College president Austin Knight, "an officer in this position needs more than instinct to see him thro."[17] Instinct alone could never produce the organization demanded by modern combat. "War has become a business," wrote marine and War College graduate George Thorpe in one of the earliest systematic volumes on the subject. "Therefore training and preparation for war is a business—vast and comprehending many departments." The new logistics argued that war preparations were "susceptible of analysis" in terms of organizing workers, materiel, and plans for operations.[18] As one officer accordingly noted, logistics was " 'scientific management' applied to the Navy."[19] Strategy might privilege genius, but the modern naval officer required diligence. "Genius," according to T. J. Cowie, paymaster general in the 1910s, "unaccompanied by logistics, invites defeat."[20]

In large part, the elevation of the status of logistics resulted from confronting the practical supply problems that emerged following the American seizure of new colonies in the Caribbean and Pacific. Through most of the 1890s, American war planning largely focused on conflicts near American shores. After 1898, American shores included Manila Bay. In the Philippines, the United States committed both to defeating a rebellion from within the islands and defending them from without, and neither goal appeared possible to achieve without a steady supply of fuel from home. To help manage these challenges, in 1900, navy secretary John Long created the General Board, an advisory body of senior officers chaired by the returning war hero George Dewey. Among the board's ex-officio members was the president of the Naval War College, providing a conduit to research there. From their earliest meetings in April of that year, the board analyzed fueling problems and the location of overseas refueling bases. It was a subject to which they repeatedly returned, revising lists of priorities. Four coaling stations in Cuba or one? Three in the Philippines? In Africa? Alaska? China? The board regarded the fueling problem as being at the center of American strategy. During one explosive bureaucratic conflict that nearly had the board disbanded, the chief of the navy's Bureau of Equipment, Royal B. Bradford, claimed exclusive jurisdiction over locating coaling stations, while his fellow chief of the Bureau of Navigation, Arent S. Crowninshield, insisted that "if this matter was not to be left in the hands of the General Board, there was hardly a necessity for the board's existence."[21]

As the General Board considered coaling and naval bases in relation to American strategy, the subject became an increasingly important element of study at the War College. By the early 1910s, Knight related that moving the Pacific Fleet from the American coast to the Philippines for even a month would create a demand for some 500,000 tons of coal. To defend the colony, Knight claimed, naval steamers needed this coal at all times, some of which should accompany the westward-steaming fleet and some of which should be deposited previously at designated coaling depots. Moreover, he noted, "all arrangements must be so coordinated that there shall be no chance of failure." With the very mobility of the navy and the success of U.S. defense and foreign policy at the mercy of the logistics of coal, Knight concluded that there was "no more important subject studied" at the War College or elsewhere.[22] Consequently, War College instructors began employing new pedagogical approaches. Responding to one officer characterizing logistics as "one of the least interesting, least studied (therefore least understood), but at the same time one of the most important subjects of study in training for war," T. J. Cowie asked, "If it is so important, how can it possibly be the least interesting?"[23]

College instructors sought to stoke that interest by assigning projects that revealed the necessity and enormity of planning for war, planning in which fuel figured largely. Lectures introduced the systematic study of coal and offered resources to naval officers. These officers helped develop logistics tables, principally for determining fuel consumption for individual ships at various speeds, as well as for quantities of ammunition. Students employed these tables when they participated in war simulations like Roosevelt's in which an American fleet had to be dispatched across the Pacific. For many years, the college required its students to produce a logistics thesis. These essays analyzed particular logistics problems, drawing on the theory presented in lectures and on statistics of engine fuel consumptions, sea routes, and war-making capabilities that were stashed away in the secret files of the college's library. Once submitted, the library preserved drafts for future study. Instructors later merged this essay requirement with a thesis in strategy, though both were discontinued in 1925, when the college adopted a new curriculum based on "the study of the strategy, logistics and tactics of actual battles and campaigns," a program that consequently led to less emphasis being put on the gritty calculations necessary for actual logistics problems. The following year, however, the college established a formal Logistics Department for specialized instruction in the problems of supply and organization.[24]

In war simulations, students practiced breaking complex operations into manageable segments. "War exists between Orange and Blue" was the introduction students received to problem 8 for the class Strategic 49. "Orange," in the not-so-secret language of naval discourse, referred to Japan, "Blue" to the United States. Problem 8 imagined a souring of diplomacy between Orange and Blue, after which Orange declared war and invaded the Philippine island of Luzon. The Orange navy, meanwhile, remained in its own waters but maintained a base in the Pescadores, the island archipelago off the western coast of Taiwan. As Blue's hold on the Philippines slipped away, it found its fleet and transport vessels stationed at Panama. The problem assumed that Pearl Harbor and Guam remained in solidly Blue hands, with both Oahu and Guam "considered secure from attack." The problem then asked students to fuel the coming war, moving Blue's fleet from Panama to the Philippine port of Polillo, passing through Pearl Harbor and Guam. Students calculated several elements of the Blue fleet's logistics plan: the path the ships ought to take (the "line of operations"), the locations of appropriate bases for fuel and supplies, sources for coal and oil, the paths of "lines of communication" between the fleet and home territory and along which fuel itself would flow, the means of fuel transportation, and, often overlooked, the source of fuel transport vessels. Detail was essential; the model solution to problem 8 comprised over forty typeset pages.[25]

To solve logistics problems like problem 8 (which was modified from year to year), students used logistics tables. Faculty detailed to the College regularly requested and collected reports on the fuel consumption of naval vessels, com-

Drawing of the Panama Canal's Cristobal Coaling Station. Completed in 1916, this station on the Atlantic terminus was the largest in the world at the time, capable of loading more than twenty-four hundred tons of coal an hour. The station contained storage areas for both commercial and naval coal and, together with the smaller Balboa coaling plant on the Pacific terminus, could store as much as seven hundred thousand tons. The station served American naval logistics interests in both the Caribbean and Pacific and figured prominently in developing war plans. "Canal Zone—Cristobal Coaling Station," folder 71-CA-76E, box 76, RG 71-CA, NARA-2.

piling this material into data-rich catalogs for war planners.[26] But if Vogelgesang supposed that tabulation translated into "simple arithmetic," Cowie asked his students to "imagine for a moment the mass of figures and the number of computations necessary to ascertain the requirements for fitting out a fleet with ordnance alone; and then consider the magnitude and intricacy of the task when it involves not only ordnance but fuel, food, clothing, tentage, camp equipment, supplies, hospital equipment and supplies, additional armament and equipment for repairs, and a multiplicity of details, all requisite to put that fleet in complete preparedness to engage in war in home waters." Tables organized this vast amount of information, offering the logistician a chance to prepare for war at any time in advance of combat while taking account of every imaginable contingency, but there was nothing simple about the mathematics that generated them.[27] Some logistics tables collated a broad spectrum of potentially useful figures and statistics, like one Cowie himself compiled in 1917, using nearly a hundred pages to provide an economic snapshot of the resources of the nation. From forests to mines, agriculture to manufacturing, imports to exports, he summarized the material assets of the nation. The object was "considering

Logistics along broad lines" and through which "many problems of vital interest to the country . . . may be solved."[28]

For the navy, the most pressing logistics questions involved coal and coaling. There were questions of quantity—how much coal was needed—but there was also the question of quality. "The areas of production of good steaming coal . . . are very restricted," Vogelgesang noted, "and while depots may and do exist all over the world where such coal is kept on hand the supply is in limited quantities in any one of them and is usually covered by government contract that will not permit of its release to an outside purchaser." The navy demanded the best coal possible for combat, introducing an additional constraint on the provision of supplies. "Here enters into the calculation, therefore, a vastly different proposition in logistics than we would have if we had the coal fields of the world to draw upon." There may have been abundant and nearby coalfields in the Pacific, for example, but the navy would only accept coal from the semibituminous mines of Appalachia.[29]

Still, even as the War College and General Board studied the coaling problem, few of their grandest requests were heeded. Congress refused to allocate funds for fortifying outlying bases. Naval colliers remained in short supply. The navy lacked reserve supplies of fuel in case of war. "If war were to come upon us under present condition," according to one 1910 assessment, "the supply of fuel would be attended with serious embarrassment at the best, and might prove a fatal handicap."[30] Aside from the ever-important lack of appropriations from Congress, part of the failure was bureaucratic. Both the War College and the General Board could only make recommendations, not mandate actions. In January 1915, Secretary Daniels proposed to the House Naval Affairs Committee the creation of a new position, a "Chief of Operations," which would include responsibility for a "logistics section." Daniels tentatively assigned the section's staff all the tasks necessary for engaging in warfare: determining expected demand for supplies; identifying their sources and availability; planning for transportation, supply vessels, and the conversion of merchant vessels for the needs of combat; and crafting plans and orders to carry out these activities. As Cowie added in his review of the proposal, "Plans for the conduct of war would be of little use if they only embraced the distribution, maneuvers, and employment of the fighting forces: they must also include arrangements for supplying that force with all the requirements necessary for carrying on the war." In modern war, success or failure in battle might well depend less on soldiers, sailors, or salvos and more on the calculations and preparations of war planners months or even years before combat. And among those preparations, Cowie noted, "the necessary fuel supply for our fleet in case of war will be the largest proposition we will have to handle." In March 1915, Congress approved the creation of the position of chief of naval operations.[31]

Unexpectedly, however, logistics planning took hold in a different, existing office, the Bureau of Supplies and Accounts, and under a dynamic new chief, Samuel McGowan. A native of South Carolina, McGowan had joined the navy as an ambitious assistant paymaster in 1894. Within a decade, he was serving in the navy's Bureau of Supplies and Accounts in Washington as assistant to the paymaster general, helping address what his outgoing supervisor called "perplexing problems of organization and administration." Rising in the ranks to pay inspector in 1906, he began representing the navy at annual meetings of the Association of American Railway Accounting Officers, the national organization of accountants for the massive industry. In 1907, he helped integrate state naval militia accounting with that of the Navy Department. A year later, McGowan joined the Great White Fleet as paymaster aboard the *Connecticut*, quickly assuming the role for the entire fleet. In this position he developed a reputation for spending long hours locating waste (and those responsible for it) and developing what he called "comprehensive, fleet-wide organization and standardization." The problem was that accounting and resource allocation—what was coming to be known as logistics—remained entirely unsystematic. "As the navy regulations are practically nil on the subject of the fleet paymaster's work and as I had no instructions or suggestions from any source and not even any precedent to go by," McGowan later recalled, "it was necessary for me to strike out on my own account and absolutely originate practically everything that was done." His work was so successful that after the cruise, he returned to the United States to help manage problems of fleet supply more generally.[32]

McGowan's star continued to rise while he was running naval pay offices in Charleston and Philadelphia, managing the accounts of the Atlantic Fleet and reforming the accounting systems of the department's many navy yards. In 1914, navy secretary Josephus Daniels appointed him paymaster general. At forty-four, he was possibly the youngest officer ever to have filled the position. In short order, he came to be known for his logistics reports, as well as his work on a newly formed permanent Logistics Committee comprised of members from each of the bureaus handling war materiel and tasked with planning how to fuel, clothe, and feed the navy during wartime. "The Bureau of Supplies and Accounts is the navy's great business office and incidentally it is one of the biggest enterprises in the United States," stated the House Committee on Naval Affairs in 1918, as well as an organization to which McGowan brought "a nation-wide reputation for business efficiency."[33]

Within his bureau, McGowan acted with unprecedented vigor in forcing bureaucratic reforms. He described entering his office during the summer of 1914 to find utter disarray: an excess of staff with few clear responsibilities, no organization—neither literally nor figuratively—and an atmosphere of loitering lawyers and lobbyists. As one of his first actions as chief, he removed the easy chairs these contractors' representatives so often occupied. He banished roll top desks, moved division chiefs' desks to the centers of their rooms and clustered the desks of attendant staff around them, and connected offices by removing the doors that separated them from their hinges. Another action, one of sixty-four intrabureau orders McGowan issued during his first six months, reformed the navy's coal operation, largely automating the contracting system and attending to detail in a way that was unprecedented. This reform of coal purchases was part of a larger transformation of fuel contracting pushed by navy secretary Daniels and his assistant secretary, Franklin D. Roosevelt, who had taken personal charge of naval fuel contracting in 1914. Their work cleared out an insiders' club of dealers who had long divided naval coal bids into equal shares at nearly equal prices. Daniels and Roosevelt considered uprooting this system a major priority in order to effect managerial reform and increase competition.[34]

World War I tested McGowan's reforms. When it comes to war mobilization, historians have focused mostly on the War Industries Board, the committee ultimately led by financier Bernard Baruch and tasked with coordinating American industry with government demands for supplies and resources. But despite some attempts to broaden its scope in the fall of 1917, the board only managed the industrial relations for the army, whose procurement agencies were what one journalist described as "hav[ing] long been in a hopeless maze of red tape and confusion." In contrast, the navy's Bureau of Supplies and Accounts, which McGowan had rebuilt over the preceding three years, was recognized both inside and outside the government as a model of modern efficiency and professionalism—its supply system was "splendid," according to Washington's most important paper, the *Evening Star*—and the department was nearly unanimously against subordinating its power to a new and untried organization. When the War Industries Board itself sought that authority, the department quickly mobilized support in Congress and by early October had prevailed in its independence. Many lawmakers voiced their confidence in the navy's independent operations, as had hundreds of vendors comfortable with the existing system. Army quartermasters, who had been blindsided by the secretary of war's decision to sign over authority to the War Industries Board, also contributed to the decision, as did McGowan's own insistence that the navy preserve as much open bidding as possible to retain public confidence and save the government needed funds. McGowan agreed, however, to consult and coordinate often with the new body.[35]

McGowan promised coordination because the war effort presented unprecedented challenges. "Logistic problems have never before been given so serious consideration as at present," McGowan wrote in 1916. The point applied to everything from structural steel to paints to lumber, but fuel introduced particular obstacles. After the European war had begun, the cost of shipping coal to the

Philippines increased nearly threefold and to the Mediterranean nearly fivefold. Americans in 1914 spoke often about preparedness but typically in an abstract way. To the navy it meant something quite concrete, like securing contracts for transporting coal and fuel oil to distant stations when commercial shippers could be contracted at low rates. By the time the United States entered the war, the department began turning to new methods. Beginning in July 1917, after failing to secure adequate fuel by competitive bidding, the navy began commandeering coal for the war effort. Using "navy orders"—over seven thousand of which were issued by early 1920, the department requisitioned fuel at prices it considered fair. Through the second half of 1917, the navy spent about $317 million on materiel, nearly a tenth of which went to purchases made under the auspices of navy orders. Fully 20 percent of these orders were for coal alone, the market prices of which the department judged excessive. Navy orders worked to ensure the delivery of a materiel when it was needed at reasonable rates; final prices could be determined later. "In these days of invention, rapid transit and quick communication," McGowan explained to Josephus Daniels, "the primary lesson learned has been that 'Things, not men, lose wars.'"[36]

The use of navy orders to procure coal followed on a clash between coal producers, who saw in war demand an opportunity to reap fat profits, and navy secretary Josephus Daniels, who believed producers ought to receive no more profit during wartime than peace. A month after Congress declared war on Germany, McGowan, his staff, and representatives of twenty-three of the navy's largest coal suppliers met in Washington. The suppliers came from the Pocahontas, New River, and George's Creek fields and represented but a small portion of the national industry—perhaps 10 percent—but they alone handled coal that met the stringent requirements of naval service. Like other collaborations between government and industry during the war, the purpose of the meeting was to organize resources for the war effort amid impossibly high demand. Between exports, increased industrial production, munitions manufacturing, and government needs, coal demand in 1917 far outstripped available supply. The suppliers who were meeting in Washington already held contracts that, if filled, would leave no fuel remaining for naval consumption. McGowan sought the greatest possible voluntary industrial cooperation and hoped that dealers would propose only fair prices, thus making it unnecessary for the navy to implement more coercive measures. The suppliers themselves wished to satisfy government needs without alienating their other customers. They suggested requiring dealers of non-naval coal to substitute their products in eastern markets harmed by naval demand and insisted that the government employ its new war powers to commandeer needed fuel, which would shift the blame for coal shortages from dealers and to the government. According to one supplier, the idea was to become "conspirators against the people who have the contracts now," a conspiracy he confessed he was happy to join. Ultimately, they unanimously agreed to establish a coal committee to allocate the equitable sharing of naval coal needs among all dealers, allowing the industry to prioritize filling contracts based on its judgment of importance.[37]

Cooperation did not lead to agreement, however. The suppliers' committee proposed tonnage rates the navy deemed impossibly high, though still less than half what the market would bear. At the end of May, Daniels met with the committee of suppliers to explain how when it came to coal, winning the war trumped the usual operations of the market. "I am persuaded that the producers have given too much weight to the prevailing very high market price and have based their figures on an abnormal demand rather than upon the real value and cost of producing the coal," Daniels wrote to the suppliers. For the coal industry, labor shortages and especially lack of railroad rolling stock made filling government contracts harder than normal, but for the government, the proof was in the profits. If contractors received higher profits than before the war, Daniels insisted that the government, the American people, and the Allied war effort itself was doomed to suffer. To the coal operators he threatened "that everyone ought to be commandeered by the Government and all you gentlemen ought to be enlisted to do the work you are doing." Indeed, Daniels predicted something along those lines would occur within six months. Daniels declared a fair price of $2.33½ per ton, later reduced by the results of a Federal Trade Commission investigation to $2.24—this at a time when in the open market, even low-grade coals sold for $6, $7, or even more. The suppliers had offered the navy coal at $2.95 per ton, not including freight, but they were forced to comply with the government's lower rate.[38]

These negotiations, which set the stage for both naval and larger Allied fuel logistics, failed to anticipate the broader, national significance coal consumption would play in the war effort. Besides ships, factories churning out munitions needed coal, as did power plants, businesses, homes, and the ever-important railroads. Like other aspects of war mobilization, early efforts to meet national fuel demands were voluntary, but by the summer of 1917, Congress decided the executive needed more authority. Under the Food and Fuel Control Act, or Lever Act, of August, Woodrow Wilson created the United States Fuel Administration to regulate prices, increase supplies, and manage the most pressing problem, distribution. Wilson appointed Harry Garfield to run the new body, and Garfield, the former president of Williams College, spent the next five months struggling to eliminate a crippling railroad car congestion that was preventing fuel from reaching desperate consumers. In a determined bid to get coal flowing again, Garfield joined William McAdoo, the treasury secretary who took over the new Railroad Administration in late December, to flush the clogged distribution network. On January 17, 1918, Garfield issued a "closing order," which banned all but the most indispensable industries from consuming coal for the week beginning January 18 and for two months of Mondays thereafter. These coal restrictions applied to office buildings and shops, theaters and brewers—a vast swath of increasingly urban America. "Yes, it's the worst order ever issued," conceded Secretary Daniels when pressed by a reporter, "but it was the worst situation that ever existed."[39]

Fuel Administration policies were never more than ad hoc and their effects limited. Far greater organization backed by far greater power kept coal flowing to the navy. Unlike with industrial planning, the war merely accelerated an existing apparatus of logistics planning. The navy did not create anything entirely new. So thoroughly had McGowan reorganized the Bureau of Supplies and Accounts before the war that the changes in naval business operations during the war were mostly in size, not method. The bureau's 23 clerks became 714. Its floor space of fewer than three thousand square feet became more than forty thousand. Disbursements rose from $8.9 million a month before the war to nearly $84 million after.[40] Six thousand contractors became eighteen thousand. Payments to some sixty thousand sailors became payments to more than five hundred thousand. The greatest prewar expenditure of $27 million for supplies in one year became $30 million in a single day. Over the course of the war, the department transported 130,000 tons of coal, 746,000 tons of fuel oil, and 12,000 tons of gasoline to Europe. Fighting the war cost the navy more than Congress had previously spent on it since its creation.[41]

Those numbers reflected the scale of mobilization. To make that mobilization possible, McGowan's bureau had to process vast quantities of information. That processing relied on what we would now call data visualization. To manage the navy's vast system of contracting, the bureau indexed its contracts—according to one journalist, for "everything the navy uses from steel, coal and wool, to eggs, butter, and beans"—and updated charts daily to nearly instantaneously reflect price changes. The charts, which the bureau called "fever sheets," displayed the volatile American wartime economy. Among the data displayed for each commodity were seven years' worth of weekly price data, allowing the bureau to estimate likely cyclical future price swings. Another set plotted the volume of coal and oil available at every depot within twelve hours. Still another traced current fuel volumes aboard every vessel in the navy, information that was constantly communicated to Washington by wireless. Mechanical accounting machines produced the visualizations, allowing the bureau to dispense with scores of clerks and to employ punch cards to automatically tabulate, calculate, and display information. The coal chart received special attention, as it traced how booming industrial consumption and shipments to nations cut off from English exports pushed domestic coal prices ever upward, though government price fixing soon nearly halved the price for the war effort.[42] Holding the once obscure title of paymaster general, Samuel McGowan came to be widely cited in the press as "the business manager of the navy."[43]

With respect to energy, the war experience catalyzed two points. First, supplying the army and navy alone with coal and oil did not encompass the complete strategic significance of fuel for the country. One coal operator acknowledged that the government had to have ways to fuel the war effort without becoming subject to price gouging but pleaded that measures like maximum prices be instituted to protect the broader coal-consuming public. The price must be low enough that consumers could afford to heat their homes, run their railroads, and fuel their factories, but not so low as to discourage production. "The fuel crisis at present is a grave one. Coal is the basic necessity of the nation," wrote one progressive Chicago organization. "Without it not a wheel can be turned in our great industries. Nor is it possible to transport either men or material by rail or water. Food cannot be shipped, cooked or refrigerated without it. In short, it is our greatest necessity."[44] The organization believed the government did too little to prevent war profiteering on the backs of workers. At the other extreme were those who agreed with the importance of coal but not the role of government. "The bituminous coal industry of the United States is, next to food, the first in importance at this time," explained one speaker before a convention of New York coal dealers, "and you may say at all times because it is basic, it is vital in importance in this industrial age." But instead of worrying about consumer prices the speaker cautioned against the prospect of creeping socialism.[45]

Second, the concept of logistics itself began to take on an even more capacious meaning than it had had earlier in the 1910s. "These plans go deeper into the matter than the application of the questions of transport and safeguard along the lines of communication by the military authorities," wrote one marine officer describing the role of logistics in war planning. "They involve the business-military organization of the entire country, its people and resources, so that the object of the military, the efficient conduct of the war, may best be consummated." On this view, logistics was more than army or navy organization; it also entailed organizing national industry, railroads, and recruitment. This new, more expansive view of logistics also inspired naval leaders to investigate if they could gain control over strategic fuel supplies themselves and thereby avoid the constraints of the private market. Seeking new solutions to fueling national security led them to the site of the greatest deposits of coal on public lands in the United States: Alaska.[46]

Logistics in the Matanuska Valley

In May 1902, the engineer Harrington Emerson surveyed the prospects for American trade in the Pacific basin, and he nodded approvingly at what he saw.

"Commerce and civilization have passed from the Mediterranean to the Atlantic" he declared, "and perhaps in turn will pass from the Atlantic to the larger ocean, the Pacific." Emerson used "perhaps" modestly, for he was confident in the westward march of civilization. At the dawn of the twentieth century, he no longer imagined this march as one of homesteaders but of modern industry with its appetite for natural resources, most importantly the energy sources supporting regional growth. "Certainly the Pacific Ocean is assuming importance," Emerson announced, "and modern commercial importance is founded on coal."[47]

Fittingly, the lands surrounding the Pacific were rich in coal. Along the western North American coast alone, geologists estimated that coal reserves rivaled the massive fields of central Appalachia. From Australia to Chile to Alaska, Emerson identified fields he believed were destined to elevate American industry, support its international trade, and secure geopolitical influence for it in the Pacific. Alaska, in particular, drew his notice. It was, of course, already an American possession and thus easier for Americans to exploit than other parts of the Pacific rim. Furthermore, the preceding six years had brought momentous changes to the territory. Gold discoveries in the Klondike in 1896 followed by additional strikes in Nome three years later brought labor and capital investment. The explosive growth of Alaskan salmon fisheries similarly attracted the attention of investors. The construction of the territory's first two railroads offered the prospect of further settlement and industrial development. The United States' acquisition of Hawaii, Guam, and the Philippines stimulated a newfound interest in the nation's older Pacific domain as well, for Alaska's long coastline and island chains lay along the shorter, northern great circle route across the ocean and offered valuable harbors for refueling en route to Asia. And, of course, Emerson saw great prospects in the exploitation of Alaska's potentially massive coal fields.[48]

Two decades later, even though Emerson's vision of global commerce and industry shifting to the Pacific had not yet been fulfilled, the movement to develop Alaska had only grown stronger, and development still meant coal. For the regional planner Benton MacKaye, coal would usher Alaska into the twentieth century, completing what he called a "big three" of extractive resources beginning with fur seals in the eighteenth century and gold in the nineteenth. Furs had first attracted Europeans to the land, and gold had offered the prospect of instant wealth. Coal, according to MacKaye, awakened Americans to the prospect of systematically developing Alaska in a permanent manner. MacKaye envisioned opening Alaska to massive colonization and the sustainable development of its resources. He further imagined replacing exploitative and temporary mining camps with permanent mining communities where miners would labor under fair conditions. Developing coal resources would stimulate copper mining, which would contribute to global electrification, and lumbering, which would support construction in Alaska and provide a product ripe for export. With this prosperous future barely visible on the horizon, MacKaye wrote optimistically of "a potential nation," "a hinterland to be opened up," and "the chance to build a nation within a nation."[49]

If Alaskan coal tantalized the promoter and planner with prospects of regional development, it also began to play a role in naval logistics planning. After 1898, the new Pacific island empire introduced new strategic questions to the United States. Could the country hold these islands in peacetime and in war? If so, how? How best could the government encourage commerce between the Americas and Asia? How could the navy maintain its vital fuel supply, especially during wartime when sea-lanes became vulnerable to attack and the ports of neutral nations were closed? Before the turn of the twentieth century, rarely had American planners thought seriously about conducting large wars halfway around the world. With a new colony in east Asia, this became a pressing problem. Considering these questions, many planners, officers, and officials looked at Alaska's proximity to central Pacific islands and mainland Asia and saw the possibilities of both its coal and its location. They began to imagine Pacific geography in a new way, and in this new conception, Alaska no longer occupied the outermost periphery of strategic thought but assumed a position several steps closer to the center. Most importantly, this reimagining of the place of Alaskan coal also transformed the government's role in providing it, all the way from mine to market to the bowels of a battleship.[50]

The matter began in the Aleutian archipelago, the northwestern tail of Pacific North America. The Aleutians are a chain of more than three hundred islands stretching roughly twelve hundred miles from the Alaska Peninsula in the east to the remote Attu Island in the west. The islands had long been home to sealing and fishing industries. By the turn of the century, the commercial significance of the Aleutians began yielding to a new, strategic one by virtue of its proximity to the great circle route. At the General Board's very first meeting in the summer of 1900, its members considered just two Pacific bases as contenders to defend their new colony in the Philippines, one in the Philippines itself and the other in the Aleutians along the great circle route. From August through October, the board sought out anyone who might have knowledge of the archipelago's weather: naval officers, the Revenue Marine, the army's transport service, the Hydrographic Office, the Weather Bureau, and merchant captains familiar with the northern Pacific. From offices in Washington and Newport, Dewey and his fellow board members concluded that five harbors in the region merited further investigation. Four clustered around the 180th meridian, near the northernmost arc of the great circle: the Bay of Waterfalls on Adakh Island, Kiska Harbor on Great Kiska Island, and Nazan Bay and the Bay of Islands on Atkha Island. All were pristine and undeveloped. The fifth, Unalaska's Dutch

Harbor, already served as a regional trading post, refueling station for commercial vessels, and port of anchor for fishing vessels, and private companies based there maintained a small commercial coaling station that served the local maritime economy. Dutch Harbor lay, however, over four hundred miles distant from the great circle route. During the next survey season in the summer of 1901, the General Board requested the USS *Concord*, a veteran of Dewey's assault on Manila Bay, to examine the Bay of Waterfalls and Kiska Harbor.[51]

Little came of these surveys, though Dewey later recalled having "urged the matter upon the Department's attention." During the summer of 1902, the navy sent the *McCulloch* of the U.S. Revenue Cutter Service to survey Atkah Island's Nazan Bay and the Bay of Waterfalls (which had already been surveyed once before). A year later, the entire Pacific Squadron steamed to the Aleutians to survey various islands. "It is desirable" wrote William Moody, the secretary of the navy, to the commander in chief of the Pacific Squadron, Rear Admiral Henry Glass, "to find some harbor near the 180th meridian that can be utilized by naval vessels crossing the Pacific Ocean to take coal from colliers, or possibly for the establishment of a permanent coal depot protected by fortifications." Moody instructed Glass to survey the Bay of Islands on Adakh Island and Kiska Harbor on Great Kiska Island; the two islands lay within about 150 miles of either side of the 180th meridian, where the great circle route swept closest to the Aleutians.[52]

The Pacific Squadron's obstacles to conducting this survey illustrated how even after six decades, steam travel remained fundamentally limited by weather and supplies. Rear Admiral Glass began his preparations in May. The voyage, he indicated to his superiors, should begin no later than July, and he advised against remaining in northern latitudes after the middle of August, after which "the liability of encountering heavy gales increases." Glass also had to plan for coaling, highlighting the very logistical limitations that prompted his survey in the first place. His smaller ships could not steam from their proposed departure port of Bremerton, on Puget Sound, to Adakh Island and back, a round trip of more than forty-five hundred miles, without refueling, and squadrons could only travel as far as the bunkers of their smallest vessels permitted. Glass proposed either sending a collier with the fleet or coaling from private shipping lines with stations at Unalaska's port of Dutch Harbor. The navy was already low on colliers and Secretary Moody vetoed sending one as "impracticable." As for Dutch Harbor, both the Alaska Commercial Company and the North American Commercial Company maintained coal depots there, the latter supplied with Comax coal from British Columbia. Yet stocks were limited and expensive, and Glass acknowledged that an expected shipment of additional coal in August was contingent "upon the settlement of labor troubles at the Comax mines," a further complication. Ultimately, despite all the preparations, however, Glass's survey, which he began on June 22, 1903, proved disappointing. The waters were exposed to harsh weather and the anchorage was too narrow for more than two ships, which convinced him that "it is entirely impracticable to establish a coaling station at this place."[53]

Kiska Harbor was a different story. Glass led the *New York*, the *Marblehead*, and the *Fortune* to Kiska on June 23, where "it was immediately apparent that this harbor offered advantages unusual in Alaskan waters." Geographically, it was ideally situated near the great circle route, a full six hundred miles west of the nearest existing coaling station at Dutch Harbor but less than two thousand miles east of Yokohama and a mere sixteen hundred miles east of the Japanese port of Hakodate. The crew surveyed the entrance to the harbor, finding it accessible from both north and south. In addition, all navigational hazards were clearly identifiable. "There is abundant room for a large coaling station," noted Glass in his report. The water was deep enough to anchor any size ship. The shore provided ample land for buildings. And the harbor boasted an "abundant" fresh water supply, needed aboard ship to generate steam. If its resources were rich, however, its labor supply was not. Kiska, like so many Aleutian Islands, was uninhabited. Glass suggested that indigenous Aleuts could be induced to move there, adding that a naval station on the island might be incentive enough to attract them to build a self-sustaining settlement.[54]

Glass's report generated contention within the Navy Department. Specifically, the Bureau of Equipment, whose portfolio then included both the navy's major coal purchases and the logistics of fueling the fleet, objected to the suggestion that Kiska alone should become the major coaling station in the Aleutians. Royal B. Bradford, the chief of the bureau who jealously defended his autonomy over coaling matters, observed that if the navy were to build only one depot in the Aleutians, it should do so at Dutch Harbor; if two, he insisted that Dutch Harbor should come first. Bradford believed that placing too great an emphasis on strictly *naval* logistics might cause the navy to lose sight of the larger commercial and maritime enterprise it ought to support and protect. Unlike Dutch Harbor, Bradford noted that Kiska lay far from popular trade routes, and he considered its lack of population a serious liability. He maintained that "in establishing fortified coaling stations the needs of the merchant marine should be considered." Bradford's support of Dutch Harbor over Kiska cut to the heart of a fundamental tension within the navy at the turn of the twentieth century: was the navy an autonomous institution whose sole task was to support national defense or was it subordinate to a larger strategy under which it was also meant to undertake projects that encouraged economic growth? The tension between these two positions was never fully resolved.[55]

The General Board did not, however, ignore the needs of commerce. Its members countered Bradford by arguing that a coaling station at Kiska would in fact encourage steamers to chose the shorter, northern course and thus contribute to the development of the resources of the Aleutians themselves. With this image of a prosperous, commercial, and above all *American* northern Pacific in mind, the navy ordered yet another detailed study of Kiska's harbor. This study would far exceed the previous three seasons of exploration. "An ordinary hydrographic or topographic survey is not sufficient," explained George Converse, Bradford's successor as chief of the Bureau of Equipment. Instead, he imagined a carrying out a detailed engineering analysis with a degree of scrutiny that required expertise and equipment not available aboard ordinary naval vessels. During the summer of 1904, the Coast and Geodetic Survey, joined by officers of the navy's Bureau of Engineering, dispatched two vessels to chart the island group.[56]

What they discovered confirmed that geographic value was more complicated than mere position. The Coast Survey found a chilly humidity clinging to Kiska. Between June and September, temperatures hovered around 45° Fahrenheit, and "numerous mists and light drizzling rains" kept the island air damp. Fog occasionally blew in from the south, carried along by the Japan stream. Although winds were erratic in strength and direction, Kiska at least remained free of williwaws, the fierce storms that plagued Dutch Harbor. The surveying was complicated, however, by the discovery that "the islands are nearly all wrongly charted." When Bureau of Equipment officers explored the harbor, they too met difficulties. The exasperated crew "nowhere found a solid foundation," and even deep in the water they discovered "only . . . an excellent variety of peat." These conditions meant that creating a functioning harbor would be harder than initially thought and would likely increase the expense. Even without spongy terrain, navy engineers estimated the cost of a major coaling station at Kiska as high as $1.6 million over several years of construction; the discoveries of the Coast Survey team only promised to increase this figure.[57]

While naval vessels explored Aleutian harbors, the federal government set in motion the legal machinery necessary for constructing naval coaling stations. President Roosevelt began this process in June 1902. Since the General Board had at first been enthusiastic about a station at Dutch Harbor, Roosevelt issued an executive order reserving a parcel of land for coaling there. Still awaiting reports from surveys of other Aleutian islands, later that month Roosevelt reserved 900 acres at Kiska and 580 acres along the Bay of Waterfalls on Adakh Island. By the end of 1902, after the surveys of these harbors arrived in Washington, Dewey and the General Board reversed their recommendation that a depot be constructed at Dutch Harbor, concluding that "the ordinary commercial facilities" at Dutch Harbor were adequate for navy needs, which were modest since the harbor lay far from the great circle route. They instead called for a massive installation at Kiska that would be garrisoned by the War

Department and maintain a hundred thousand tons of coal. President Roosevelt concurred and issued another executive order to reserve the entire Kiska island group—Kiska, Little Kiska, nearby islets—all to support future naval construction and to hedge against "squatters and speculators" looking to profit from proximity to a new base.[58]

While considering Aleutian islands along the great circle route for a coaling station, one question the navy did not explicitly address was where the tens or even hundreds of thousands of tons of coal that would fill it might come from. They did not have to. At the turn of the twentieth century, no coal known in the Pacific region matched "navy standard" coals of Appalachia in energy content, smokeless combustion, and ease of handling. The navy anticipated supplying the proposed Aleutian station the same way it supplied its other West Coast and Pacific stations, by shipping Appalachian coal, much of it marketed as "Pocahontas," around Cape Horn. In early 1905, however, a new possibility emerged. Writing to the secretary of the navy, the chief engineer of the struggling Alaska Central Railway inquired whether the department had interest in a coaling station around Resurrection Bay, on mainland Alaska's Kenai Peninsula. Although such a station would no doubt serve the financial interests of the ailing railway (it would go bankrupt in 1908), the General Board maintained that Resurrection Bay was too far removed to serve the strategic needs of the navy. It was too remote, too undeveloped, too impractical for serious consideration. But the railway drew its coal from the Matanuska Valley, near the camp that would become known as Anchorage, whose rich deposits of bituminous, semibituminous, and anthracite coal were of the highest quality anywhere then known in the Pacific region. Declining the pursuit of a coaling station at Resurrection Bay, the General Board instead expressed interest in Matanuska coal for the proposed station at Kiska.[59]

As late as 1910, however, the Kiska coaling station remained more an idea than a concrete policy. The General Board affirmed the importance of fortified bases en route to Asia and declared that "the war combinations in Eastern seas held the attention of the world that summer of 1900 no less than now." But developing the base was hampered by the geographical and climatic challenges of Aleutian navigation and construction, the General Board's awareness of the likely difficulty of defending these islands in an emergency, the board's perception that conflict in the central Pacific was more likely than in the north, and finally the navy's already strained budgets. These factors finally led the board to focus instead on bases at Pearl Harbor, Guam, and Manila. Ultimately, the surveys of the Aleutians never resulted in a major base along the great circle route to Asia. But the navy's persistent interest in the region for over a decade brought it a much greater involvement in and knowledge of Alaskan affairs. At no time before 1898 had the department been concerned with Alaska from a strategic, defensive, or resource perspective. That was changing. With a greater presence in Pacific waters, with increased American commerce with Asia, with a greater integration of Alaska into the trade and defense of the Pacific, and with developing Alaskan coal fields, the navy had begun a venture with Alaska that would only grow over the following decade.[60]

The navy accelerated its investigations into Alaskan coal when the armored cruiser *Maryland* arrived at Controller Bay on July 31, 1913. Its mission was to collect and examine coal from Alaska's Bering River field, miles from known and traveled seaways. The captain, John M. Ellicott, had anticipated this voyage for more than a decade. His interest in Alaskan waters had been sparked when he studied "ocean highways" during a stint at the Naval War College, when he "happened one day to stretch an elastic across a globe from Puget Sound to China and Japan" and recognized the proximity of the Alaskan coast to the great circle route in the north Pacific. Unaware of the General Board's interest in this topic, he mused about exploring the harbors of Alaska for a way station en route to Asia. He later requested, and received, a commission as inspector of the Thirteenth Lighthouse District, which held jurisdiction over the expanse of American coastline between California and the Arctic Ocean. With this command, Ellicott developed a deep familiarity with Alaskan waters. When naval planners sought an officer to scout Alaskan harbors and test samples of coal, Ellicott was the obvious choice.[61]

Geologists believed that two coalfields in Alaska contained great quantities of soft, semibituminous steaming coal: the Bering River field, some twenty miles northeast of Controller Bay where the *Maryland* anchored, and the Matanuska field, two hundred miles to the northwest beyond the Chugach Mountains. Estimates of the size of the Bering River field ranged widely, from $1 billion to $6 billion. Mining engineers familiar with the region quoted the figure of 500,000,000 tons worth of coal, a number frequently cited by the national press. This estimate took into account only the coal lying above the water level. U.S. Geological Survey estimates considered all coal to a depth of three thousand feet and its figures were consequently much higher. One survey held that both Bering River and Matanuska semibituminous coals were "better than anything that is being mined in the West" and compared them to the navy standard coals from the Pocahontas, New River, and Georges Creek fields in Appalachia. This report also noted that these Alaskan coals were "eminently adapted for use on warships" by virtue of their "smokeless" properties. These coals were expected to drive competition—whether from eastern coals or coking coals produced in Washington or Vancouver—straight out of the Pacific market.[62]

More sober accounts of the Bering River field acknowledged that its geology was complex, that its seams folded and faulted in unpredictable ways, and that much of its coal crushed into a fine, sooty powder that made it difficult to mine, transport, and consume. These circumstances made mining laborious and expensive. Alaskan coal nevertheless impressed even usually dispassionate observers. According to Alfred Brooks, the U.S. Geological Survey's leading expert in Alaskan geology (and a scientist quick to point out the exaggerations of others), the development of this coal was destiny, the instrument of modernity that would support commerce and industry along the Pacific rim, furthering the march of civilization ever westward. There were obstacles to overcome first—transportation, markets, competition with Californian oil—but it was inevitable, he argued, that Americans would exploit the fuel. "What is the future of Alaskan coal?" he rhetorically asked members of the American Mining Congress in 1911. "The answer is simple enough—it will be burned."[63]

The Bering River field rose to national prominence in 1909 as the focal point of a national political spectacle. That year, Gifford Pinchot, the chief forester of the United States, and his allies launched a public broadside against interior secretary Richard Ballinger for allegedly mishandling coal leases there to Clarence Cunningham, leases that were then pursued by the Guggenheim-Morgan Alaska Syndicate. Pinchot and Ballinger had long been at odds over conservation policy, and the forester capitalized on accusations against Ballinger by a Land Office employee that the secretary had illegally helped monopolists gain private control over public coal lands along the Bering River. As investors and politicians wrestled for a share of the field's seemingly fabulous profits, an incendiary article in *McClure's* magazine breathlessly called the coal there "the greatest single prize ever played for in this country."[64]

In the wake of the Ballinger-Pinchot controversy, Ballinger's Interior Department successor rescinded the disputed leases. For a time, Alaskan coalfields remained effectively closed, and the prize appeared forfeited. But not for long. In July 1913, Washington senator Miles Poindexter submitted a bill in Congress to open Alaskan coalfields and construct a regional transportation network. Poindexter, along with his Washington colleague James W. Bryan in the House, based the bill on a proposal by James MacKaye, brother of regional planner Benton. The bill (S.R. 2714) proposed dividing Alaskan coalfields into two halves, one for private companies, the other for a government mining agency. The plan gained significant press coverage during the summer of 1913 for its proposal to distribute the anticipated 10 percent profit from the government's fields equally between miners and consumers. The *Seattle Star*, expecting an economic windfall for its city pending the passage of the bill, threw the weight of the Scripps newspaper chain behind what it called "the greatest and the most vital project it has ever launched for the upbuilding of this city," asking its readers in a massive front-page editorial, "ARE YOU WITH US ON THIS GREAT PROJECT? **IT'S TO HELP SEATTLE**." Other Scripps papers printed supporting articles. Supporters of the plan pointed to the high costs of naval steaming coal along the west coast—nearly all of it from West Virginia—as ample evidence that competition from Alaskan coal would lower fuel costs. The navy paid $7 per ton for its coal in Pacific ports, observed Alaskan boosters, some $5 of which went for transportation from Appalachia and around Cape Horn. Alfred Brooks of the Geological Survey observed that these costs might be halved with an Alaskan coal supply.[65]

Thus, when the *Maryland* anchored in Controller Bay in late July 1913, it was with much anticipation. A successful test might lead to large-scale mining in the region and simultaneously guarantee the navy enormous quantities of coal. The tests, however, did not produce the expected results. First, the coal samples were wet. "By picking up a handful it was possible to squeeze water out of it" reported the testing board. Chemical experiments revealed 5 percent moisture, meaning that a ship carrying 2,000 tons of coal also lugged fully 100 tons of useless water, weighing down the ship while costing the vessel a full day of steaming. Additional tests revealed that even when dried, Bering River coal could not compare to navy standard Pocahontas. By one measure, it was less than half as efficient, meaning a ship burning Pocahontas would have no trouble voyaging between San Francisco and Yokohama by the great circle, yet while burning Bering River coal that same ship would just barely reach Honolulu. These results were so unexpectedly dismal and diverged so widely from expectations that rumors circulated among Alaskan miners that the navy had received an intentionally inferior coal sample, supposedly the result of a deliberate intervention by a former interior secretary. The poor performance did not result from sabotage, however. They were instead a consequence of sloppy mining practices along the frontier, a punishing climate, and lack of transportation infrastructure, all of which contributed to degrading the samples once they were brought to the surface.[66]

The poor Bering River tests did not lead to a loss of interest in Alaskan coal. The following summer, in 1914, the *Maryland* returned to Alaska and received another shipment, this time from the Matanuska field. Performing identical steaming tests, the experimenters found strikingly different results. Matanuska coal burned nearly as well as Appalachian navy standard. The question was only whether the coal could be mined and transported at rates competitive with those prevailing in the Pacific. A lack of railroad infrastructure and then war in Europe interrupted action on Alaskan coal development, but in 1919, a naval commission returned for renewed geological exploration. The commission's senior member, Captain Sumner Kittelle, predicted they would find more than enough coal to justify mining the field. He estimated that in the entire Matanuska Valley, there was 46 million tons, and in the five most promising leasing units of the Chickaloon district alone he predicted there was some 19 million, all adequate for naval use. He recommended that the navy assemble a mining expedition, advising Secretary Daniels that "the whole Matanuska region should be thoroughly and scientifically investigated by means of diamond drilling, shaft sinking, tunnel and slope driving" to positively determine the volume of coal available and the techniques to mine it most effectively.[67]

From the perspective of Josephus Daniels, the strategic advantages of an Alaskan coal supply were obvious. He had long been preoccupied with the length of time involved in shipping coal from the Eastern Seaboard to the West Coast.[68] But would proximity of supply compensate for the costs of development? Shipping eastern coal to the Pacific made it expensive, but it remained unclear whether the navy's development of an Alaskan coal supply would reduce the total expense. While officers could insist that no price was too high to secure the highest quality fuels, Daniels ultimately expected to turn Alaskan coal mining over to private industry after government mining proved and developed the fields. If the mines could not profit, they would perish along with the navy's new coal supply. Nevertheless, as early as 1917, Daniels imagined Matanuska coal eventually competing in world markets, and he prepared to commit the navy to purchasing 150,000 tons per year through the mid-1920s.[69]

It was with these considerations in mind that the navy's Alaska Coal Commission entered the Matanuska Valley camp of Chickaloon in August 1920. The commission came to mine. Admiral Hugh Rodman, the commander in chief of the Pacific Fleet, noted that it was known more or less for certain that there were at least 400,000 tons of coal, but he expected to find much more. Nevertheless, early geological reconnaissance indicated that the field's structure would be difficult to decipher. Geologists estimated that the Chickaloon formation, an alternating series of shale and sandstone, was some two thousand feet thick, but individual coal seams embedded within the earthy matrix were not "persistent," appearing irregularly and terminating abruptly. Igneous masses intruded into the coal beds, forming dikes and sills. Deformations in the landscape brought about by intense underground forces further complicated the field's geology. A report to Secretary Daniels cautioned that estimating the coal buried in the region was impossible and noted that "the steep dips and complex folding and faulting of the coal areas calls for careful investigation and development of the structural conditions of each individual tract before the development of a mine is attempted."[70]

Careful investigation and mining lasted nearly two years but was beset by obstacles. Shortly after exploration began, Commander Otto Dowling, the naval officer in charge of the mining, reported to Washington that the Coal Creek field showed "unusual conditions." There, diamond drilling and tunneling revealed a variety of impediments: coked coal beds in the south; faulted or "dirty and bony" beds in the north. Fair coal seams gradually shaded into shale, and even the high-grade coal required extensive and expensive rail lines to transport it. At nearby Kings River, the complicated geology of the field made predictions impossible. The commission found satisfactory coal but judged it "totally valueless" nonetheless, "as the beds cannot be correlated and traced for sufficient extent to warrant driving upon them." As with Coal Creek, even these fair beds were difficult to reach and therefore promised to be expensive to mine. There were even tensions within management, as the navy's supervisors found it difficult "to maintain friendly relations" with the chief mining engineer from the Department of the Interior, whose "ideas of expenditures," according to one report, "tend very closely to extravagance."[71]

Suspicion within the navy that Matanuska coal would never compete economically with Appalachian coal arose during the summer of 1921. Geological reconnaissance indicated irregular beds, making high production costs likely, if not inevitable. Even if coal could be mined for as little as $7.00 per ton, the combined costs of freight, washing, and handling would easily exceed the cost of eastern coals. "I doubt whether the coal can ever be laid down for the same price as eastern coal," admitted Dowling, "certainly not for less."[72] Markets for other coal consumers held little promise. In the summer, only canneries along the coast consumed large quantities of coal, but they subsisted on cheap and dirty lignite. In the winter, even growing towns like Anchorage still demanded very little—about 1,000 tons per month—and they opted for a lower grades of commercial bituminous coal mined near Chickaloon.[73]

These market constraints remained unknown to the majority of Alaskans, who grew increasingly discontented over the progress of government mining in the Matanuska Valley in late 1921. One Alaskan paper editorialized the situation by demanding to know why the navy's project in Chickaloon had not resulted in any obvious benefit to the residents of Alaska. "Where is the Coal?" it asked, "and why is there not more coal available for commercial use in Alaska towns?" Other papers reproduced the piece, to which the navy eventually responded. Explaining the progress of its exploration and mining, the department noted that difficulties in marketing Alaskan coal were not limited to the government, as even the four private companies operating in the Matanuska Valley likewise struggled for consumers. It was simply difficult—logistically difficult—to transport coal from remote fields to markets in Alaska or elsewhere along in the Pacific. The navy concluded its defense by diffusing expectations that Alaskan coal might ever be able to compete with eastern coals and their "present ridiculously low freight rates."[74]

Then there was labor, a subject that tied this remote government mine to a national debate over wages and power in post–World War I America. Postwar wage reductions, strikes, and, in many places, violence affected nearly every American industry, but these years were particularly volatile for coal in particular. During the war, heavy coal demand and increased prices had stimulated

Coal camp and mines at the U.S. Navy's mine in Chickaloon, Alaska, circa 1920. The Navy Alaskan Coal Commission attempted to open the vast coal resources of this territory that was nestled in Alaska's Matanuska Valley for both the U.S. Navy as well as the growing markets of the vast Pacific basin. A government coal mine was the culmination of decades of logistics planning for fueling American power in the Pacific ocean. "Chickaloon Coal Mines," report, December 1920, box 1, Reports from the Navy Alaskan Coal Commission, December 1920 and May 1, 1922, General Records, RG 80, NARA-1.

a nearly 30 percent increase in new mines between 1917 and 1919. The war brought inflation, too. In response, miners fought for, and received, a series of wage increases—around 50 percent above the 1914 scale—in return for promises to abstain from striking. With postwar demobilization, the nation slid into an industrial recession; in the coal industry, utility and industrial coal orders slackened and production plummeted. Faced with a precipitous decline in demand, mine operators closed mines, cut hours, and dismissed miners. A nationwide coal strike of six hundred thousand miners in November 1919 shut down more than 60 percent of the country's bituminous mines along with the entire anthracite region. Only government arbitration brought a resolution, when miners and operators in most of the country accepted wage increases of between 20 and 27 percent in March 1920.[75]

These labor struggles spilled over into the navy's operation in Alaska. Miners in Alaska were frequently drawn from Washington State, and private operators up north typically enticed these workers with higher wages, typically a premium of 10 percent over Washington rates. Washington was not governed by the March 1920 national labor agreement; because its labor and production conditions varied so greatly, government arbitrators thought it best that an agreement there be effected by a subsequent regional negotiation, the conclusion of which raised rates modestly but ultimately pleased no one. By the summer of 1920, however, with rising inflation, the wage premium in Alaska had all but disappeared. Miners struck for higher wages at the government's Alaska Engineering Commission mine that provided coal for the under-construction Alaska Railroad. They threatened another strike when the Washington arbitration led to higher wages, and by the end of 1920, successfully secured an increase. It was barely 4 percent over Washington rates, but it was something.[76]

For the navy, however, it was also too much. No sooner had the government agreed to this increase than it began deliberating over how to rescind it. In 1921, the cost of living in Alaska began to decrease. With miners receiving regular work and unusually good lodging, food, and medical care, navy and interior officials argued that the government wage was "manifestly too high and out of all reason." High wages were "a handicap" and "seriously embarrassed" private coal operators in Alaska, who were unable to offer such rates and remain solvent. The navy could ill afford to ignore its effect on private operators, for the exploratory mining at Chickaloon was designed precisely to entice capital to invest in the high-grade coals of the Matanuska district. Capital would only invest, reasoned navy planners, if investors believed that Alaskan coals could compete in Pacific markets with coals of the Eastern Seaboard. The department concluded that industry's perception that labor costs were excessive acted as a deterrent against this investment, with high production costs "tending to discredit" the field.[77]

From the perspective of the miners, however, the discredit came from how they were treated. In March 1921, operators in Washington announced a return to prewar wages (and prewar coal prices) in return for providing miners with food, rent, and supplies at 1917 rates. At first, miners answered with a strike that shut down 90 percent of the state's mines, but by August, they buckled under operator demands. The next month, the navy used this reduction as an impetus to reduce its own wages and expenses, announcing a cut of nearly 25 percent that lowered the pay scale from $8.60 to $6.50 per day. The navy's miners protested, organizing an impromptu campaign for navy secretary Edwin Denby and interior secretary Albert Fall in Washington, pressuring them to not go through with the cut or to at least postpone putting it into effect. For their part, the officers who ran the mine acknowledged "that if any man on the globe deserves utmost consideration, it is the citizen who breathes the coal dust in dark, damp, dusty, close coal mines." But they also believed that the demands of economy must be considered first.[78]

The response from the Navy Department surprised the miners. Mining Alaskan coal was not the only way government officials thought about the geopolitics of energy in the Pacific. In Chickaloon, the dispute over miners' wages was central to the mining operation. In Washington, D.C., it was just one aspect of the struggle toward the larger goal: developing secure and economical fuel supplies for national defense. When Denby and Fall learned in September 1921 of the extent of miners' dissatisfaction with a proposed wage reduction, they agreed to investigate the situation themselves and quickly postponed any changes until the following spring. The two departments admitted that a wage reduction on the cusp of winter appeared cruel. But more, in fact, was going on. To Dowling in Chickaloon, Denby could only wire instructions to delay the wage reduction and exclaim ambiguously, "Situation more complicated than appears to you."[79]

What exactly he meant remains unclear. But at precisely the moment the navy was struggling to mine coal in Alaska, it was engaged in policy changes that would see the sun set on the strategic value of coal far faster than anyone had previously anticipated. In February 1922, Dowling received a terse telegram from Washington: "Due to lack of funds all expenditures in Matanuska Field will cease April one period Advise all concerned period Letter follows." After nearly two years of digging, drilling, strikes, and strategizing, the leadership in the navy decided to terminate its direct involvement in mining Alaskan coal. The decision was not unanimous in Washington, where the chiefs of both the Bureau of Engineering and the Bureau of Yard and Docks protested to Denby that the closure of the Chickaloon mine must be accompanied by further development of an emergency coal supply in the Pacific. From the field, officers reported that they had never dug deeply or widely enough to be able to state with certainty how much coal the field actually contained—perhaps another 12 or 15 million tons. But the exploration never happened. Instead, a small group of officers and officials had decided to abandon coal and pursue oil instead.[80]

The Other Teapot Dome Scandal

John Keeler Robison was one of the naval officers shaped by the new science of logistics. Before his appointment to head the Bureau of Engineering in 1921, Robison had served in the office of the chief of operations, the bureaucracy that effectively served as the navy's chief of staff. Since its creation in 1915, the office had managed the fleet and planned for war. There, Robison coordinated American war plans for the department's many shore stations and served as secretary to an interbureau committee studying the "problem of the Pacific," or, how to wage war against Japan. Between 1920 and 1921, Robison came to recognize the logistical challenge of fueling a future war. Every year, the navy consumed more oil, yet facilities for transportation and storage remained inadequate, and after the armistice, appropriations for fuel had shrunk, not grown. Coal remained a critical consideration, but oil appeared to be the future. When Robison heard rumors in early 1921 that the incoming Harding administration sought to transfer government land set aside a decade earlier as naval oil reserves from the Navy Department to Interior in order to lease them, he readily endorsed the plan. The oil reserves were called Elk Hills and Teapot Dome.[81]

Teapot Dome was one of the great scandals of American history. It is best understood, however, as two separate scandals. The first is more widely known. In the lead-up to the 1920 Republican presidential nomination, a few wealthy oil barons helped support the nomination of Warren Harding in exchange for his promise to appoint a secretary of the interior willing and eager to lease lands set aside as an oil reserve for the navy. After Harding's election, Albert Fall, an influential New Mexico senator, secured the helm of the Interior Department, collaborated with the navy secretary to turn management of the reserves over to him, and then secretly leased the reserves without competitive bidding to those same oil barons who had helped nominate Harding. Edward Doheny obtained a lease to California's Elk Hills reserve, while Harry Sinclair acquired one for Wyoming's Teapot Dome. Geological estimates at the time suggested that Teapot Dome alone contained as many as 135 million barrels of oil, suggesting profits for Sinclair upwards of $400 million (or $5 billion in today's dollars). Elk Hills in California was thought to hold some 300 million barrels and perhaps be worth $1 billion (or nearly $13 billion today) to whomever produced its oil. To secure the leases, Sinclair, Doheny, and other, shadier figures in the oil business showered Fall with hundreds of thousands of dollars in cash and Liberty Bonds, which they later claimed were simply unsecured and undocumented loans. Along the way, in a complicated set of dubious transactions, Sinclair and three other oil men bilked their own companies to siphon proceeds and redirect them as illegal campaign contributions to the Republican party. By the end of the 1920s, Fall had become the first cabinet secretary to go to prison for crimes committed while in office, a jury had convicted Harry Sinclair for jury tampering, and the courts (and public pressure) had returned the naval petroleum reserves securely to the Navy Department.[82]

Gross political corruption was at work in this scandal, but it only explains a part of the story. Fall, Doheny, and Sinclair may have been the prime movers on leasing the reserves, but why did the navy appear to acquiesce to it? The 1925 district court opinion over annulling the leases noted that of the many government participants, only Fall was (rightly, as it turned out) accused of corruption.

Other participants all approved of the new policy and pursued it diligently without any awareness of Fall's schemes or prospect of benefit to themselves. The list of participants included navy secretary Denby; his assistant secretary, Theodore Roosevelt Jr. (son of the late president and known as Ted); Bureau of Engineering chief John K. Robison; assistant secretary of the interior Edward Finney; Bureau of Mines director H. Foster Bain; and Arthur Ambrose, Bain's chief petroleum technologist. Moreover, oil operators consulted during the trial acknowledged that the benefits for the government from Fall's leases were so extensive that they would not have agreed to the leasing terms themselves.[83]

What had changed that encouraged government officials to develop the reserves? Before World War I and even before the strategic value of oil was as widely understood, navy administrators had considered the reserves essential national assets. "The very life and future existence of the United States Navy is at stake," Franklin D. Roosevelt had exclaimed in 1916 without a sense of hyperbole. Already, the navy had begun designing new scouts, destroyers, and battle cruisers to burn oil fuel exclusively, and while coal use continued, no one imagined a wholesale return to exclusive coal consumption. Yet Roosevelt, along with most geologists unaffiliated with the oil industry, believed that existing American oil fields had reached or would soon reach their peaks "and that there is not much probability of discovering fields of the same magnitude as those already opened." Increasing domestic use in automobiles and trucks brought additional strains. The only answer, Roosevelt argued, was preserving a protected domestic supply for national defense.[84] That same year, the Naval Consulting Board, headed by Thomas Edison, announced that any action to deprive the navy of its oil reserves would "seriously weaken the navy and imperil the national defense." This was a year in which peacetime naval oil consumption reached a mere 842,000 barrels; within a decade the board predicted an increase of more than an order of magnitude. A state of war might produce another threefold increase over that, at least.[85]

The view of naval counsels, from within which people like John Keeler Robison studied the logistics of war, was that preserving a distant future oil supply for the navy was indeed essential, so long as the navy possessed the infrastructure to make use of it. This tension led to the other Teapot Dome scandal—how and why naval leadership not only acquiesced to Albert Fall's designs but actively encouraged them. Naval officers led by Robison tendentiously interpreted ambiguities in the law so as to justify circumventing congressional restrictions they believed fatally limited preparations for national defense. Between 1921 and 1922, Robison and others devised a scheme not only to turn petroleum reserves into a range of refined oil products like fuel oil, gasoline, and lubricating oils but also to establish a network of storage facilities, all outside any specific congressional appropriation. It is unlikely that those involved had any hint of Fall's motives or the machinations of the reserve-leasing oil barons. Put simply, naval officers dealing with fuel logistics had their own reasons to see leasing the reserves as in the national interest. It was a view that in part reflected knowledge about the peculiar characteristics of oil and in part the way the new logistics introduced a kind of resource planning around energy into security debates.

Before the development of logistics, oil had long been considered a potentially strategic fuel before being rejected in favor of coal. In 1864 naval engineers investigated petroleum as a coal substitute, and two years later Congress allocated $5,000 for additional studies. Results, however, were disappointing. It was then impossible to prevent explosive fumes from leaking into the bowels of a ship. Despite prospective advantages in greater density and ease of refueling, "the use of petroleum as fuel for steamers is hopeless," wrote engineer Benjamin Isherwood in 1867. "Convenience is against it, comfort is against it, health is against it, economy is against it, and safety is against it." With these technological constraints, Isherwood dismissed the possible advantages of petroleum. Although some engineers continued to support a fuel switch, Isherwood's Bureau of Engineering returned to studying coal boilers and the properties of various coals.[86]

Yet by the turn of the twentieth century, the picture looked very different. Between 1897 and 1901, spectacular petroleum discoveries in Texas and California made the country appear awash in oil, just as market instability and the enormous 1902 anthracite coal strike in Pennsylvania led navy planners to worry about the security of their existing fuel supply. Coal was expensive and relatively scarce in the Pacific, while western states were becoming large oil producers. A wartime appropriation in 1898 granted the navy $15,000 to study liquid fuel. After promising results, in 1901, the Bureau of Steam Engineering ranked the study of oil its highest of eleven research priorities. This request attracted the ear (and purse) of Congress, which allocated an additional $20,000 in 1902 for more comprehensive tests. Thus began a two-year investigation into the feasibility of oil fuel. Chaired by engineer and lawyer Commander John R. Edwards, the U.S. Naval "Liquid Fuel" Board presented the results of its extensive analysis in 1904. The report concluded that "the engineering or mechanical feature of the liquid-fuel problem . . . [was] practically and satisfactorily solved." Only "financial and supply features" needed to convert the fleet remained to be addressed. These features included the "serious difficulty" of ensuring an adequate oil supply. This final point proved to be the fundamental stumbling block for naval fuel policy for more than two decades to come.[87]

Even amid new discoveries, the future of American oil fields were far less certain than the country's mines of coal. In Teddy Roosevelt's White House, conservation became a national political issue and a major plank for progressive politics. "Conservation of our resources is the fundamental question before this nation," Roosevelt declared in 1909, as "our first and greatest task is to set our house in order and begin to live within our means." To set the national house in order, Roosevelt convened a conservation conference for governors in 1908 and then commissioned a national inventory of mineral wealth. Appointing forestry chief Gifford Pinchot to chair the National Conservation Commission, the president charged him with assessing the use and longevity of all economically significant natural resources. The commission was as much a political maneuver as a scientific project, for Roosevelt had already advocated conservation policies and now sought to rally support for them. The commission produced the first estimate of national oil resources conducted by a federal agency, petroleum being one resource among many assessed under the four headings of "waters," "forests," "lands," and "minerals."[88]

That estimate proved bleak. While acknowledging an intrinsic uncertainty, David T. Day of the Geological Survey predicted that between 10 and 24 billion barrels remained in the U.S. oil supply—enough to last at then current rates of increasing consumption until at most 1935. This study seemingly confirmed speculation about waning resources. Day noted that the U.S. would never actually run dry—market forces would simply push prices up to unaffordable levels—but with the growing national need for automobile fuel, lubrication, and now the navy's supply, Day feared that petroleum could not be treated like other resources. "In the preparation of the tables of production of petroleum and the comparison of these with the estimated supply," Day later wrote, "it has become manifest that the necessity for conserving our supply stands in a class by itself."[89]

As federal geologists observed the few remaining known oil deposits on government land passing into private hands, they successfully lobbied President Roosevelt to remove these lands from private entry and reserve them for future naval use. The director of the Geological Survey fretted that without immediate action, "the Government will be obliged to repurchase the very oil that it has practically given away" in order to meet coming naval needs. First the government reserved a handful of townships in California's Fresno County and Kings County in August 1907. After oil companies continued grabbing public oil-bearings lands under spurious claims for gypsum, in June 1909, President Taft's interior secretary withdrew 430,340 acres of public oil lands from entry across the state's McKittrick-Sunset region. Taft approved and three months later authorized an additional withdrawal totaling over three million acres, from which he designated in 1912 the Naval Petroleum Reserve No. 1 and No. 2—Elk Hills and Buena Vista Hills. Three years later, Woodrow Wilson withdrew a third reserve at Wyoming's Teapot Dome.[90]

The status of the reserves remained uncertain, however. Despite the withdrawal orders, several oil producers continued to operate under their earlier claims, action considered by the navy tantamount to trespassing. For the remainder of the 1910s, navy secretary Daniels pushed Congress to give his department complete control over the reserves, arguing that he had committed the navy to consuming oil in 1913 only after assurances from the Interior Department that the land would provide decades of security to the nation's fuel supply.[91] What Daniels precisely sought by demanding naval control over the reserves remained uncertain. At the outset of his term, he anticipated turning the navy into a producer and refiner of oil itself. As a southern progressive, Daniels imagined that an integrated government oil operation would not only ensure a dependable supply for the fleet but also loosen the grip of big oil companies. By 1915, he had concluded instead that simply keeping naval oil in the ground would best ensure fuel in a future emergency.[92]

Even this approach, however, did not guarantee a secure reserve. Neither Elk Hills nor Buena Vista Hills entirely enclosed underground oil deposits, and these deposits did not follow the boundaries of property maps. Oil producers on neighboring properties could extract oil on their land, siphoning oil from under the naval reserves. The challenge of the so-called rule of capture—oil belonged to the land on which it was extracted, even if that oil originated under a neighboring property—pointed to the difficulty of melding legal concepts of land ownership with the stubborn geology of oil pools. Owing to these complex legal issues and continuous litigation with oil producers within the reserves, the navy struggled to gain full control.[93]

To address these problems, Daniels established the Fuel Oil Board in May 1916. Comprised of technical experts from the navy, the board investigated the state of commercial oil production and assessed the future fuel needs for the department. Fitting with the navy's broader focus on logistics, the board's charge included investigating the economics of fuel oil purchasing, how this fuel should be purchased, and most importantly, from where the navy might expect its future supply to come. As it conducted its study, the board painted an unsettling picture of oil and American security. "Oil fuel is vitally necessary to the Navy because its use contributes directly to the fighting efficiency of the fleet," noted one report in September 1916. "Although this nation has, for a number of years, produced the greater part of the world's supply of petroleum, our deposits are being exploited at a rate which, if continued, will exhaust them within thirty years." It was an assessment drawn from experts at the Geological Survey. In response, the board called for "assuring an adequate, dependable supply of petroleum for the future at prices which will not be prohibitive." Its members gave a clear endorsement of naval control over the petroleum reserves and argued for prioritizing the exploitation of Mexican oil, examining the possibility of requisitioning oil-bearing Osage Indian lands in Oklahoma, and adding a further fail-safe by acquiring a vast reserve of oil shale. As they starkly explained, "It is no longer safe for us to depend upon the commercial petroleum market."[94]

This planning was interrupted by the entrance of the United States into World War I, where for the first time oil joined coal as an energy source possessing strategic and logistic significance. "No other power is in sight" wrote Daniels about fuel oil, "and we can not assume that other power suitable for ship propulsion will be discovered." This was, of course, not entirely true—the navy still burned the same coal it had for decades. Yet even during war, Daniels's concern was less for the present conflict than for naval fuel a generation hence in wars that nearly all assumed would be fueled by petroleum. According to Paymaster General McGowan, with the commitment to oil-burning vessels having been made, "there will be too much of the nation's capital invested in the floating armed forces to follow any such short-sighted policy as has been advanced by the advocates of tapping the naval fuel reserves." Draining the reserves for present needs, McGowan believed, would be catastrophic, but so too would allowing its fuel to remain in the ground without providing an adequate network of above-ground tank storage. In his annual reports, Daniels repeatedly admonished Congress to plan for the future.[95]

As it did with coal, war mobilization gave government planners new experiences with the management of oil. Yet within Harry Garfield's Fuel Administration, oil never received the same focus as coal. Consumed with managing industrial and railroad coal demand, Garfield only belatedly established an oil division at the end of 1917, when domestic price increases, lagging production, and uneven refining of petroleum products made the industry's problems inescapable. Garfield appointed Mark Requa to lead the Fuel Administration's Oil Division. Requa, a protégé of Herbert Hoover from the Food Administration, tried using the power of the Lever Act to foster industrial cooperation to coordinate wartime needs. Requa brought with him the conviction that since the "war cannot be won without the products of petroleum," he had the duty to deploy whatever legal means were available to support the war effort. Speaking in July 1918 before the National Petroleum War Service Committee—the industrial board he had reorganized to coordinate with producers and oil servicers—Requa declared "that individualism is for the time submerged." As for increasing production, he pushed wildcatters to drill more, drillers to develop fields with an eye toward conservation, and consumers to burn natural gas instead of oil whenever possible. In the face of opposition from the navy, the Federal Trade Commission, the Geological Survey, and the Council of National Defense, he joined western drillers in calling for production on the naval petroleum reserves. Yet while Requa spoke to producers about his power under the Lever Act "to requisition necessaries for the army and navy or any public use connected with the common defense," Garfield and Wilson never supported using these kinds of coercive measures to procure oil, even though they were invoked to secure coal. Historian John G. Clark has noted the difference was most likely due to the greater political power commanded by the more concentrated oil industry. Whatever the reasons, Requa was also unable to open Elk Hills and Teapot Dome to production.[96]

After the armistice, Congress returned to peacetime legislating, and among its first orders of business was updating the country's public lands laws. The issue had been debated to an unsuccessful conclusion for a decade. When it came to oil production on western federal land, existing laws encouraged producers to drill on prospective oil land in order to stake claims for future production, necessary steps for ultimately patenting the land into private hands. The process led, however, to market oversaturation with oil and chronically depressed prices, not to mention wasted oil. One faction of conservationists argued that the government ought to seize control of production itself; a more moderate one proposed introducing the leasing of mineral land under conditions Washington could closely regulate. After intense debate and negotiation, in February 1920, Congress passed a leasing law that set rules for prospecting and claiming land, resolved various outstanding conflicting land claims, and provided for a 37.5 percent royalty payment on oil produced to the state containing the leased federal land. Just over three months later and under intense pressure from the soon-to-be ex-navy secretary Daniels, Congress added an amendment to the yearly naval appropriation bill that finally granted the navy broad control over the petroleum reserves. With formal "possession" of the reserves, the secretary could now "conserve, develop, use, and operate the same in his discretion, directly or by contract, lease or otherwise, and use, store, exchange, or sell the oil and gas products thereof, and those from all royalty oil from lands in the naval reserves for the benefit of the United States." Daniels left office believing that he had finally secured the future supply of naval oil and thus guaranteed that the needs of American defense would be met.[97]

Defenders like Daniels of keeping navy oil in the ground represented their view, which they characterized as in the national interest, as widely shared among members of the service. When Thomas Walsh, the assiduous Montana Democrat who spearheaded the Senate's Teapot Dome investigation, later called navy witnesses, nearly all had experience managing and protecting the reserves and had opposed transferring them from the navy to the Interior Department. At one of Walsh's hearings on Teapot Dome, Commander H. A. Stuart revealed during testimony that when he had learned of the reserves' proposed transfer to Interior, he had appealed to John Keeler Robison in the chief of operations' office and pleaded with him to do whatever he could to prevent Interior from gaining control. Stuart's chief, Robert Griffin, the admiral and former head of the Bureau of Steam Engineering, had worked for years to secure the petroleum reserves for the navy and had tried to maintain some measure of naval control after their transfer. After Walsh began his investigation, an outraged (and out of office) Josephus Daniels connected the senator to Griffin, who began quietly providing Walsh with the names of officers who were knowledgeable on the subject and sympathetic to his view, mostly fellow engineers who had served under him and with firsthand experience with the reserves, among them Stuart.[98]

Unlike these officers, those working with war planning and logistics saw the value of the reserves very differently. Just as Walsh queried Griffin for experts on the reserves, he likewise asked the General Board to provide a list of potential witnesses, which returned an entirely different list, this time officials from the Bureau of Supplies and Accounts, which managed fuel logistics, the Bureau of Yards and Docks, which superintended fuel storage, the War Plans Division of the Office of Operations, and Robison's Bureau of Steam Engineering. Of the six officials recommended, Walsh called but one, the chief of Bureau of Yards and Docks, but under Walsh's questioning, this official only answered questions about contracts for dredging Pearl Harbor. It was hardly clear from Walsh's extensive hearings, but different branches of the navy held different views about the relative merits of leaving oil underground or producing it for storage in tanks above ground. According to Frank Schofield of the General Board, naval officers mostly supported undertaking immediate production from the petroleum reserves. He reported that no one had contemplated asking Congress for a specific appropriation for additional fuel storage; rather, everyone expected "royalty oil"—the government's share of oil produced on the reserves it received as lessor of the land—to fund what naval planners on the General Board, in the bureaus, and around the chief of naval operations believed to be needed construction. Yet few of the officers holding this view ever received invitations to testify on Capitol Hill.[99]

To the dismay of conservationists, these high-ranking officers supported turning the fields over to Interior for leasing. To the disappointment of H. A. Stuart, the architect of this plan was John Keeler Robison, who had left his post with the chief of operations to become chief of the Bureau of Engineering in the fall of 1921. In his new position, Robison received an admonishment from navy secretary Denby that the navy had entered a period of retrenchment and not a dollar was available for anything but the urgent needs of national defense. In fact, Congress had just passed the Budget and Accounting Act of 1921. Among its provisions, the act created the Bureau of the Budget, requiring the president to coordinate spending requests from across the executive branch and send an annual budget proposal to Congress. From the perspective of the navy, the new process upended its traditional way of doing business with Congress. Previously, officers could present their wish lists directly to lawmakers and hope for authorization acts that approved and funded ship construction or base fortification. Now, funding requests passed first through the Bureau of the Budget, and the Harding administration (which eagerly embraced the new system) was resolved against funding any program not already separately authorized by Congress, a policy of which the navy's war planners were all too well aware.[100]

Among the planners, Robison was acutely aware of the department's deficiencies—the "insistent and immediate" need for oil, as he put it—yet believed that there was still one untapped naval fund. Under the leasing act passed in February 1920, the government had settled with oil drillers it argued operated on Naval Petroleum Reserve No. 2 (Buena Vista Hills) illegally. This field was a checkerboard of public and private ownership. As part of the settlement, the navy agreed to receive royalties from continued production there. Additionally, the government drilled some twenty wells in Naval Petroleum Reserve No. 1 (Elk Hills) to offset oil drainage from production outside the reserve. Millions of dollars flowed into the Treasury from the sale of this oil, yet the revenues were not credited to specifically naval accounts. Robison believe they could be, provided the revenue was used only to fuel the fleet and fund construction of fuel storage tanks needed to do so. Under later questioning by Tom Walsh in the Senate, Robison insisted that despite the agreement with the Interior Department, the navy retained as much control over the reserves as it had previously, as it maintained veto power over any changes to oil leasing policy. When asked by an incredulous Walsh what Interior received if not control, Robison had a simple reply. "Work," he said, "and nothing else."[101]

For Robison, the work was an attempt to solve the problem of perceived fuel shortages using the bureaucratic and material tools he believed he had available. Under the terms of the original contracts, the government leased land in the California naval reserves exclusively to Ed Doheny's Pan American Petroleum and Transport Company and in Wyoming's Teapot Dome reserve to Harry Sinclair's Mammoth Oil. In return, the government claimed up to 50 percent of the oil produced as a royalty (with lower royalty percentages for lower production). Then, in an unusual provision, instead of taking physical possession of the royalty oil, the government received from the producers a specially designed "oil certificate," which the government could redeem from Doheny and Sinclair for a range of services. The navy anticipated initially redeeming the certificates to fund construction of a network of naval oil storage facilities. Once the companies completed the storage tanks, the navy would redeem additional certificates to fill them with fuel oil for naval ships or with other refined petroleum products like gasoline, kerosene, or lubricating oils. When Robison presented the proposal to the counsel of bureau chiefs, it was largely endorsed, with the chief of naval operations, Robert Coontz, and his assistant chief, William Cole, among the supporters. These officers handled war plans and studied fleet logistics and as naval leaders, felt deeply insecure about their fuel.[102]

Though the policy took shape over the course of 1921, by the time the Sinclair contract for leasing Teapot Dome was signed in April 1922, navy logisticians had an additional reason for worry. In February 1922, representatives of the United States, Britain, Japan, France, and Italy met in a conference to limit naval armaments. The most consequential result was the Five-Power Treaty. Headlines trumpeted the treaty's restrictions on battleship construction. Less noticed was its article 19, which placed limitations on colonial fortifications in the Pacific. Contrary to the wishes of the navy, the treaty froze base construction at all American Pacific territories except along the West Coast and continental Alaska, by the Panama Canal Zone, and in Hawaii. All the rest—the Aleutians, Guam, the Philippines—were subject to treaty restrictions of no new fortifications, including no additional fuel storage facilities. Under the treaty, naval planners witnessed the upending of two decades of war planning intended to establish fuel reserves to reach Asia and operate there unhindered. The treaty forced navy strategists and logisticians to concentrate the entirety of their planning for fueling war at the one remaining western outpost: Pearl Harbor.[103]

Ultimately, the treaty posed a paradox for American naval planning. It was designed to prevent an arms race in capital ships from escalating and to decrease the likelihood of war, yet by facilitating international cooperation and thus increased global trade, it encouraged precisely the kinds of economic rivalries and competitions that many Americans in the 1920s presumed led to war. Signatories, including the United States, did not look for loopholes in the restrictions merely from cynical or militaristic motives but also because they had convinced themselves that preparedness would temper these second-order consequences. As the treaty limited the construction of new fuel storage in outlying territories, Americans immediately set out to accelerate the conversion of their remaining coal burners to oil, thus increasing their steaming distances by 50 percent. From the planners' perspectives, the limitation on fortifications introduced a tremendous obstacle to preparedness. "If anyone asks why fortified outlying naval bases expedite naval action," explained navy secretary Edwin Denby, "the answer lies in the extent of the logistic support a great naval expedition requires." Denby noted the vast quantity of oil tankers likely required for war in east Asia and the vulnerability of stored fuel in the absence of fortifications. "The permanent and secure defense of suitable outlying positions and the accumulation there of suitable fuel reserves will put more speed into the ships of the fleet advancing to war than any other act of which I know," he observed. Yet with the naval treaty, the only outlying Pacific base available for fortified fuel reserves was Pearl Harbor.[104]

After the Teapot Dome scandal broke, the extent of American logistic weakness became widely known. Curtis Wilbur, Edwin Denby's successor heading the Navy Department, explained to Congress in 1924 that while it was likely that no other great naval power had more than a three-year reserve fuel oil supply, the American navy had fuel that would last barely six months. Worse, these

August 1919 photograph of the coaling station at Pearl Harbor. Though the station was always smaller than naval logisticians desired, in the two decades after 1898, fueling in Hawaii took on an increasingly important role in American naval strategic planning. Note the oil tanks in the lower left; between 1940 and 1943, the navy would replace these vulnerable aboveground tanks with a series of massive, underground ones at Honolulu's Red Hill. Folder 71-CA-160B, box 160, RG 71-CA, NARA-2.

projections accounted for only peacetime operations. During war, fuel supplies might last less than a single month.[105] Ted Roosevelt, still assistant secretary, reported that on the Pacific coast, naval reserves could barely meet a third of anticipated needs, while things appeared even worse in the Atlantic. Oil reserves at Pearl Harbor measured about 1.75 million barrels; then current war projections for the Pacific estimated future consumption at 70 million barrels for the first year of fighting alone.[106]

As Walsh's Senate committee uncovered the details of the deals, the country learned how navy planners had tried to manage their fuel problem within the new statutory and diplomatic constraints. The contract with Sinclair's Mammoth oil specified twenty-seven ports where the navy might desire oil storage facilities, from Machias, Maine, to Houston, Texas, to Guantanamo Bay, Cuba, for an estimated total construction cost of $25 million (or over $300 million in contemporary dollars). A subsequent contract identified the first four locations as Portsmouth, New Hampshire; Melville, Rhode Island; Boston; and Yorktown, Virginia; when the scandal broke, only work on the Portsmouth storage facility had begun.[107] The deal with Doheny's Pan American similarly provided for fuel storage construction in the Pacific for an estimated cost of $15 million (or nearly $200 million today). On top of construction costs, the deals provided for an estimated $60 million worth of refined fuel oil, diesel fuel, lubricating oils, and other products to fill those storage tanks, bringing the total expense to over $100 million ($1.3 billion today). The navy anticipated meeting these vast expenses ultimately from the sale of oil from Elk Hills and Teapot Dome. Congress, however, had never appropriated anything beyond a mere $500,000 toward fuel in 1920.[108]

For supporters within the navy, the contracts helped in many ways. They served to test the petroleum reserves for exactly how much oil they contained (a significant advantage if this time-consuming work had to take place during an emergency). If the reserves did prove to be oil rich, the plan anticipated taking crude oil from where the navy could not use it (like landlocked Wyoming) and turning it into refined products that could be used in key navy yards. The contracts additionally helped modernize the navy's underfunded logistical network of storage tanks, which planners long believed insufficient under various war-planning scenarios. Indeed, in his justification for the new policy, Secretary Denby cited logistic studies for war operations, though the supporters rarely focused on this sensitive detail of defense vulnerability, preferring instead to rely on anxiety over the uncertain volume of oil contained in each reserve and fears about drainage by neighboring properties. Of course, the terms of the contract also sidestepped Congress's role in appropriating specific amounts for specific purposes. The navy essentially began generating its own revenue to use for its own purposes—a circumvention of Congress that infuriated even those House and Senate Republicans most inclined to protect Albert Fall and the administration.[109]

In many ways, an observer's view of national security shaped reaction to the scandal. As the story broke in 1923 and 1924, the leading conservationists in government service, like the director of the Geological Survey, George Otis Smith, focused less on the corruption evident in the leases than on the damage it did to national security. As a leading proponent of the gospel of efficiency, Smith insisted it was the "un-business-like" approach to government oil that was the true crime. Smith reserved particular scorn for the deal to exchange royalty oil for storage tanks, as the price of oil then was relatively cheap and that of construction relatively dear. "In the name of good business," Smith caustically explained, "the Navy's oil has been 'saved' by spending 92 barrels out of every 100 barrels extracted from the California reserves in order to put less than 8 barrels into storage. That kind of liquidation of an irreplaceable asset must suggest to you business men only the desperate effort of a landowner facing bankruptcy, surely not the deliberate policy of a great nation planning for a long future."[110]

Perhaps, but Smith was a geologist, and Robison was the head of the navy's Bureau of Engineering. When questioned by Irvine Lenroot, the Republican senator who chaired the Public Lands Committee charged with investigating the matter, Robison insisted that Congress's action in 1920 delivering control of the reserves to the Navy Department authorized *protection of the oil*, not merely preservation of it in the ground (as most in Congress believed they had intended). For Robison, opening up the reserves was itself intended to defend national security. Even members of Congress conceded that drilling to prevent drainage by neighboring landowners was necessary to protect the naval asset; for Robison, this was only a small part of what was really needed to protect national security. When Lenroot asked him if he thought Congress would have allocated the huge $100 million sum needed for constructing a fuel storage network, Robison replied yes. When the skeptical Lenroot further inquired whether he thought Congress would have consented to the navy's deal to exchange two barrels of oil in the ground for only one in royalty, Robison again replied yes. Robison's explanation to the incredulous Lenroot suggests the global dimensions of this attempted security strategy: "If we had where we needed sufficient quantities of oil," Robison explained, "war would never come within our coasts. That is the object of the navy: To make it impossible that any war in which this Nation engages shall be one of invasion of our country, and with oil where we need it we can accomplish that mission. That is why I want oil where we need it rather than in the ground anywhere." Through these arrangements, Robison anticipated ultimately setting aside some 40 million barrels of oil for future emergency use.[111]

As the scandal came to light, Ted Roosevelt conceded that transferring the reserves from the navy to Interior was probably illegal, that the secret leases were a mistake, that opening production on the reserves not merely to prevent drainage was wrong, and that Albert Fall had been corrupt all along. Still, even as the government expanded its investigations in the summer of 1924, Roosevelt maintained, surprisingly, that the actual terms of the contracts, especially with Sinclair, "were good business propositions." Good, because they helped advance the construction of strategically placed oil storage facilities the navy had already decided it needed.[112] Roosevelt conceded that naval reserve oil, which he called "an important part of the national insurance," ideally should remain underground but insisted that when oil must be drilled, it must be stored above ground. Roosevelt argued that even if all leases to Doheny and Sinclair were invalidated by the courts, many of the wells then under production could not be shut down, in large part to protect against drainage. Further production would accrue to the navy's account—some 100,000 barrels per month from Naval Petroleum Reserve No. 2 alone—and would need to be stored somewhere. At the same time, the navy's barely 4 million barrels worth of existing storage was nearly full. The facilities at Pearl Harbor, already by 1924 some 70 percent complete, would add over 2.5 million barrels of storage; the 90 percent complete system in Portsmouth would add another 300,000 barrels of storage. Roosevelt implored Congress to permit the two companies under the Teapot Dome cloud, Ed Doheny's Pan American Petroleum and Transport in Pearl Harbor and Harry Sinclair's Mammoth Oil Company in New Hampshire, to complete the work they had begun for an additional $2.5 million.[113] The government's special counsels conceded that the almost completed work should be finished (so as not to be wasted), but they, and Congress, refused to go along with any further entertaining of oil leasing for fuel storage. As a result, with only a few exceptions, there was little subsequent development of the naval fueling network until World War II. By then, logistics had gone from marginal tool of mid-level bureaucrats to a principal lens through which American war planners conceived of national security.

The New Science

World War I changed the calculus of American interests with respect to fuel. For oil especially, the needs of the navy and commercial shipping remained central, but the American interest now extended to the already vast and growing domestic market for powering cars, trucks, and ships. Looking to the postwar period, a joint statement by the heads of the Bureau of Mines, Geological Survey, and Fuel Administration anticipated skyrocketing oil consumption alongside diminishing volumes of harder-to-reach petroleum. The fuel experts pleaded for "sympathetic Government cooperation in acquiring additional foreign sources of supply and by protection of properties already acquired," a proposal they couched in terms of a broad, national interest in oil. "This means a worldwide exploration, development, and producing petroleum company financed with American capital, guided by American engineering, and supervised in its international relations by the United States Government. In its foreign expansion, American business needs this Governmental partnership, and through it the interests of the public can best be safeguarded." It was a proposal that would be raised repeatedly over the following quarter century.[114]

World War I also produced among administrators and broad swaths of industry and labor alike an increased desire for government supervision of fuel industries. A 1919 symposium between coal operators, the United Mine Workers, and the wartime Fuel Administration produced a call for continued, permanent cooperation. "The public interest is violated by the existence of any such conditions," they wrote in a joint statement, warning of a return to prewar overpro-

Scene from a ship collision at the Brooklyn Navy Yard's coal pier, August 1920. This ship is positioned to receive coal from the bunker above. The soot on the seamen's faces underscores one reason naval officers and crew alike welcomed the transition to oil fuel, a process that accelerated over the coming decade. "New York—Navy Yard—Coaling Plant," folder 71-CA-271A, box 271, RG 71-CA, NARA-2.

duction, ill health and job insecurity among miners, and poor coal quality for consumers. "Some form of governmental regulation must be exercised in order to avert disaster." The groups called for federally licensing coal operators, strictly regulating prices and wages, exempting the industry from antitrust laws, and making the Fuel Administration a permanent agency that could promote conservation, trade regulation, and the settlement of labor disputes. The head of the Fuel Administration, Harry Garfield, went so far as to propose reorganizing the executive branch: a political cabinet would handle affairs of state and an industrial one would supervise agriculture, commerce, labor, transportation, and fuels. To aid the industrial cabinet, he suggested collecting all the government's technical bureaus from chemistry to the National Academy of Sciences to the Interstate Commerce Commission into a single "Commission of Science and Statistics."[115]

Garfield's dramatic plan never took hold, but the significance of coal and oil to national security remained permanently transformed. With respect to the navy, a consensus gradually emerged both within and without the government that the problems of the naval oil supply were inseparable from the problems of the national oil supply. For many, the challenges of logistics planning and the ever growing need for oil demanded a new role for the federal government. Oilman Henry Doherty of Cities Services entreated the government to abandon the rule of capture and asked it to take a less myopic view of the security issue. To the chief of the Geological Survey, George Otis Smith, Doherty observed that "you can not solve the problem of a supply of oil for our navy without bringing about radical changes in the production of oil."[116] In 1924, Doherty contacted President Coolidge, urging conservation in oil fields. Smith concurred, observing that roughly 30 percent of the yearly production of domestic oil came from newly drilled fields and that a full half of the oil in new fields was pumped within its first two years of production. Smith noted that "from the point of view of national security" the nation was obligated to preserve its oil supply when the burning of coal would suffice.[117] Coolidge was encouraged by interior secretary Hubert Work to invite Doherty to express his concerns in person to the four cabinet secretaries most directly affected by the oil situation—those of the interior, commerce, war, and the navy.[118]

With Smith's support, Doherty's arguments succeeded. On December 19, 1924, Coolidge notified the four secretaries of their appointment to the new Federal Oil Conservation Board. National security was foremost on the president's mind, an issue that now concerned more than the fuel demands of naval vessels alone. "Developing aircrafts indicate that our national defense must be supplemented, if not dominated, by aviation," he wrote. "It is even probable that the supremacy of nations may be determined by the possession of available petroleum and its products." Chaired by Secretary Work, the board was charged by the president with studying the state of the entire petroleum industry to determine the volume of waste and what conservation measures might be implemented by the federal government. Thus by early 1925, the question of the security of the naval fuel supply was expanded to address the conservation needs of the nation as a whole. The Federal Oil Conservation Board extensively studied oil issues affecting the nation until 1934, serving as the de facto conduit of petroleum information and policy until New Deal policies further altered the terms of political debate.[119]

Within the navy, too, the war brought about a transformation in planning. According to one officer, C. S. Baker, from the war experience "there has arisen a modern aspect of logistics, far broader than that of the past." Mobilization ceased to be the province of strictly military planning, let alone an academic exercise. Its scope had expanded, too, for after the war, writers and lecturers at the Naval War College went beyond the narrow conception of logistics as enabling naval strategy and announced instead "its national aspect." Logistics became a topic of *national* reach. The new conception embraced the mobilization of financial resources through taxation and borrowing, the alliance with industry to produce war materiel like ships, munitions, and fuels, and the provision of vast amounts of food, fuel, and other resources to Americans and their allies. Baker called the world war "the greatest problem in logistics ever given to a warring power." But even that problem would pale before whatever challenge came next. "Success in the future," Baker declared, "more than ever, will depend on a greater task, that of the mobilization of a nation's finances, resources, materials and man power and their employment in the most effective way."[120]

Logistics once meant managing the resources of war. Now, it meant managing the resources of the entire country. By the 1920s, according to Baker, logistics involved "the mobilization of a nation's energy." It embraced what he called "industrial strategy, national efficiency." Logistics was "professional war." In the 1890s, the subject was subordinate to strategy. In the 1910s, naval thinkers began considering logistics and strategy as conceptual equals. After the war, the study of mobilization fully inverted the old order. Put another way, by the 1920s, logistics *was* war.[121] What was more, logistics in the 1920s not only embraced "the study of the processes and sufficiency of production, storage, transport, and distribution, from the standpoint of their dependency upon the country's internal facilities," according to Robert Coontz, who had recently stepped down from his post as the chief of naval operations, but also went "far afield into the and to the root of the country's external policies."[122]

Instruction on logistics at the War College also changed. Unlike many of his predecessors lecturing on this topic over the previous fifteen years, the new head of the college's Logistics Section, Reuben E. Bakenhus emphasized the novelty, not historical continuity, of logistical problems. Bakenhus rejected appeals to dictionaries or prominent nineteenth-century military thinkers for the proper boundaries of the subject, insisting instead that to understand the meaning of logistics "the dictionary writer should come to the War College and not the War College to the dictionary." The war and industrial revolution had so transformed the field that "we may suffer if we adopt a dictionary definition . . . or take the viewpoint of some authoritative writer of the past." Moreover, he explained, "we may not suffer from the limitations of previous thought on the subject while taking full advantage, at the same time, of all that has been written." In 1926, Bakenhus twice repeated for his students a statement from the college's sibling institution, the Army War College: "The greatest difficulty in executing all phases of the War operations lies in logistics." When he recited it the second time, Bakenhus emphasized "greatest," "all phases," and "logistics."[123]

While early logistics study at the college emphasized supplies and provisioning—fueling the fleet representing the most studied and most important example—logistics after the war embraced what Bakenhus called "a broader subject." This "broader" conception moved beyond the navy itself to include both the larger industrial and material activities that made naval warfare possible. In addition, it embraced the navy's role in maintaining the economic life of the nation whether at war or at peace. Postwar logistics thus included not only coal, oil, and materiel for the navy alone but the broad spectrum of "strategic raw materials" available only from overseas from metals like antimony, chromium, and manganese to food products like coffee and sugar. The details of these materials were kept secret from the broader public, and students were advised to consult the college's archives for specifics. Materials questions also included sources of supply, the trade routes these supplies followed, and the effects of possible war on them. Fuel shaped industrial productivity and economic mobilization. National wealth shaped the economic limits of peace and war.[124]

In the classroom, War College instructors approached the subject from both theoretical and practical levels. Theoretically, students studied constraints like the physical characteristics of landscapes and weather, supplies, and finances. They learned how to address increasing scales of complexity, from ensuring the mobility of individual warships and the larger fleet of repair ships, merchant vessels, and other auxiliaries that facilitated the activities of warships to managing the global network of naval bases, themselves connected to the vast "natural resources of the nation and its mercantile and industrial facilities."[125] In terms of practical calculations, they studied innovations like the "logistic allowance" for coal or oil, a figure devised by logistics planners to make future fuel consumption more predictable.[126] Students also studied and helped develop war plans, most importantly War Plan Orange for defeating Japan, which included increasingly elaborate procedures for logistics, especially with regard to fueling.[127]

As the conception of naval logistics broadened to include strategic raw materials, the implications on naval strategy likewise broadened. "If we must have a detailed knowledge of our own strategic raw materials and their sources of supply and rates of trade," Bakenhus lectured, "then we must also have the same information as to the enemy's strategic raw materials." This subject became one of widespread study in the 1920s, and this sort of thinking in terms of strategic raw materials could also influence students' perception of the character of international rivalries, at times in a way that led them to erroneous conclusions. Several students of the logistical aspect of economic growth concluded that Japan's lack of certain materials in both its home islands and its colonies in nearby mainland Asia necessarily implied that it could "never become a first class industrial nation." If this analysis proved conclusive, observed Bakenhus, "it would have a profound effect on the feeling of security which the United States might have."[128]

At its center, the emergence of logistics was an effort to increase this feeling of security in an inherently and increasingly insecure world. Despite America's newfound wealth, material abundance, and technological innovations, the country's war planners always saw the next war on the horizon and they never felt prepared. Logistics brought order to the chaos. It made planning for ever more awful wars a business proposition. It made officers question war plans but not the national policies that contributed to or helped prevent war. It offered the illusion that securing particular quantities of fuel or planning wartime operations would mean certain victory. It promoted the idea that ensuring supplies of fuel brought security, but it did not consider the geopolitical costs of those supplies. The study of logistics may have helped link energy to American national security, but it did little to explain what American interests were worth securing.

ENERGY AND SECURITY
IN PERSPECTIVE

We undoubtedly have more oil than any other industrial nation, but certainly far from enough.

George Otis Smith, "A World of Power," (1925)

In March 1941, members of the House of Representatives worried about petroleum pipelines. As war expanded across Europe and Asia, federal agencies like the National Resources Planning Board and the Office of Production Management had begun surveying the materiel that would likely prove essential should the United States join the conflict. Unsurprisingly, petroleum ranked among the most essential commodities they examined, and though the oil industry continued to earn its reputation for waste and mismanagement, government bureaucrats expressed reasonable optimism that enough unused capacity existed in the ground, in refineries, and in transportation networks to fuel a war effort successfully.

There was only one problem. Some 95 percent of the oil consumed along the Atlantic arrived by a fleet of some 260 tankers, mostly from the Gulf of Mexico. By early 1941, a host of pressures strained this transportation route. After twenty years of service, many tankers were approaching the end of their useful lives. Others had been discharged from the service to aid Britain, replacing ships sunk at sea. As for the fleet's newest and largest vessels, they had already been drafted into emergency naval service, a prospect looming for additional ships as well. Rough winter weather conditions delayed the voyages of the vessels remaining. Should the United States enter the war, worried the Roosevelt administration, this already-taxed infrastructure could prove especially vulnerable to disruption. A disruption in the flow of tankers would mean cutting off the densely populated Northeast from vital supplies of gasoline, lubricants, and heating oil. "It is in the interest of national defense to augment currently these facilities," Roosevelt wrote of the Atlantic oil corridor in an appeal to Congress.

The obvious solution to the transportation problem would be to construct domestic pipelines to carry both crude and refined oil products directly from the Gulf of Mexico to markets along the Atlantic. Most oil companies endorsed the measure, but pipelines meant politics. Lines from gulf refiners to Atlantic consumers would have to cross the southeastern states, and railroad corporations there pressured state officials, especially in geographically critical Georgia, to prevent any future competitors from gaining rights-of-way at their expense. In Congress, the response of the House Committee on Interstate and Foreign Commerce was H.R. 4816, a bill that would allow the government to declare certain pipelines critical to national defense and ensure that any desired route could be built under eminent domain. With the president's backing, the committee opened a series of hearings to air the issue and assemble support for a national defense pipeline bill.[1]

To the surprise of the committee, however, even the fact that the president had designated these pipelines as critical to national defense did not persuade logisticians from either the army or the navy that this assessment was correct. Despite Roosevelt's call and the endorsements of the secretaries of war and the navy, the officers dispatched to testify to Congress had a different understanding of what energy for national defense meant. Major Clifford V. Morgan of the army astonished the committee by explaining "that from a strictly military point of view, the War Department has very little interest in the bill." Rear Admiral H. A. Stuart of the navy, acutely aware of his audience, called the pipelines "desirable at least" for national defense, "generally speaking" and "as a broad picture," but demurred from calling them "necessary." He further insisted that private industry should construct and pay for them, not the government. What then was the national interest in energy? What were the proper energy needs for national defense or national security? The House hearings revealed that the terms remained contested.[2]

Both army and navy war planners believed they had a good handle on prospective wartime needs. Since the end of World War I, neither service had ignored the details of future mobilization. Ever since the Wilson administration, the services had devoted sustained attention to war planning and logistics calculations. The navy expanded and refined its war plans for various scenarios, building on the methods and bureaucracy it first put in place when Franklin Roosevelt had been assistant secretary. By the beginning of World War II, the navy's Bureau of Supplies and Accounts boasted that it had performed some five million calculations to plan for the mobilization of fourteen thousand separate articles, chief among them petroleum. Its intelligence operatives also kept a close eye on the fuel strategies of prospective rival powers. Outside of naval offices, however, a general lack of urgency or enthusiasm in Congress about defense spending meant fueling facilities vigorously discussed since the 1920s were built only slowly, like the underground storage tanks belatedly begun at Pearl Harbor in 1940 and only finished between September 1942 and July 1943.[3]

The army came later to the planning game than the navy. Prior to World War I, both political aversion to excessive militarism and bureaucratic infighting within the War Department prevented it from engaging in excessive speculation about future large wars. War planning efforts, as conducted by the Army War College after its creation in 1903, instead focused on small-scale operations, like occupying Cuba and campaigning for Pancho Villa in Mexico between 1916 and 1917. After a chaotic administrative experience during World War I, however, in 1920 Congress charged the army's General Staff more explicitly with a duty to plan for war. Two years later, fearful of overlap and jealous of their own authorities, the two services created the joint army and navy Munitions Board to coordinate mobilization plans. For several years, the services still largely planned separately but over time came to coordinate more and more. Planners in the army feared their work would be wasted if war brought a prioritization of naval needs, while planners in the navy feared an emergency would give the army authority over at least some of the navy's perceived domain and create needless delays. Their cooperation was never complete—the navy continued to insist on different war plans for specific contingencies, the army always on a full mobilization—but at least they had begun talking.[4]

By the time the House Committee on Interstate and Foreign Commerce met in 1941, logistics plans for both services anticipated that a complete war mobilization would create a demand for some 125 million barrels of crude oil per year between the army and the navy. This was a large volume but still less than 10 percent of total national output. Of that 10 percent, planners estimated that the navy would consume an overwhelming share—some 80 percent—mostly as fuel oil. As for the army, it conveyed the impression that its strategic dependence on oil was limited. When House committee members worried that their hearing might divulge classified information, the army's witness shrugged, remarking that "our requirements are so small that if everybody knew them, it would not affect us strategically." In light of the central importance oil would play in World War II, this statement sounds astoundingly naive, but at the time it reflected certain common assumptions about fueling modern war: that then current domestic oil production would be sufficient to meet wartime needs, that no operation could conceivably require more than a tenth of that amount, that oil production was sufficiently large that prioritizing civilian uses would be unnecessary, that even if some kind of civilian curtailment *were* necessary, especially of gasoline, this consumption could be effectively halved without any deleterious consequences for the war effort. Both the army and the navy, in short, saw the energy aspect of national defense as fueling the army and the navy alone.[5]

Increasingly, the view of the services appeared too narrow. While the War and Navy departments refined their war plans, the American economy itself grew more dependent on fossil fuels. Americans' 2.4 million registered cars and trucks of 1915 became 31 million by 1939. Over those same years, the nation's 25,000 service stations became 246,000. On farms, tractor use more than doubled between 1928 and 1940 alone to nearly 1.8 million. Outside of gasoline use, petroleum product consumption more generally expanded as well. As new home construction brought on new demand for roofing and the New Deal helped pave old dirt roads, the 895,000 short tons of asphalt consumed in 1926 became 2.3 million tons by 1939. Home construction trends also turned the 12,500 oil-heating furnaces of 1921 into over 2.16 million twenty years later. Oil consumption increased in every category of petroleum derivatives at rates that far exceeded even the growth of coal in the nineteenth century. Petroleum was everywhere: as diesel oil in motorboats, fuel oil in steamships, and aviation fuel for airplanes. In the space of a few decades, oil had become central to American life, wealth, and strength. "In this power age," read a 1934 report on mineral policy by National Resources Planning Board, "petroleum is of paramount importance to our national welfare and security."[6]

With the expansion of oil, something had changed. The rhetoric of energy for national defense had slipped beyond the control of the designated defenders of the nation in the army and navy. The two services had spent over twenty years writing and revising logistics plans, sketching out precisely what resources they would need to serve the nation. Suddenly, they learned that members of Congress, and even the commander in chief, held much more expansive views of a national interest in energy. Despite the services' tepid endorsement of the pipeline plan, the bill passed easily in July.[7]

Seen in a longer perspective, however, raising economic considerations with respect to energy in the context of national security also revived much older nineteenth-century precedents. From mail steamers in the 1840s to Chiriquí colonizationists in the 1860s to economic imperialists in the 1880s, Americans had conceived of fuel as valuable or even essential to national security precisely because it supported certain forms of economic and social activity. Encouraging trade, improving communication, even enforcing a racial order—these motivations framed security not strictly in terms of protection from assault from without but instead in terms of order, growth, and prosperity from within. These issues were never solely under the purview of military officers. By 1941, both Congress and the president perceived that ensuring oil for transporting goods, lubricating factories, and heating homes was as essential an aspect of national security as fueling the army and navy.

A Century of Figuring Things Out

This book has traced a century of Americans learning to think about energy in terms of national security and the national interest, a complex, halting, and circuitous process that nevertheless illuminates important themes about the emergence of the United States as a global power. Understanding how Americans fueled themselves is a subject of importance in its own right, but exploring the development of ideas about energy and power also reveals aspects of American foreign relations, the practice of warfare, and the politics of technology that are more obscure otherwise. Returning to the broad, thematic questions first posed in the introduction, let us examine these aspects.

What was the national interest in energy? The familiar way of thinking about the development of American energy policy, and especially energy security policy, has been to regard the federal government's focus on oil as growing directly from increased consumption beginning in the early twentieth century. In this view, Americans reacted to the voracious appetite for petroleum exhibited by automobiles, ships, aircraft, and tanks. As the country grew more dependent on oil, Congress held committee hearings and passed conservation laws, presidents withdrew valuable oil land from public entry, the navy pushed for oil reserves, and the State Department sponsored American corporations seeking oil fields abroad. These responses were contested and of varying degrees of effectiveness (a national interest did not mean a settled national policy, after all), but by around World War I, the subject of energy had become a subject of national concern, a concern based on the evident connection between oil, new technology, and national security. The national *interest* in energy, then, followed a national *interestedness*.[8]

This view is not wrong, but it is incomplete. As the concept of a national interest in energy evolved, it came to reflect not only changing ideas about proper policy proscriptions but changing ideas about the American state itself. The emergence of a recognizable American interest in energy predated the emergence of oil as a strategic commodity and reflected a changing conception of the place of the United States in the wider world. With the acquisition of a new island empire after 1898, questions that had appeared settled had to be asked anew. What was the purpose of government? How did the country stand in relation to other states in an ever-changing international system? How did technology provide or undermine security? What rules or norms constrained action? In a way, it was the global problems of *empire* that created a *national* interest in energy.[9]

The problems Americans faced in managing an empire made the world after 1898 appear very different from what had come before, not least with regard to energy.[10] During the antebellum era, the federal government had turned its focus to coal largely for economic reasons, as individual states and their citizens vied for commercial opportunity abroad. Through chemical experimentation, engineering innovation, geological exploration, and diplomatic missions, Americans sought to harness government for their economic benefit. Support for the navy and its fuel needs likewise followed from its role in stimulating industry (supporting domestic steam engine manufacturers and coal interests, for example) and protecting commerce itself (the traditional role of the navy). As Thomas

Butler King found in the 1840s mail steamer debates, concern over domestic security alone remained an insufficient impetus to push Congress to take on a more aggressive role in developing the new infrastructure of steam or expanding the steam fleet. Americans took a long time to conclude that sources of power were unambiguously subjects of vital security interests.

Even the Civil War—a conflict powered in essential though often unreliable ways by coal—proved insufficient to persuade Americans to make coal a matter of pressing national concern. After the war, calls to annex various islands or harbors for use as coaling stations in the future could hardly ever overcome domestic opposition. Fueling American security with coal and steam power presented real obstacles, but could still be addressed by familiar means. The United States was a weaker state compared with the great powers in Europe but no less committed to technological innovation. When this innovation led to larger steamships and increased foreign trade, Americans turned not to seizing foreign coaling stations but still more innovation. Technological innovation maximized fuel economy aboard ships, leading to new battleships that steamed close to home and speedy cruisers with vast coal endurances that went great distances without need for frequent refueling. It also brought a studied interest in inventions for coaling at sea to make harbors on land unnecessary. Mathematical innovation promised more direct navigation by great circle routes, stretching the steaming distances possible for naval and merchant vessels alike. Legal and diplomatic innovation produced international agreements on the laws of war, neutrality, and contraband, creating international norms with regard to coal that all powers anticipated relying on during times of conflict.

But no unambiguous national interest in energy—meaning coal, essentially—appeared before 1898, when the United States seized its island empire of Hawaii, the Philippines, Guam, and Puerto Rico. In the years before 1898, at the height of European colonial island expansion in the Pacific, writers like Alfred Thayer Mahan had popularized new, strategic arguments for why the United States ought to acquire at least a few island coaling stations, but these arguments were essentially defensive, limited to a handful of nearby Caribbean ports and especially Hawaii. Mahan's national interest in energy was effectively to prevent potential Pacific rivals from occupying this strategically located island group and ensure the navy had access to enough fuel to defend an isthmian canal. The importance of energy to national security changed after 1898, as the boundaries of American sovereignty expanded into the Caribbean and across the Pacific. Empire produced new questions about American security—questions about protecting these islands from external attack or internal rebellion, about integrating them into the national polity, and about adjusting to the constantly changing foreign policy decisions of other powers. None of these questions could be adequately answered without ensuring American ships could move.

As Americans, especially in the navy, sought to answer these questions, they helped build the intellectual, material, and bureaucratic infrastructure around energy that would largely be in place well before oil became a central subject in conceptions of American security policy.

Would the adoption of new technology enhance or constrain Americans' opportunities in the world? For the United States, steam power simultaneously brought advantages and disadvantages. New routes for faster, regular trade and communication also meant the potential for a handful of nations to monopolize the global infrastructure of maritime commerce, thus dictating costs and access to markets. The perception that steam power liberated steamships from wind and waves masked the fact that these vessels could not do everything their civilian masters expected of them. Advantages in wartime tactics meant new challenges to strategy and new subjects for diplomatic disputes. There were even two sides to the unquenchable appetite for fuel that kept ships closely tethered to shore. On the one hand, potential European and Asian rivals seemed less threatening, since their naval operations in American waters became more difficult further from their own home ports. On the other, American naval vessels were similarly limited by precarious supply lines and sparse fueling stations.

But new steam technology did not force particular choices; instead, it only posed questions. The ways Americans argued about and then ultimately answered these questions reflected their changing views of the place of the United States in the world. These questions were "about technology" only in the most trivial sense. Coal-fired steam power introduced new problems to commerce and warfare, but the most consequential constraints to American action that followed were largely rooted in law, politics, and political culture. Domestic politics during the Gilded Age proved more powerful than post–Civil War assertions about the need for Caribbean coaling stations in places like the Danish West Indies, Santo Domingo, Chiriquí, or for many decades even Hawaii. Americans' fundamental distrust of permanent alliances prevented diplomatic agreements that promised access to coaling infrastructure in return for military support. Agreements in international law at the first arbitration conference in Geneva redefined the meaning of neutrality for the machine age and established a legal category for coal in wartime, thus creating the guiding framework for fueling naval strategy for decades to come. The constraints each of these choices placed on the United States were not a consequence of technology but a consequence of Americans' many and varied responses to technology.

Would the infrastructure necessary to support steam power come from international cooperation or unilateral action? Avoiding binding alliances was a precept of American foreign policy into the twentieth century, but the world's functional energy infrastructure was, in practice, mostly collaborative. Despite repeated calls for American coaling stations abroad, the vast majority of American commercial and naval vessels made use of foreign coaling stations in ports around the world. Though Matthew Perry could argue that an American station in the Bonin Islands was essential to future commerce with China, or Alfred Thayer Mahan could point with trepidation at the growing number of British coaling stations across the Pacific, most of the time, coal remained available at market prices in seaports around the world.

Of course, the exceptions mattered. British colonies in the Caribbean made refueling the Union Navy exceedingly difficult during the American Civil War. Similarly, the British decision not to refuel the Spanish fleet en route to the Philippines in 1898 effectively ended the prospect of beating back the American assault. But even after Americans secured territory ostensibly for fortified coaling stations, they typically did not make much use of it. Since the 1880s, Samoa had remained more of an idea of a coaling station than a reality. After 1898, Guam, Puerto Rico, Hawaii, and the Philippines were all envisioned as links in the American defensive chain, but again, the visions of naval planners far outpaced facts on the ground. Guam and Puerto Rico remained poorly stocked outposts. Congressional purse strings long prevented construction at Pearl Harbor. After the Washington Naval Conference in 1921, Americans agreed to abandon even the possibility of building fortified coaling stations west of Hawaii. Still, regardless of peacetime planning, when wars did come, both international cooperation and unilateral action increased simultaneously based on the contingencies of the moment.

Would technology drive choices in foreign affairs, or would the desires of American policy makers, merchants, and naval officers catalyze the development of new technologies? The intersection of technology and politics is at the heart of this history. The development of steam power, after all, both allowed Americans to imagine new possibilities for global trade and naval strategy while simultaneously forcing them to confront the new limitations and vulnerabilities that a dependence on coal would impose. These changes took place in a political realm. Engineers designed engine components to improve fuel economy with government consumers in mind; politicians subsidized commercial boiler manufacturers. Coal dealers relied on government contracts to fund and market their operations; the Navy Department leveraged war emergency powers to literally force fuel industry executives to the table to ensure the American war machine would remain well supplied with coal and oil.

Yet a distinctively American answer to the problems introduced by the new technology of steam power was often more technology. Before the Civil War, both the navy and Congress looked to inventions like Charles Grafton Page's electromagnetic power or John Ericsson's caloric engine to make the coal question obsolete. After the war, Americans experimented with a range of technical innovations, from the use of petroleum to devices for coaling at sea to the briefly popular Gamgee zeromotor. More broadly, Americans settled on a New Navy of steel and steam that was premised on a naval strategy of close-to-home defense by battleship augmented by swift, fuel-economical cruisers protecting American commerce everywhere else. Technology, naval strategy, and foreign relations all evolved together. Moreover, throughout this period, other powers likewise confronted fuel problems, and the choices they made—from seizing territories as coaling stations to introducing policies of naval buildup—further shaped the questions Americans had to answer.

Would particular choices of motive power predetermine particular geographies of expansion? To explain the overseas territorial acquisitions of 1898, historians have long asserted that steam power necessitated acquiring coaling stations and that coaling stations meant empire. In this book I have argued the opposite, that the post-1898 acquisitions created a new strategic need for coal, not the other way around. Until 1898, the "need for coaling stations" was an argument, not a fact, and one whose justifications changed over time. With these different justifications came different geographies of expansion.

When Americans first began building oceangoing steamships in the 1840s and 1850s, they already had places to go. At first, the new steam lines crossed the Atlantic for European ports like Bremerhaven and Liverpool. From New York, they connected the East Coast to California through the Isthmus of Panama. Almost immediately, the lure of the China market directed Americans to potential sources of coal and refueling ports in east Asia like Brunei, Japan, and the Bonin Islands. This geography reflected the interests of antebellum commerce but also the way new perceptions of space were shaped by new perceptions of time. It was not steam per se that drove Americans to pursue coal in the Far East but the worry that new British steam networks would quicken the flow of Chinese information, goods, and wealth away from the United States. The construction of steam networks in space was always a race over the relative rate of communication in time.

As a conflict fueled by coal and powered by steam, the Civil War and its aftermath directed American attention to new geographies. For the Union, the Civil War produced a geography that focused on maintaining a coastal blockade from the South Atlantic to the Gulf of Mexico and pursuing blockaderunners throughout the Caribbean and beyond. The difficulty in fueling these operations led Lincoln and his cabinet to consider colonization plans in Chiriquí as well as to redouble diplomatic engagement with Britain over contraband and the laws of war. After the war, Americans saw potential coaling stations everywhere. Put simply, the *idea* of needing these stations outpaced their practical *value*. The Civil War had demonstrated the importance of coal to naval operations, and afterward, every speculator found it useful to tout how every island in the Caribbean offered incalculable benefits to American trade and security.

Others saw American coaling stations as more general aids to expanding American commerce overseas in places like Africa, east Asia, and South America. But commerce was not security. The Gilded Age use of security arguments to bolster overseas expansion was usually the rhetorical dressing for speculative schemes or efforts to expand American commerce and investments in places like Santo Domingo, Samoa, and Hawaii. Supporters of expansion, usually for commercial reasons, touted the national need for coaling stations to introduce a gravity to foreign empire building that commercialism alone could not command. As Alfred Mahan showed, there were, in fact, security arguments to make for constructing coaling stations. Most boosters of coaling stations, however, did not really make those arguments. Instead, they engaged in entrepreneurial diplomacy.

Many Americans refused to see geography as destiny, however. For them, the reaction to coal and steam power in the nineteenth and early twentieth centuries focused on attempting to transcend geography altogether. A common impulse linked the pursuit of engineering fuel economy in the 1850s, the partial return to sail after the Civil War, the adoption of devices for coaling at sea around the turn of the twentieth century, and the development of the science of logistics around World War I. Each effort challenged the geographical limitations imposed by a dependence on coal and sought to make them irrelevant. These efforts—technological, organizational, computational—all reflected an American desire to cultivate commerce and project naval power in the world without the inconvenience of securing distant territories. Even the development of new mathematical techniques for great circle navigation, a subject inherently rooted in specific geographies, was an effort to make the physical realities of land, wind, and currents less important than a simple minimization of distance (and with distance, time, fuel, and money, as well). A complete triumph over geography was, of course, impossible, but making sense of American overseas imperialism must take into account the efforts Americans made to reach their objectives without resorting to seizing land.

Would policy for shipping, postal communication, and naval defense be organized by markets, politics, or technocratic experts? Before the adoption of steam power, the federal government had little reason to make policy with specific reference to coal or any other fuel. The advent of naval and mail steamers in the 1840s made the fuel question one for the federal government to confront. Its responses ranged from funding chemical and physical investigations to soliciting engineering experimentation and new ship designs to sponsoring geological and diplomatic missions. Importantly, these efforts were never considered to be opposed to the domestic coal industry but to supplement and complement it. Federal action did not counter the market but worked to create successful, functional markets in the first place.

By the early twentieth century, war planners had concluded that while a re-liance on private corporations for both fuel and technical expertise was essen-tial, it was not enough. The creation of the short-lived navy coal mine in Alas-ka's Matanuska Valley and the longer-lasting naval petroleum reserves in California, Wyoming, and Alaska were not designed to supplant private indus-try but to provide an insurance policy against catastrophic market failure. Sim-ilarly, the actions of the Fuel Administration and the use of navy orders to set coal sales at reasonable prices during World War I represented emergency mea-sures to force the industry to comply with military needs during times of acute crisis. The government held vast potential power to shape policy around energy security, but rarely did American politics favor the exercise of this power.

Yet as the story of southeastern pipelines on the eve of World War II suggests, by 1940 Americans did not conceive of the security dimensions of coal and oil merely in terms of their capacity to contribute to national defense but also in terms of their ability to ensure that fuel markets themselves—for the national industrial economy and the residential comforts of ordinary American citizens alike—could operate unimpeded. Market and government thus intertwined.

Would American fuel needs be met by domestic supplies or foreign dependence? When steamships first crossed the ocean, it remained unclear whether the United States possessed the kind of coal that was suited to transatlantic travel. That question was quickly resolved (it did), and East Coast anthracite and bi-tuminous operators spent two decades jockeying for dominance over the steam fuel market. Even with abundant American fuel, however, distant voyages re-mained difficult to supply, since added transportation costs could rarely com-pete with less expensive, often inferior local coals or British exports. Despite these challenges, until the end of the nineteenth century, American government attempts to secure overseas sources of coal—from Labuan, Formosa, Chiriquí, and elsewhere—remained haphazard, half hearted, or ineffectual.

While American-mined coal could more than satisfy domestic consumption, the problem of the second half of the nineteenth century remained fueling Americans far from home. This was the problem faced by both Union and Con-federate navies in the Caribbean during the Civil War and by American naval vessels in the Pacific in the 1890s. The outstanding question, however, was whether a guaranteed means for refueling in distant ports really mattered all that much. In times of peace, local markets stocked with either local or imported fuels allowed American naval and merchant vessels to steam nearly anywhere. In times of war, these markets would likely close—at least George Dewey seems to have thought as much, since he made a determined rush to secure British coal and colliers before a declaration of war with Spain. But prior to Dewey's cam-paign in the Philippines, few Americans thought much about significant Amer-ican naval operations that far from home waters. The aftermath of the war, of course, made fueling, defeating, and defending the Philippines an American problem. As oil came gradually to displace coal as the principal strategic fuel, the work of naval logistics and broader American strategy focused on organizing domestic resources, most notably fuel resources, to defend all American territories both continental and colonial.

Yet between 1840 and 1940, the problem of fueling American security was never simply about the desirability of domestic versus foreign sources of fuel. Instead, the sources of supply were but one part of the larger difficulty of engaging strategically and commercially in a world whose spatial relationships constantly changed with new technologies of transportation. Controlling territory to facilitate refueling was one way, but as Charles Francis Adams insightfully observed at the arbitration conference in Geneva in 1872, if every great power created a global network of fortified coaling stations, the consequence would invariably be ruinous expense and a dangerous escalation in the risk of conflict. Even with the annexation of Hawaii and the Philippines, Americans preferred to resolve issues of steam power through agreements in international law and spatially liberating new technologies.

Would and should guaranteeing the global infrastructures of energy for trade and defense become a function of national authority? And if so, how? John G. Clark concludes his comprehensive account of American energy policy during the first half of the twentieth century by stating that "federal policies toward the mineral fuels from 1900 to 1946 can be characterized as unsystematic, vague, and eminently minimal." Over this period, Clark argues, energy became a subject of political importance, but with limited exceptions, the federal government failed to assume a decisive role in shaping the political economy of energy. It largely reacted to the lobbying power of already entrenched coal, oil, gas, and electricity producers. These interests were too powerful, in Clark's view, for the development of an incontrovertible "public interest" in energy that would vest power in the government to shape patterns of production and consumption. World War I's powerful Fuel Administration was quickly dismantled after the war. The investigations of the Federal Oil Conservation Board of the 1920s went nowhere. Interior secretary Harold Ickes's Petroleum Reserves Corporation during World War II failed to persuade any major domestic constituency that the American government ought to enter the foreign oil business—not the large, integrated oil firms, who feared losing markets and not the small independents, who shuddered at the prospect of competition from more foreign oil imports.[11]

This analysis works for the larger, domestic, civilian political economy but misses the way state capacity expanded in the late nineteenth and early twentieth centuries to make possible the extraordinary growth of the American military and American naval operations around the globe. Before 1898, government officials from Matthew Perry and Robert Shufeldt to James Blaine and Benjamin

Harrison had thought the national government had an obligation to build new energy infrastructure around the globe to support American commerce and security. These opinions rarely resulted in any clear or consistent federal authority, but they did create precedents for American actions abroad to secure fueling infrastructure for security. The search for coal helped bring Matthew Perry to Japan in the 1850s. It facilitated Lincoln's pursuit of black colonization in Chiriquí in the 1860s. After the Civil War, it created a rhetoric around fuel and security that helped make the annexation of Hawaii in the 1890s fulfill a national security imperative.

After 1898, federal authority vis-à-vis coal took on a different character. The navy placed coal at the center of its new responsibilities of planning for war and defending an overseas empire. But if these responsibilities were new, the fact that energy was a subject that drew Americans into tangled diplomatic and economic relationships around the globe was not. More than anything else, this history shows how it was not oil that forced Americans into a geopolitics of energy, but the United States' very engagement with the rest of the world in the age of fossil fuels itself that did. This engagement, and the geopolitics of energy which is a part of it, lasted through the twentieth century and is not likely to go away any time soon.

The Geography of Energy Independence

Every source of power has its own geography and its own politics. This point is especially relevant in the context of American debates since the 1970s about the future of oil consumption. By 1973, U.S. domestic oil production had already peaked, petroleum consumption was increasing, and the Arab nations of OPEC had instituted an embargo, leading to the quadrupling of U.S. oil prices in just a few months. In response, President Nixon announced an ambitious energy program he called "Project Independence," invoking the greatest technological successes in American history and calling upon the nation to "set as our national goal, in the spirit of Apollo and with the determination of the Manhattan Project" developing "the potential to meet our own energy needs without depending on any foreign energy sources."[12] Nixon exhorted the country to commit tremendous resources, administrative acumen, and the best scientific and engineering minds in the world. Yet despite Nixon's entreaties, Project Independence was a failure. For thirty more years, oil imports continued to climb, and even after declining from a peak in 2005, the United States of 2014 still imports about one-third of its total oil consumption.[13]

But while Project Independence died, the rhetoric of energy independence persisted. Subsequent presidential administrations after Nixon have all given at least lip service to versions of energy independence. Jimmy Carter spoke of rising energy demand as constraining the nation's "independence of economic and political action." Ronald Reagan called energy independence the nation's "proper goal."[14] Presidents George H. W. Bush, Bill Clinton, George W. Bush, and Barack Obama all similarly endorsed the concept, at least rhetorically, even as they differed in how to get there.[15]

When unmoored from specific policy proposals or policy caveats, however, the idea of energy independence can be filled with all sorts of utopian hopes. Among the highest-profile cheerleaders for an energy-independence panacea has been *New York Times* columnist Tom Friedman. "If President Bush is looking for a similar legacy project [to Kennedy's mission to the moon]," Friedman wrote in his 2006 bestseller, *The World Is Flat*, "there is one just crying out—a national science initiative that would be our generation's moon shot: a crash program for alternative energy and conservation to make America energy-independent in ten years." For Friedman, channeling Nixon's own historical analogies, energy independence would be a mechanism for solving a whole host of problems. "If President Bush made energy independence his moon shot," he continued, "in one fell swoop he would dry up revenue for terrorism, force Iran, Russia, Venezuela, and Saudi Arabia onto the path of reform—which they will never do with $60-a-barrel oil—strengthen the dollar, and improve his own standing in Europe by doing something huge to reduce global warming."[16]

As a policy goal, energy independence has been critiqued by economists, business leaders, editorialists, and most definitely the oil industry. They've argued at one time or another that the global energy supply is just that, global, with prices set on international markets; that oil is a fungible commodity—if the U.S. stopped buying oil from Saudi Arabia, that same oil would be purchased (as most of it already is) by Europe, Japan, and China. But proponents and critics tacitly agree on one thing, that there once was a time when the United States *was* energy independent, a time when the United States was safely isolated from reliance on foreign sources of energy, and that at some point between the beginning of the twentieth century and the 1970s, the country became so entangled economically and geopolitically with places in the Middle East, Latin America, and elsewhere that it lost its energy independence.[17]

Examining how Americans came to think about energy, national security, and the national interest between 1840 and 1940 helps reframe the premises of energy independence. Before oil, coal had already become a major subject for diplomacy and naval strategy. Even when coal supplies were largely domestic, the challenges of distribution and storage—what came to be called logistics—hardly insulated the United States from vulnerabilities to its fuel supply. National aspirations and national capacities were not always in alignment. New technologies could solve old problems but also introduce new ones. At no time could new technologies make difficult political choices disappear.

Coal brought new questions in the nineteenth century, and oil brought more questions in the twentieth. In the twenty-first century, should the United States become a net exporter of energy again or move to new sources of power altogether, it will not likely find independence from the rest of the world just new questions. We cannot yet know what these questions will be, but remembering how Americans came to think about energy in terms of the national interest in the first place will leave us better prepared to answer them when they do inevitably arise. What was true for coal and oil will remain true for whatever comes next. First, technology is only part of the story, a vital component of making the modern world run but, more generally, merely a foundation on which to build political, social, ecological, and economic relationships. Those relationships, whatever they may look like, will matter regardless of the particularities of technology. Second, at least since the early nineteenth century, fueling the United States has always connected the country with the rest of the world. Simply moving away from a reliance on oil imports will not absolve the United States from the necessity of facing difficult international challenges, and we should not expect it to. Finally, the "national interest" itself is not a material fact but a contested concept whose significance changes over time. An unthinking reliance on a fixed concept of a national interest in energy may risk mistaking the means to better policy for policy ends themselves. New energy questions may yield new obstacles, but with better understanding (and a little luck), they may also yield new hopes and new opportunities.

In the notes, most publications from the United States Senate and House of Representatives only appear in abbreviated form. Full citations for these documents appear below, sorted by year of publication, then by Congress and session, and finally by document number. In a handful of cases, the dates of publication are later than the Congress that produced the document. Cited congressional hearings and other state and federal publications outside the serial set may be found with complete citations in the notes.

H.R. Ex. Doc. No. 396, 25th Cong., 2nd sess. (1838): Walter R. Johnson, *Establishment of an Institution for Experiments in Physical Science*

S. Ex. Doc. No. 229, 26th Cong., 1st sess. (1840): Robert Strange for the Committee on Naval Affairs, *Report [To Accompany Bill S. No. 240]*

H.R. Rep. No. 3, 27th Cong., 1st sess. (1841): Thomas Butler King for the Committee on Naval Affairs, *Home Squadron*

H.R. Ex. Doc. No. 2/7, 28th Cong., 1st sess. (1843): *Schedule of Papers Accompanying the Report of the Secretary of the Navy*

S. Ex. Doc. No. 386, 28th Cong., 1st sess. (1843): Walter R. Johnson, *A Report to the Navy Department of the United States on American Coals Applicable to Steam Navigation, and to Other Purposes*

S. Ex. Doc. No. 167, 28th Cong., 1st sess. (1844): James Alfred Pearce for the Committee on Naval Affairs, *Report [To Accompany Bill S. 96]*

H.R. Ex. Doc. No. 2/8, 28th Cong., 2nd sess. (1844): Charles A. Wickliffe, *Report of the Postmaster General*

H.R. Ex. Doc. No. 13, 29th Cong., 1st sess. (1845): Robert J. Walker, *Letter from the Secretary of the Treasury Transmitting the Annual Report of Commerce and Navigation, &c.*

H.R. Rep. No. 476, 29th Cong., 1st sess. (1846): Henry Hilliard for the Committee on Post Office and Post Roads, *Atlantic Mail Steamers*

H.R. Rep. No. 685, 29th Cong., 1st sess. (1846): Thomas Butler King for the Committee on Naval Affairs, *Ocean Steamers*

H.R. Ex. Doc. No. 8/7, 30th Cong., 1st sess. (1847): *List of Papers Accompanying the Report of the Secretary of the Navy*

H.R. Ex. Doc. No. 8/8, 30th Cong., 1st sess. (1847): Cave Johnson, *Report of the Postmaster General*

H.R. Rep. No. 275, 30th Cong., 1st sess. (1848): Hugh White for the Committee on Naval Affairs, *Captain John Percival*

H.R. Rep. No. 596, 30th Cong., 1st sess. (1848): Thomas Butler King for the Committee on Naval Affairs, *Steam Communication with China, and the Sandwich Islands*

H.R. Ex. Doc. No. 1/8, 30th Cong., 2nd sess. (1848): *List of Papers Accompanying the Report of the Secretary of the Navy*

H.R. Ex. Doc. No. 7, 30th Cong., 2nd sess. (1848): Robert J. Walker, *Letter from the Secretary of the Treasury, Transmitting His Annual Report on the State of the Finances*

S. Ex. Doc. No. 1/8, 31st Cong., 1st sess. (1849): *List of Papers Accompanying the Report of the Secretary of the Navy*

H.R. Ex. Doc. No. 5/8, 31st Cong., 1st sess. (1849): *List of Papers Accompanying the Report of the Secretary of the Navy*

H.R. Ex. Doc. No. 20, 31st Cong., 1st sess. (1850): Thomas Ewbank, *Report of the Commissioner of Patents, for the Year 1849*

S. Misc. No. 117, 31st Cong., 1st sess. (1850): *Memorial of Citizens of Massachusetts, Praying That Provision Be Made for Continuing the Experiments on American Coal Commenced by Professor Johnson in 1843*

S. Ex. Doc. No. 1/5, 31st Cong., 2nd sess. (1850): William A. Graham, *Report of the Secretary of the Navy*

S. Ex. Doc. No. 1/6, 31st Cong., 2nd sess. (1850): *List of Papers Accompanying the Report of the Secretary of the Navy*

S. Ex. Doc. No. 1/7, 32nd Cong., 1st sess. (1851): N.K. Hall, *Report of the Postmaster General*

H.R. Rep. No. 34, 31st Cong., 2nd sess. (1851): Frederick P. Stanton for the Committee on Naval Affairs, *Steamers between California and China*

S. Ex. Doc. No. 50, 32nd Cong., 1st sess. (1852): William A. Graham, *Report of the Secretary of the Navy, Communicating, in compliance with a resolution of the Senate, Information in relation to Contracts for the Transportation of the Mails by Steamships between New York and California*

S. Ex. Doc. No. 74, 32nd Cong., 1st sess. (1852): Charles M. Conrad, *Report of the Secretary of the Navy, Communicating, in Compliance with a Resolution of the Senate, a Report of the Engineer-in-chief of the Navy, on the Comparative Value of Anthracite and Bituminous Coals*

S. Ex. Doc. No. 2, 33rd Cong., special sess. (1853): James C. Dobbin, *Report of the Secretary of the Navy, Communicating, in Compliance with a Resolution of the Senate, Information in relation to the Contract with Howland & Aspinwall for Supplying the Japan Squadron with Coal*

H.R. Ex. Doc. No. 147, 34th Cong., 1st sess. (1855): James Guthrie, *Report of the Secretary of the Treasury, Transmitting a Report from the Register of the Treasury, of the Commerce and Navigation of the United States for the Year Ending June 30, 1855*

S. Ex. Doc. No. 34, 33rd Cong, 2nd sess. (1855): Franklin Pierce, *Message of the President of the United States, Transmitting a Report of the Secretary of the Navy, in Compliance with a Resolution of the Senate of December 6, 1854, Calling for Correspondence, &c., Relative to the Naval Expedition to Japan*

S. Ex. Doc. No. 79, 33rd Cong., 2nd sess. (1856): Francis L. Hawks, ed., *Narrative of the Expedition of an American Squadron to the China Seas and Japan, Performed in the Years 1852, 1853, and 1854, under the Command of Commodore M. C. Perry*, 3 vols.

S. Rep. No. 447, 34th Cong., 3rd sess. (1857): Stephen Mallory for the Committee on Naval Affairs, *Report [on the Memorial of George T. Parry]*

S. Misc. Doc. No. 144, 35th Cong., 1st sess. (1858): U.S. Court of Claims, *Report . . . in the case of Benjamin H. Springer vs. the United States*

S. Rep. No. 317, 35th Cong., 1st sess. (1858): Stephen Mallory for the Committee on Naval Affairs, *Report [on the Memorial of George T. Parry]*

H.R. Ex. Doc. No. 82, 35th Cong., 2nd sess. (1858): Isaac Toucey, *Coal–United States Navy: Letter from the Secretary of the Navy, Transmitting a Statement of the Quantity of Coal Used since the Introduction of It into the Navy; Names of the Agents for the Purchase of the Coal, and the Amount Paid Them as Commissions*

S. Ex. Doc. No. 2/15, 36th Cong., 1st sess. (1859): Isaac Toucey, *Report of the Secretary of the Navy, December, 1859*

H.R. Rep. No. 184, 35th Cong., 2nd sess. (1859): Thomas S. Bocock for the Select Committee on Naval Contracts and Expenditures, *Naval Contracts and Expenditures*

S. Ex. Doc. No. 2, 36th Cong., 1st sess. (1860): James Buchanan, *Message from the President of the United States to the Two Houses of Congress [Accompanying Documents]*

H.R. Rep. No. 568, 36th Cong., 1st sess. (1860): Freeman H. Morse for the Committee on Naval Affairs, *Contract for Coal*

H.R. Rep. No. 568, 36th Cong. 1st sess. (1860): Charles B. Sedgwick for the minority of the Committee on Naval Affairs, *Contract for the Purchase of Coal: Minority Report*

S. Ex. Doc. No. 1, 36th Cong., 2nd sess. (1860): Isaac Toucey, *Annual Report of the Secretary of the Navy*

S. Ex. Doc. No. 20, 36th Cong., 2nd sess. (1860): Howell Cobb, *Report of the Secretary of the Treasury, Transmitting a Report from the Register of the Treasury of the Commerce and Navigation of the United States*

H.R. Ex. Doc. No. 41, 36th Cong., 2nd sess. (1861): James Buchanan, *Chiriqui Commission: Message of the President of the United States, Transmitting Reports from the Chiriqui Commission*

H.R. Rep. No. 87, 36th Cong., 2nd sess. (1861): Henry L. Dawes for the Select Committee on the Special Message of the President of January 8, 1861, *Naval Force of the United States—Where Ships Are Now Stationed, Etc.*

H.R. Ex. Doc. No. 1/2, 37th Cong., 3rd sess. (1862): *Papers Relating to Foreign Affairs*

H.R. Ex. Doc. No. 1/15, 38th Cong., 1st sess. (1863): Gideon Welles, *Annual Report of Secretary of Navy*

H.R. Ex. Doc. No. 1, 38th Cong., 2nd sess. (1864): Gideon Welles, *Report of the Secretary of the Navy*

H.R. Ex. Doc. No. 83, 38th Cong., 2nd sess. (1865): Abraham Lincoln [transmitting report of Edwin M. Stanton], *Annual Report of the Secretary of War*

H.R. Ex. Doc. No. 1/20, 39th Cong., 1st sess. (1865): Edwin M. Stanton, *Annual Report of the Secretary of War*

H.R. Ex. Doc. No. 1/25, 39th Cong., 1st sess. (1865): Gideon Welles, *Report of the Secretary of the Navy*

S. Ex. Doc. No. 55, 39th Cong., 1st sess. (1866): Andrew Johnson, *Message from the President of the United States, Communicating . . . Information Touching the Transactions of the Executive Branch of the Government Respecting the Transportation, Settlement, and Colonization of Persons of the African Race*

S. Rep. No. 116, 39th Cong., 1st sess. (1866): John Conness for the Committee on Post Offices and Post Roads, *Report to Accompany Joint Resolution S.R. No. 98*

H.R. Ex. Doc. No. 1/19, 39th Cong., 2nd sess. (1866): Alexander W. Randall, *Report of the Postmaster General*

H.R. Ex. Doc. No. 1, 40th Cong., 2nd sess. (1867): Gideon Welles, *Report of the Secretary of the Navy*

H.R. Ex. Doc. No. 12, 40th Cong., 2nd sess. (1867): Hugh McCulloch, *Appropriations Post Office Department: Letter from the Secretary of the Treasury, Transmitting a Letter from the Postmaster General, Relative to Appropriations for the Service of His Department*

S. Ex. Doc. No. 79, 40th Cong., 2nd sess. (1868): Gideon Welles, *Letter of the Secretary of the Navy, Communicating . . . Information in Relation to the Discovery, Occupation, and Character of the Midway Islands, in the Pacific Ocean*

H.R. Ex. Doc. No. 275, 40th Cong., 2nd sess. (1868): Hugh McCulloch, for Benjamin Peirce, *Report of the Superintendent of the United States Coast Survey, Showing the Progress of the Survey during the Year 1867*

Confidential Ex. Doc. No. W, 40th Cong., 2nd sess. (1868): Andrew Johnson for William H. Seward, *Message Transmitting a Report of the Secretary of State Relating to the Vote of St. Thomas on the Question of Accepting the Cession of that Island to the United States*

Confidential Ex. Doc. No. AA, 40th Cong., 2nd sess. (1868): Andrew Johnson for William H. Seward, *Message of the President of the United States Transmitting a Report from the Secretary of State, with Accompanying Papers, on the Subject of a Transfer of the Peninsula and Bay of Samana to the United States*

S. Rep. No. 194, 40th Cong., 3rd sess. (1869): James W. Nye for the Committee on Naval Affairs, *Report [on Deepening the Harbor at Midway]*

Confidential Ex. Doc. No. K1, 40th Cong., 3rd sess. (1869): Charles Sumner, *Papers Relative to the Negotiation with Denmark for the Purchase of St. Thomas and St. John*

H.R. Ex. Doc. No. 1, Part 3, 42nd Cong., 3rd sess. (1872): George M. Robeson, *Report of the Secretary of the Navy*

H.R. Ex. Doc. No. 1, Part 3, 43rd Cong., 1st sess. (1873): George M. Robeson, *Report of the Secretary of the Navy*

H.R. Misc. Doc. No. 113, 42nd Cong, 3rd sess. (1874): George M. Robeson, *Reports of Explorations and Surveys to Ascertain the Practicability of a Ship-Canal between the Atlantic and Pacific Oceans by the Way of the Isthmus of Darien*

H.R. Ex. Doc. No. 1, Part 3, 46th Cong., 3rd sess. (1880): Richard W. Thompson, *Report of the Secretary of the Navy*

H.R. Ex. Doc. No. 1, Part 3, 47th Cong., 1st sess. (1881): William H. Hunt, *Report of the Secretary of the Navy*

H.R. Ex. Doc. No. 46, 47th Cong., 1st sess. (1882): William H. Hunt, *Chiriqui Grant: Letter from the Secretary of the Navy, in Response to Resolutions of the House of Representatives Relative to Certain Lands and Harbors Known as the Chiriqui Grant*

H.R. Ex. Doc. No. 1, Part 3, 47th Cong., 2nd sess. (1882): William E. Chandler, *Report of the Secretary of the Navy*

H.R. Ex. Doc. No. 1, Part 3, 48th Cong., 1st sess. (1883): William E. Chandler, *Report of the Secretary of the Navy*

H.R. Ex. Doc. No. 1, Part 3, 48th Cong., 2nd sess. (1884): William E. Chandler, *Report of the Secretary of the Navy*

H.R. Ex. Doc. No. 1, Part 3, 49th Cong., 1st sess. (1885): William C. Whitney, *Report of the Secretary of the Navy*

H.R. Ex. Doc. No. 238, 50th Cong., 1st sess. (1888): Grover Cleveland, *American Rights in Samoa*

H.R. Ex. Doc. No. 1, Part 3, 51st Cong., 2nd sess. (1890): Benjamin F. Tracy, *Report of the Secretary of the Navy*

H.R. Misc. Doc. No. 239, 51st Cong., 1st sess. (1891): A. J. Bentley, ed., *Official Opinions of the Attorneys-General of the United States, Advising the President and Heads of Departments in Relation to Their Official Duties*, vol. 19

H.R. Ex. Doc. No. 1, Part 3, 52nd Cong., 1st sess. (1891): Benjamin F. Tracy, *Report of the Secretary of the Navy*

S. Ex. Doc. No. 77, 52nd Cong., 2nd sess. (1893): Benjamin Harrison, *Correspondence respecting Relations between the United States and the Hawaiian Islands from September, 1820, to January, 1893*

H.R. Ex. Doc. No. 1, Part 3, 53rd Cong., 3rd sess. (1894): Hilary A. Herbert, *Report of the Secretary of the Navy*

H.R. Ex. Doc. No. 1, pt. 1, 53rd Cong., 3rd sess. (1895): *Foreign Relations of the United States, 1894: Affairs in Hawaii*

H.R. Ex. Doc. No. 1, Part 1, 53rd Cong., 3rd sess. (1895): Benjamin Harrison, *Appendix II: Foreign Relations of the United States, 1894: Affairs in Hawaii*

S. Ex. Doc. No. 62, 55th Cong., 2nd sess. (1898): John T. Morgan and John M. Schofield, *Annexation of the Hawaiian Islands*

S. Ex. Doc. No. 82, 55th Cong., 2nd sess. (1898): Richard F. Pettigrew and Frederick T. Dubois, *Against the Annexation of Hawaii*

S. Ex. Doc. No. 188, 55th Cong., 2nd sess. (1898): George W. Melville, *Views of Commodore George W. Melville, Chief Engineer of the Navy, as to the Strategic and Commercial Value of the Nicaraguan Canal, the Future Control of the Pacific Ocean, the Strategic Value of Hawaii, and Its Annexation to the United States*

S. Ex. Doc. No. 315, 55th Cong., 2nd sess. (1898): John T. Morgan and George W. Melville, *Value of the Hawaiian Islands*

S. Rep. No. 681, 55th Cong., 2nd sess. (1898): Cushman K. Davis for the Committee on Foreign Relations, *Annexation of Hawaii*

H.R. Ex. Doc. No. 3, 55th Cong., 3rd sess. (1898): John D. Long, *Annual Reports of the Navy Department for the Year 1898*

H.R. Ex. Doc. No. 3, 57th Cong., 1st sess. (1901): John D. Long, *Annual Reports of the Navy Department*

S. Ex. Doc. No. 231, Vol. 8, 56th Cong., 2nd sess. (1901): *Compilation of Reports of the Committee on Foreign Relations, United States Senate, 1789–1901*

H.R. Ex. Doc. No. 3, 57th Cong., 2nd sess. (1902): William H. Moody, *Annual Reports of the Navy Department*

H.R. Ex. Doc. No. 681, 63rd Cong., 2nd sess. (1913): Josephus Daniels, *Annual Reports of the Navy Department for the Fiscal Year 1913*

H.R. Ex. Doc. No. 876, 63rd Cong, 2nd sess. (1914): Josephus Daniels, *Report on Coal in Alaska for Use in United States Navy*

H.R. Ex. Doc. No. 1484, 63rd Cong., 3rd sess. (1914): Josephus Daniels, *Annual Reports of the Navy Department for the Fiscal Year 1914*

H.R. Ex. Doc. No. 20, 64th Cong., 1st sess. (1915): Josephus Daniels, *Annual Reports of the Navy Department for the Fiscal Year 1915*

S. Ex. Doc. No. 26, 64th Cong., 1st sess. (1915): Josephus Daniels, *Experimental Tests of Matanuska Coal for Naval Ships*

H.R. Ex. Doc. No. 1480, 64th Cong., 2nd sess. (1916): Josephus Daniels, *Annual Reports of the Navy Department for the Fiscal Year 1916*

H.R. Ex. Doc. No. 618, 65th Cong., 2nd sess. (1917): Josephus Daniels, *Annual Reports of the Navy Department for the Fiscal Year 1917*

H.R. Ex. Doc. No. 1450, 65th Cong., 3rd sess. (1918): Josephus Daniels, *Annual Report of the Secretary of the Navy for the Fiscal Year 1918*

H.R. Ex. Doc. No. 994, 66th Cong., 3rd sess. (1920): Josephus Daniels, *Annual Reports of the Navy Department for the Fiscal Year 1920*

Abbreviations

ALP	Abraham Lincoln Papers, Library of Congress, Washington, DC
AWT	Ambrose W. Thompson Papers, Library of Congress, Washington, DC
BHL	Benjamin H. Latrobe Photostats, Maryland Historical Society, Baltimore
BSA	General Correspondence, Planning Division, Advance Base Section, 1918–1942, Records of the Bureau of Supplies and Accounts (Navy), RG 143, NARA-1
Captains' Letters	M125, Letters Received by the Secretary of the Navy from Captains, RG 45, NARA-1
CFGP	Caspar F. Goodrich Papers, New-York Historical Society, New York
CIC-AL	Chiriqui Improvement Company Papers, Abraham Lincoln Presidential Library, Springfield, IL
CJMP	Colin J. McRae Papers, Alabama Department of Archives and History, Montgomery
DDPFP	David D. Porter Family Papers, Library of Congress, Washington, DC
DDS	Despatches to the Department of State from the U.S. minister in London, RG 84, NARA-2
DIDS	M77, Diplomatic Instructions of the Department of State, 1801–1906, RG 59, NARA-2
DSADS	M37, Despatches From Special Agents of the Department of State, 1794–1906, RG 59, NARA-2
DUSMG	M44, Despatches from U.S. Ministers to German states and Germany, 1799–1906, RG 59, NARA-2
EDP	Edwin Denby Papers, Bentley Historical Library, University of Michigan, Ann Arbor
FDRL	Papers as Assistant Secretary of the Navy, Franklin D. Roosevelt Collections, Franklin D. Roosevelt Presidential Library and Museum, Hyde Park, NY
FOCB	Records of the Federal Oil Conservation Board, RG 232, NARA-2
GBP	George Bancroft Papers, box 2, New York Public Library, New York
GBSF	General Board Subject File, RG 80, NARA-1
GCSN	General Correspondence of the Secretary of the Navy, RG 80, NARA-1
GDP	George Dewey Papers, Library of Congress, Washington, DC

GPWP	George P. Welsh Papers, Library of Congress, Washington, DC
HCNA	House Committee on Naval Affairs, RG 233, NARA-1
HCPO	House Committee on Post Office and Post Roads, RG 233, NARA-1
HL	Huntington Library, San Marino, CA
JDP	Josephus Daniels Papers, Library of Congress, Washington, DC
JEP	Microfilm Edition of the John Ericsson Papers, American Swedish Historical Museum, Philadelphia, PA
JMSP	John McAllister Schofield Papers, Library of Congress, Washington, DC
JPKP	John Pendleton Kennedy Papers, Enoch Pratt Free Library, Baltimore, MD
Leahy Papers	William D. Leahy Papers, Library of Congress, Washington, DC
M124	M124, Letters Received by the Secretary of the Navy: Miscellaneous Letters, 1801–1884, RG 45, NARA-1
M1493	M1493, Proceedings and Hearings of the General Board of the U.S. Navy, 1900–1950, RG 80, NARA-1
MHS	Massachusetts Historical Society, Boston
NACC-Alaska	Navy Alaskan Coal Commission, RG 80, NARA-Alaska
NARA-Alaska	National Archives and Records Administration, Anchorage
NARA-1	National Archives and Records Administration 1, Washington, DC
NARA-2	National Archives and Records Administration 2, College Park, MD
NFOB	Naval Fuel Oil Board Letters Sent, RG 80, NARA-1
ORUCN 1	U.S. Naval War Records Office, *Official Records of the Union and Confederate Navies in the War of the Rebellion.* Ser. 1. 27 vols. Washington, DC: Government Printing Office, 1897.
ORUCN 2	U.S. Naval War Records Office, *Official Records of the Union and Confederate Navies in the War of the Rebellion.* Ser. 2. 3 vols. Washington, DC: Government Printing Office, 1897.
PMF	Papers of MacKaye Family, Dartmouth College Library, Hanover, NH
PMSC	Pacific Mail Steamship Company: Letters to Alfred Robinson, Bancroft Library, University of California, Berkeley
RG	Record Group
Roberts Letter Book	Letter Book of Marshall O. Roberts, New York Agent, 3 vols., U.S. Mail Steam Ship Company Papers, New-York Historical Society, New York
ROCNO	Intelligence Division: Naval Attaché Reports, 1886–1939, RG 38, NARA-1
RWT	R. W. Thompson Collection, Rutherford B. Hayes Presidential Center, Fremont, OH
RWT 1	R. W. Thompson Correspondence, Indiana State Library, Indianapolis
SCFR	Senate Committee on Foreign Relations, RG 46, NARA-1
SCNA	Senate Committee on Naval Affairs, RG 46, NARA-1
SCPO	Senate Committee on Post Office and Post Roads, RG 46, NARA-1
SLMB	Samuel Latham Mitchill Barlow Papers, HL

SMP	Samuel McGowan Papers, Library of Congress, Washington, DC
T33	T33, Despatches from U.S. Ministers to Colombia, 1820–1906, RG 59, NARA-2
TBKP	T. Butler King Papers, University of North Carolina at Chapel Hill
TJRP	Thomas Jefferson Rusk Papers, University of Texas at Austin
TJWP	Thomas J. Walsh Papers, Library of Congress, Washington, DC
TRJP	Theodore Roosevelt Jr. Papers, Library of Congress, Washington, DC
USBM	Special Files, Records of the U.S. Bureau of Mines, RG 70, NARA-2
USGS	Records concerning Naval Oil Reserves, RG 57, NARA-2
USNWC	U.S. Naval War College Archives, Newport, RI
WHSP	William Henry Seward Papers Microfilm, University of Rochester, Rochester, NY
WMGP	William McKendree Gwin Papers, Bancroft Library, University of California, Berkeley
WPBFP	William Phineas Browne Family Papers, Alabama Department of Archives and History, Montgomery

Introduction · *In Which the President Seeks an Audience with the King*

1. Michael F. Reilly, as told to William J. Slocum, *Reilly of the White House* (New York: Simon and Schuster, 1947), 216.

2. William D. Leahy, *I Was There: The Personal Story of the Chief of Staff to Presidents Roosevelt and Truman Based on His Notes and Diaries Made at the Time* (New York: McGraw-Hill, 1950), 326. Leahy does not mention the presence of the king's brother (Amir Abdullah) and two sons (Amir Muhammad and Amir Mansur, the minister of defense) or the foreign minister (Yusuf Yassin), the minister of finance (Abdullah Sulaiman), and the minister of state (Hafiz Wahba). Nor does he mention that along with "sabers and daggers," the king's guards also carried modern machine guns. See William A. Eddy, *F.D.R. Meets Ibn Saud* (New York: American Friends of the Middle East, 1954), 19–32, and Reilly, *Reilly of the White House*, 222.

3. John S. Keating, "Mission to Mecca: The Cruise of the Murphy," *U.S. Naval Institute Proceedings* 102 (1976): 56.

4. Ibid, 59; Reilly, *Reilly of the White House*, 222–23; Eddy, *F.D.R. Meets Ibn Saud*, 19–20; W. Barry McCarthy, "Ibn Saud's Voyage: U.S. Destroyer Creates Naval History with Deck Full of Royalty, Sheep and Coffee-Makers," *Life*, March 19, 1945, 60; Waveney Ann Moore, "Sailor Was the Piper of History: Thomas Hilliard Was an Eyewitness When the U.S. and Saudi Arabia Joined Destinies in 1945," *St. Petersburg (FL) Times*, February 12, 2005.

5. Reilly, *Reilly of the White House*, 221–22.

6. William Leahy is one of the few witnesses to claim that Roosevelt and ibn Saud even discussed oil (see his 1950 memoir, *I Was There*), though his contemporaneous diary records only a discussion of Palestine; see William D. Leahy, diary, 1945, 36, reel 4, Leahy Papers. On Eddy, see William A. Eddy and Yusuf Yassin, "Memorandum of Conversation Between the King of Saudi Arabia (Abdul Aziz Al Saud) and President Roosevelt, February 14, 1945, Aboard the U.S.S. 'Quincy,'" in *Foreign Relations of the United States, Diplomatic Papers, 1945*, vol. 8: *The Near East and Africa* (Washington, DC: Government Printing Office, 1969), 2–3, and Eddy, *F.D.R. Meets Ibn Saud*. The most thorough historical account of the meeting

may be found in Ross Gregory, "America and Saudi Arabia, Act I: The Conference of Franklin D. Roosevelt and King Ibn Saud in February 1945," in *Presidents, Diplomats, and Other Mortals: Essays Honoring Robert H. Ferrell,* ed. J. Garry Clifford and Theodore A. Wilson (Columbia: University of Missouri Press, 2007), 116–33.

7. In 1933, the concession originally went to Standard Oil of California (SOCAL) alone, which transferred it to a subsidiary, the California Arabian Standard Oil Company (CASOC). In 1936, to ensure access to markets for future oil production, SOCAL invited the Texas Company to share ownership of CASOC, which they renamed ARAMCO in 1944. The company's $6.8 million advances were reportedly made between 1939 and 1940 alone and did not include the roughly $27.5 million the company had already invested in oil exploration and production since the concession had begun. In 1941, CASOC promised an additional advance of $3 million, but when the Saudi government asked for a further $1.5 million, CASOC turned to the U.S. State Department, which shored up Saudi finances, thus stabilizing the Saudi government and, consequently, CASOC's own investments. Political and legal obstacles made direct American financing unavailable, but Americans secured indirect aid from Britain until 1943, when a direct lend-lease assistance program was created. See Aaron D. Miller, *Search for Security: Saudi Arabian Oil and American Foreign Policy, 1939–1949* (Chapel Hill: University of North Carolina Press, 1980), xii n. 7, 35–37. On Saudi-American relations in the early twentieth century, see also Rachel Bronson, *Thicker than Oil: America's Uneasy Partnership with Saudi Arabia* (New York: Oxford University Press, 2006) and Robert Vitalis, *America's Kingdom: Mythmaking on the Saudi Oil Frontier* (Stanford, CA: Stanford University Press, 2007), which emphasizes ARAMCO's reproduction of exploitative labor conditions.

8. Though sent by Edward Stettinius, the secretary of state, this document reflected the consensus policy views of the departments of War, Navy, and especially State, the document having originally been drafted by the Office of Near Eastern and African Affairs. See "Memorandum by the Secretary of State to President Roosevelt," *Foreign Relations of the United States, Diplomatic Papers, 1944,* vol. 5: *The Near East, South Asia, and Africa, the Far East* (Washington, DC: Government Printing Office, 1965), 757.

9. Bronson, *Thicker than Oil,* 42; Miller, *Search for Security,* x–xii, 128–31. See also Daniel Yergin, *The Prize: The Epic Quest for Oil, Money, and Power* (New York: Simon & Schuster, 1991), 403–5, and for broader analysis, John A. DeNovo, *American Interests and Policies in the Middle East, 1900–1939* (Minneapolis: University of Minnesota Press, 1963), David S. Painter, *Oil and the American Century: The Political Economy of U.S. Foreign Oil Policy, 1941–1954* (Baltimore, MD: Johns Hopkins University Press, 1986), and Stephen J. Randall, *United States Foreign Oil Policy Since World War I: For Profits and Security,* 2nd ed. (Montreal: McGill-Queen's University Press, 2005).

10. The contemporary connection between energy and national security is evident in the seemingly endless array of books exploring one aspect of the subject or another. For a (very) recent sampling, see Daniel Yergin, *The Quest: Energy, Security, and the Remaking of the Modern World* (New York: Penguin, 2011), John M. Deutch, *The Crisis in Energy Policy* (Cambridge: Harvard University Press, 2011), Marilyn A. Brown and Benjamin K. Sovacool, *Climate Change and Global Energy Security: Technology and Policy Options* (Cambridge, MA: MIT Press, 2011), Michael J. Graetz, *The End of Energy: The Unmaking of America's Environment, Security, and Independence* (Cambridge, MA: MIT Press, 2011), Doug Stokes and Sam Raphael, *Global Energy Security and American Hegemony* (Baltimore, MD: Johns

Hopkins University Press, 2010), Daniel Moran and James A. Russell, eds., *Energy Security and Global Politics: The Militarization of Resources Management* (New York: Routledge, 2009), Brenda Shaffer, *Energy Politics* (Philadelphia: University of Pennsylvania Press, 2009), and Michael T. Klare, *Blood and Oil: The Dangers and Consequences of America's Growing Petroleum Dependency* (New York: Metropolitan Books, 2004).

11. Here, and elsewhere in this book, an imperfect yet suggestive barometer of usage frequencies for key phrases may be found from the Google Books Ngram viewer at books.google.com/ngrams. On the emergence of the vocabulary of "interests," see Daniel Rodgers, *Contested Truths: Keywords in American Politics since Independence* (New York: Basic Books, 1987), 176–211.

12. Charles A. Beard with G. H. E. Smith, *The Idea of National Interest: An Analytical Study in American Foreign Policy* (New York: Macmillan, 1934), 3.

13. On the security dimension of the national interest, see J. Fred Rippy, "A Reconsideration of the Monroe Doctrine," *World Affairs* 97 (1934): 104.

14. On the origins of the oil-security nexus, see Gerald D. Nash, *United States Oil Policy, 1890–1964: Business and Government in Twentieth Century America* (Pittsburgh, PA: University of Pittsburgh Press, 1968), 9–11, 23–97, John G. Clark, *Energy and the Federal Government: Fossil Fuel Policies, 1900–1946* (Urbana: University of Illinois Press, 1987), 156–64, 385–86, Yergin, *The Prize*, 13–14ff., Martin V. Melosi, *Coping with Abundance: Energy and Environment in Industrial America* (New York: Knopf, 1985), 97–102, 147–50, 160–95, Harold F. Williamson et al., eds., *The American Petroleum Industry*, vol. 2 (Evanston, IL: Northwestern University Press, 1963), 181–84, 261–64, John A. DeNovo, "Petroleum and the United States Navy before World War I," *Mississippi Valley Historical Review* 41, no. 4 (1955): 641–56, John A. DeNovo, "The Movement for an Aggressive American Oil Policy Abroad, 1918–1920," *American Historical Review* 61, no. 4 (1956): 854–76, John A. DeNovo, *American Interests and Policies in the Middle East, 1900–1939* (Minneapolis: University of Minnesota Press, 1963), esp. 167–209, and Leonard Bates, *The Origins of Teapot Dome: Progressive Parties and Petroleum, 1909–1921* (Urbana: University of Illinois Press, 1963). Stephen Randall begins with the close of World War I to argue that the United States first pursued a coherent foreign oil policy "premised on the twin objectives of strategic security and economic strength and a private-public partnership based on shared values and goals throughout the interwar and war years" (*United States Foreign Oil Policy Since World War I*, 1).

15. Williamson, *The American Petroleum Industry*, 184.

16. On consumption figures, see Joseph E. Pogue, *The Economics of Petroleum* (New York: John Wiley and Sons, 1921), 61. On refined petroleum, see Raymond Foss Bacon and William Allen Hamor, *The American Petroleum Industry*, vol. 2 (New York: McGraw-Hill, 1916), 447–526.

17. "The President's Commission on Oil Reserves," *Science*, May 2, 1924, 392.

18. House Committee on Armed Services, Special Subcommittee on Petroleum, *Petroleum for National Defense*, 80th Cong., 2nd sess., (1948), 2–6.

19. The rise of fossil fuels is one of the defining characteristics of the modern world. For the United States, see Louis C. Hunter, *A History of Industrial Power in the United States, 1780–1930*, 3 vols. (Charlottesville: University Press of Virginia and Cambridge, MA: MIT Press, 1979–91), David Nye, *Consuming Power: Questions to Live With* (Cambridge: MIT Press, 2006), Melosi, *Coping with Abundance*, Brian Black, *Petrolia: The Landscape of America's First Oil Boom* (Baltimore, MD: Johns Hopkins University Press, 2000), Sean Patrick

Adams, *Old Dominion, Industrial Commonwealth: Coal, Politics, and Economy in Antebellum America* (Baltimore, MD: Johns Hopkins University Press, 2004), and Paul Lucier, *Scientists and Swindlers: Consulting on Coal and Oil in America, 1820–1890* (Baltimore, MD: Johns Hopkins University Press, 2008). For longer term and global accounts, see Valcav Smil, *Energy in Nature and Society: General Energetics of Complex Systems* (Cambridge, MA: MIT Press, 2008), Valcav Smil, *Energy in World History* (Boulder, CO: Westview, 1994), Alfred Crosby, *Children of the Sun: A History of Humanity's Unappeasable Appetite for Energy* (New York: Norton, 2006), and J. R. McNeill, *Something New Under the Sun: An Environmental History of the Twentieth-Century World* (New York: Norton, 2000), 50–117, 296–306.

20. Recently, historians have questioned the premise of an inherently weak American state, especially in the nineteenth century, arguing instead "that the United States," in Balogh's formulation, "governed *differently* from other industrialized contemporaries, but did not necessarily govern *less*." Americans welcomed direct federal authority when certain functions, like defense, foreign relations, or the governance of new territories could not be accomplished through the mediation of private institutions or state and local governments (*A Government Out of Sight: The Mystery of National Authority in Nineteenth-Century America* [New York: Cambridge University Press, 200], 2). On the institutional activity of the nineteenth-century American state, see Richard R. John, *Spreading the News: The American Postal System from Franklin to Morse* (Cambridge, MA: Harvard University Press, 1995), John Lauritz Larson, *Internal Improvement: National Pubic Works and the Promise of Popular Government in the Early United States* (Chapel Hill: University of North Carolina Press, 2000), Robert G. Angevine, *The Railroad and the State: War, Politics, and Technology in Nineteenth-Century America* (Stanford, CA: Stanford University Press, 2004), William J. Novak, *The People's Welfare: Law and Regulation in Nineteenth-Century America* (Chapel Hill: University of North Carolina Press, 1996), Stephen Skowronek, *Building a New American State: The Expansion of National Administrative Capacities, 1877–1920* (New York: Cambridge University Press, 1982), Richard Bensel, *Yankee Leviathan: The Origins of Central State Authority in America, 1859–1877* (New York: Cambridge University Press, 1990), Theda Skocpol, *Protecting Soldiers and Mothers: The Political Origins of Social Policy in the United States* (Cambridge, MA: Harvard University Press, 1992), and William R. Brock, *Investigation and Responsibility: Public Responsibility in the United States, 1865–1900* (New York: Cambridge University Press, 1984). For a succinct challenge to the weak state thesis, see William J. Novak, "The Myth of the 'Weak' American State," *American Historical Review* 113, no. 3 (2008): 752–72.

21. The fundamental question implicit here—whether technology determines social or political outcomes or is instead malleable in the face of a range of alternatives—is discussed in Merritt Roe Smith and Leo Marx, eds., *Does Technology Drive History? The Dilemma of Technological Determinism* (Cambridge, MA: MIT Press, 1994), and Nye, *Technology Matters*, 17–31.

22. The place of technology in U.S. foreign relations has received substantially less treatment than has technology in European imperialism. A plea for recognizing technology's place in American foreign relations may be found in Walter LaFeber, "Presidential Address: Technology and U.S. Foreign Relations," *Diplomatic History* 24, no. 1 (2000): 1–19. Among the best explicit accounts of technology and American foreign relations in the nineteenth and early twentieth centuries may be found in David Paull Nickles, *Under the Wire: How the Telegraph Changed Diplomacy* (Cambridge, MA: Harvard University Press, 2003), Michael

Adas, *Dominance by Design: Technological Imperatives and America's Civilizing Mission* (Cambridge, MA: Belknap Press of Harvard University Press, 2006), and Jonathan Reed Winkler, *Nexus: Strategic Communications and American Security in World War I* (Cambridge, MA: Harvard University Press, 2008). A larger literature explores technology and European imperialism: Michael Adas, *Machines as the Measure of Men: Science, Technology, and Ideologies of Western Dominance* (Ithaca, NY: Cornell University Press, 1989), Daniel R. Headrick, *The Tools of Empire: Technology and European Imperialism in the Nineteenth Century* (New York: Oxford University Press, 1981), Daniel R. Headrick *The Tentacles of Progress: Technology Transfer in the Age of Imperialism, 1850–1940* (New York: Oxford University Press, 1988), and Daniel R. Headrick, *Power over Peoples: Technology, Environments, and Western Imperialism, 1400 to the Present* (Princeton, NJ: Princeton University Press, 2010). For the experience of the English and Indians when North America was still on the periphery of colonizing Europe, see Joyce E. Chaplin, *Subject Matter: Technology, the Body, and Science on the Anglo-American Frontier, 1500–1676* (Cambridge, MA: Harvard University Press, 2001). For the intersection of firearms and imperial racial politics, see William Kelleher Story, *Guns, Race, and Power in Colonial South Africa* (New York: Cambridge University Press, 2008).

23. Bernard Brodie, *Sea Power in the Machine Age* (Princeton, NJ: Princeton University Press, 1941), 108; James A. Field Jr., "American Imperialism: The Worst Chapter in Almost Any Book," *American Historical Review* 83, no. 3 (1978): 653–56. Broader accounts of American diplomacy discussing prospective coaling stations in the nineteenth century may be found in, for example, John H. Schroeder, *Shaping a Maritime Empire: The Commercial and Diplomatic Role of the American Navy, 1829–1861* (Westport, CT: Greenwood Press, 1985), Walter LaFeber, *The American Search for Opportunity, 1865–1913*, vol. 2 of *The Cambridge History of American Foreign Relations*, ed. Warren I. Cohen (New York: Cambridge University Press, 1993), Charles S. Campbell, *The Transformation of American Foreign Relations, 1865–1900* (New York: Harper Colophon, 1976), Walter Nugent, *Habits of Empire: A History of American Expansion* (New York: Knopf, 2008), Kenneth J. Hagan, *This People's Navy: The Making of American Sea Power* (New York: Free Press, 1991), 109, Ernest N. Paolino, *The Foundations of the American Empire: William Henry Seward and U.S. Foreign Policy* (Ithaca, NY: Cornell University Press, 1973), 105, and Walter R. Herrick, Jr., *The American Naval Revolution* (Baton Rouge: Louisiana State University Press, 1966), 90. Eric T. L. Love mentions coaling stations to show how racial anxieties inhibited American acquisitions before 1898; see *Race Over Empire Racism and U.S. Imperialism, 1865–1900* (Chapel Hill: University of North Carolina Press, 2004).

24. French E. Chadwick, "Coal." 1901 lecture, folder 1, box 2, RG 14, USNWC.

25. Seward, quoted in *Cong. Globe*, April 27, 1852, 1200.

Chapter 1 · Empire and the Politics of Information

1. "Notes and Commentaries, on a Voyage to China," pt. 21, *Southern Literary Messenger* 19, no. 8 (1853): 474, 476.

2. Mira Wilkins, "Impacts of American Multinational Enterprise on American-Chinese Economic Relations," in *America's China Trade in Historical Perspective: The Chinese and American Performance*, ed. Ernest R. May and John K. Fairbank (Cambridge, MA: Committee on American-East Asian Relations of the Department of History in collaboration with the Council on East Asian Studies, 1986), 259–62.

3. "Notes and Commentaries, on a Voyage to China," 476.

4. Westray, Gibbes, and Hardcastle, memorial, June 8, 1858, HR 35A-G16.3, HCPO. On the opening of Japan, China, and Siam to American trade, these memorialists noted that "without the establishment of other ocean mails upon the Pacific, these great interests may be retarded, defeated, lost. We can have no great commerce with the countries mentioned, unless letters of advice, remittances, and official communication—diplomatic and naval— are protected by the direct assistance and power of the Government."

5. John G. B. Hutchins, *The American Maritime Industries and Public Policy, 1789–1914* (Cambridge, MA: Harvard University Press, 1941), 348–68; George Rogers Taylor, *The Transportation Revolution, 1815–1860* (New York: Rinehart, 1951), 112–31; Stephen Fox, *Transatlantic: Samuel Cunard, Isambard Brunel, and the Great Atlantic Steamships* (New York: Perennial, 2003). Those emphasizing the failure of antebellum steamship subsidies rarely note that they were revived after the Civil War. Seventy-five years after embarking on its experiment with mail steamer subsidies, the U.S. government began a remarkably similar, and ultimately more successful, program of airmail contracts to subsidize the nascent aviation industry; see Roger E. Bilstein, *Flight Patterns: Trends in Aeronautical Development in the United States, 1918–1929* (Athens: University of Georgia Press, 1983), 29–56, Roger E. Bilstein, *Flight in America: From the Wrights to the Astronauts*, 3rd ed. (Baltimore, MD: Johns Hopkins University Press, 2001), 48–59, and David D. Lee, "Herbert Hoover and Commercial Aviation Policy, 1921–1933," in *Reconsidering a Century of Flight*, ed. Roger D. Launius and Janet R. Daly Bednarek (Chapel Hill: University of North Carolina Press, 2003), 89–117.

6. *Cong. Globe*, April 27, 1852, 1200. "International postal communication and foreign commerce are as important as domestic mails and traffic," Seward continued. "Equality with other nations in respect to those interests is as important as freedom from restriction upon them among ourselves. Except Rome—which substituted conquests and spoliation for commerce—no nation was ever highly prosperous, really great, or even truly independent, whose foreign communications and traffics were conducted by other States; while Tyre, and Egypt, and Venice, and the Netherlands, and Great Britain, successively becoming the merchants, became thereby the masters of the world."

7. See, for example, Manuel Castells, *The Rise of the Network Society*, 2nd ed., vol. 1 of *The Information Age: Economy, Society, and Culture* (Malden, MA: Wiley-Blackwell, 2010). Debates over mail steamers in the 1840s and 1850s provided a preview for many of the ideas and arguments deployed around global telegraph networks in the 1860s and beyond; for those, see the insightful studies of Simone M. Müller-Pohl, *Wiring the World: The Social and Cultural Creation of Global Telegraph Networks* (New York: Columbia University Press, forthcoming), John A. Britton, *Cables, Crises, and the Press: The Geopolitics of the New Information System in the Americas, 1866–1903* (Albuquerque: University of New Mexico Press, 2013), and David Paull Nickles, *Under the Wire: How the Telegraph Changed Diplomacy* (Cambridge, MA: Harvard University Press, 2003).

8. David M. Henkin, *The Postal Age: The Emergence of Modern Communications in Nineteenth-Century America* (Chicago: University of Chicago Press, 2006), 3–5.

9. Richard R. John, *Spreading the News: The American Postal System from Franklin to Morse* (Cambridge, MA: Harvard University Press, 1995). The specifically moral dimensions of postal correspondence is explored in Wayne E. Fuller, *Morality and the Mail in Nineteenth-Century America* (Urbana: University of Illinois Press, 2003).

10. Richard B. Kielbowicz, *News in the Mail: The Press, Post Office, and Public Information, 1700–1860s* (New York: Greenwood Press, 1989); Menahem Blondheim, *News over the Wires: The Telegraph and the Flow of Public Information in America, 1844–1897* (Cambridge, MA: Harvard University Press, 1994); Richard D. Brown, *Knowledge is Power: The Diffusion of Information in Early America, 1700–1865* (New York: Oxford University Press, 1989). See also Alfred Chandler Jr. and James W. Cortada, *A Nation Transformed by Information: How Information Has Shaped the United States from Colonial Times to the Present* (New York: Oxford University Press, 2000), and Wayne E. Fuller, *The American Mail: Enlarger of the Common Life* (Chicago: University of Chicago Press, 1972).

11. Robert Greenhalgh Albion, *Square-Riggers on Schedule: The New York Sailing Packets to England, France, and the Cotton Ports* (Princeton, NJ: Princeton University Press, 1938), vii, 20–21, 35. See also George E. Hargest, *History of Letter Post Communication between the United States and Europe, 1845–1875*, 2nd ed. (Lawrence, MA: Quarterman Publications, 1975), and Frank Staff, *The Transatlantic Mail* (London: Adlard Coles, 1956), 62.

12. Staff, *The Transatlantic Mail*, 62–63; John A. Butler, *Atlantic Kingdom: America's Contest with Cunard in the Age of Sail and Steam* (Washington, DC: Brassey's, 2001), 83. See also J. C. Arnell, *Steam and the North Atlantic Mails: The Impact of the Cunard Line and Subsequent Steamship Companies on the Carriage of Transatlantic Mails* (Toronto: Unitrade Press, 1986), and Fox, *Transatlantic*.

13. John P. Heiss et al., petition, August 16, 1850, 9–10, 14–15, HR 31A-G12.8, HCNA.

14. H.R. Rep. No. 685, 29th Cong., 1st sess. (1846), 9. The unreliable practice of delivering mail at sea is recounted in *Moby-Dick*, in which Melville describes the chance encounter of the *Pequod* and another ill-fated Nantucket whaler, the *Jeroboam*. Whalers, notes the narrator Ishmael, carry mail for the crews of ships they might encounter during their years at sea. The fortunate few receive their letters two or three years late; the rest never do. To the captain of the *Jeroboam* Ahab delivers a letter for a crewmember, a wife's message to her voyaging husband, "sorely tumbled, damp, and covered with a dull, spotted, green mould, in consequence of being kept in a dark locker of the cabin." Ishmael observes that "of such a letter, Death himself might well have been the post-boy." And indeed, he was, for the addressee had already passed away at sea; see Herman Melville, *Moby-Dick; or, The Whale*, in *Redburn, White-Jacket, Moby-Dick* (New York: Library of America, 1983), 1132–33.

15. Thomas Rainey, petition, February 8, 1858; and Philadelphia Board of Trade, memorial, February 15, 1858, HR 35A-G16.3, HCPO.

16. "Theodore G. Schomburg, memorial, January 13, 1852 HR 32A-G15.3, HCPO.

17. John P. Heiss et al., petition, August 16, 1850, 9–10, 14–15, HR 31A-G12.8, HCNA.

18. Edward Everett to Daniel Webster, March 1, 1842, vol. 7, DDS; Edward Everett to secretary of state ad interim, March 22, 1844, vol. 8, DDS; Edward Everett to John C. Calhoun, October 21, 1844, and Edward Everett to John C. Calhoun, March 27, 1845, vol. 9, DDS; Louis McLane to James Buchanan, May 21, 1846, vol. 10, DDS.

19. Edward Everett to Daniel Webster, February 28, 1843, vol. 7, DDS; H.R. Ex. Doc. No. 2/8, 28th Cong., 2nd sess. (1844), 690–93.

20. Hershel Parker, ed., *Gansevoort Melville's 1846 London Journal and Letters from England, 1845* (New York: New York Public Library, 1966), 19, 53.

21. John Tyler, "Fourth Annual Message," December 3, 1844, www.presidency.ucsb.edu /ws/?pid=29485, accessed March 25, 2011.

22. H.R. Rep. No. 476, 29th Cong., 1st sess. (1846), 1; "Trial Trip of the U.S. Mail Steamer Washington," *New York Evangelist* 18, no. 21 (1847): 1 (reproduced from New York's *Journal of Commerce*); "Steam and Sailing Lines," *Scientific American* 2, no. 43 (1847): 342.

23. Henry Wheaton to Department of State, December 31, 1845, roll 4, DUSMG; Andrew J. Donelson to Karl Ernst Wilhelm von Canitz und Dallwitz, August 26, 1847, and Andrew J. Donelson to James Buchanan, November 16, 1846, roll 5, DUSMG. I thank Scott Lillard for sharing these microfilmed letters with me.

24. Initially failing to attract enough investors, Edward Mills had reorganized the Ocean Steam Navigation Company's board and ceded his control to others. The new investors continued working with Mills as a company agent, but when he tried using a loophole in his contract to construct ships of his own to surreptitiously compete with the company he founded, the board severed their relationship with him and publicly blamed the failures of their ships on his mechanical ineptitude. See A.W.J., "The United States' Mail Steamers," *Times* (London), May 16, 1848, 6, *Remonstrance of the Ocean Steam Navigation Company Against the Petition of Edward Mills* (New York: Daniel Fanshaw, 1848), a copy of which can be found in Sen 30A-H14.2, SCPO, as well as the *Anglo American* (New York), June 5, 1847, 164, and "Arrival of the Washington Steamer in England," *Scientific American*, July 17, 1847, 342.

25. George Bancroft to James Buchanan, June 17, 1847, and George Bancroft to James Buchanan, September 1, 1847, vol. 10, DDS.

26. George Bancroft to Cave Johnson, November 3, 1847, GBP.

27. George Bancroft to Cave Johnson, October ?, 1847, GBP. Bancroft was not merely projecting his view of justice onto the British public. The *Times* of London ("The New Postal Arrangements with Canada and the United States," December 14, 1847, 6) opposed the British post office's levying additional postage on American mail after the American post office had been more than generous in its own policies.

28. George Bancroft to Cave Johnson, October ?, 1847, GBP.

29. George Bancroft to James Buchanan, October 2, 1847, George Bancroft to James Buchanan, October 9, 1847, George Bancroft to James Buchanan, October 23, 1847, and George Bancroft to James Buchanan, December 27, 1847, vol. 10, DDS; "Mails Between England and North America," *Times* (London), December 16, 1847, 3.

30. James K. Polk: "Third Annual Message," December 7, 1847, www.presidency.ucsb .edu/ws/?pid=29488, accessed September 26, 2011.

31. H.R. Ex. Doc. No. 8/8, 30th Cong., 1st sess. (1847), 1326–27.

32. George Bancroft to Cave Johnson, January 28 1848, GBP.

33. Cave Johnson to George Bancroft, February 18, 1848 and March 15, 1848, GBP. After the Whig House of Representatives passed the retaliatory measure, it remained stuck in the Democratic Senate, which frustrated Bancroft. Lacking additional leverage, he refrained from beginning negotiations again, grumbling that "now they can quote against me the apparent acquiescence of Congress." See George Bancroft to Cave Johnson, May 25, 1848, as well as Cave Johnson to George Bancroft, Washington, May 3, 1848, GBP. Finding no opposition to the measure, Cave Johnson became convinced that the Senate's dithering could only be motivated by the personal animus between himself and John M. Niles, the chairman of the Committee on the Post Office and Post Roads (Cave Johnson to George Bancroft, May 29, 1848, GBP).

34. H.R. Ex. Doc. No. 8/8, 1327; Cave Johnson to George Bancroft, May 3, 1848, box 2, GBP.

35. Act of June 27, 1848, ch. 79, 9 *Stat.* 241, *Cong. Globe*, April 12, 1848, 619–22 and May 29, 1848, 793. Somewhat incongruously, after months of frustration, Bancroft observed that "people here [in London] did not think we [illegible] act so promptly & decidedly" (George Bancroft to Cave Johnson, July 21, 1848, GBP).

36. George Bancroft to Cave Johnson, September 8, 1848, GBP.

37. "The New Postal Arrangements with Canada and the United States," *Times* (London), December 14, 1847, 6; see also George Bancroft to Cave Johnson, May 12, 1848, and June 22, 1848, GBP.

38. "The American Postage Question," *Times* (London), August 8, 1848, 2; George Bancroft to James Buchanan, August 11, 1848, vol. 11, DDS.

39. Ibid.

40. George Bancroft to James Buchanan, November 24, 1848, George Bancroft to James Buchanan, December 1, 1848, George Bancroft to James Buchanan, December 12, 1848, George Bancroft to James Buchanan, December 14, 1848, George Bancroft to James Buchanan, December 15, 1848, and George Bancroft to James Buchanan, February 7, 1849, vol. 11, DDS. One additional consequence of the new American postal treaty was the decision of Britain and the United States to approach France for a tripartite agreement (in 1843, Britain and France had adopted a postal convention of their own) to reduce postal rates among all three parties. Before returning to the United States in September 1849, George Bancroft traveled to Paris to open the negotiations.

41. Tocqueville, *Democracy in America*, trans. Gerald E. Bevan (New York: Penguin, 2003), 324; H.R. Rep. No. 3, 27th Cong., 1st sess. (1841), 1–4. Even a decade later, Americans' sense of vulnerability to British mail steamers was evident. Writing of American efforts to support a transatlantic mail line, navy secretary John Y. Mason wrote in 1852 that the line "if well managed will drive the British line off the ocean." Fear of foreign encroachment was manifest. "It is not desirable," Mason concluded, "that the British should have those heavy steamers so frequently on our coast. This line, and one from San Francisco to China, will complete the circle, and produce the finest results on all the public interests" (John Y. Mason to Thomas Jefferson Rusk, February 21, 1852, box 2K148, TJRP).

42. H.R. Rep. No. 685, 29th Cong., 1st sess. (1846), 2–3.

43. Reel 17, subseries 4.1, TBKP.

44. George A. Magruder to M. F. Maury, January 24, 1846, enclosed in letter to T. B. King, January 26, 1846, reel 3, TBKP. Younger officers, however, like William D. Porter, regarded their superior's reluctance as evidence of the need to empower midshipmen and lieutenants to take charge of steam vessels (William D. Porter to D. [James?] Relf[e], January 12, 1846, HR 27A-D13.6, HCNA). Yet the reticence of the older officers proved prescient. A few years later, when Porter's brother David Dixon found himself in command of a mail steamer, the U.S. Mail Steamship *Georgia* on the route from New York to New Orleans and Havana, he regularly reported to the Navy Department that while his ship was fast and reliable, he often depended on auxiliary sails and could not push the ship's engines too hard because coal supplies were so hard to come by and rapidly exhausted (David D. Porter to M. C. Perry, March 9, 1850, in *Reports of the Secretary of the Navy and the Postmaster General, Communicating, in Compliance with a Resolution of the Senate, Information in Relation to the Contracts for the Transportation of the Mails, by Steamships, between New York and California* [Washington, DC: A. Boyd Hamilton, 1852], 114–18).

45. See, for example, Matthew Perry to William Ballard Preston, January 2, 1850, roll 347, Captains' Letters.

46. According to the influential Washington journal *Niles' National Register*, "The question is—and it is a question not to be blinked or postponed,—shall the United States abandon the mail monopoly—the transport of passengers—the freight of merchandise—the advantages of foreign markets, to the British—or to the French—or to any power upon earth? Or shall she enter the ocean of contest for her own fair share in all these advantages?" ("Steamers," *Niles' National Register*, June 13, 1846, 226–27).

47. H.R. Rep. No. 476, 1.

48. An Act Providing for the Building and Equipment of Four Naval Steamships, S. 128, 29th Cong., 2nd sess. (1847).

49. John Whitehead to T. Butler King, April 6, 1841, enclosing Andrew C. Armstrong to John Whitehead, April 6, 1841 [?], to T. Butler King, March 17, 1842, roll 2, TBKP.

50. Andrew. L. King to T. Butler King, January 28, 1841, and Andrew L. King to T. Butler King, September 16, 1842, roll 2, TBKP.

51. Mayor of Baltimore et al., memorial, February 17, 1851, HR 31A-G12.7, HCNA.

52. William C. Barney et al., memorial, February 6, 1857, Sen 34A-H16.1, SCPO.

53. T. Butler King to Frances Granger, August 7, 1841, R. R. Cuylor to ?, August 2[4?], 1841, and James Macqueen to T. Butler King, February 12, 1842, roll 2, TBKP.

54. Edward Padelford to T. Butler King, December 2, 1845, roll 2, TBKP.

55. S. Ex. Doc No. 1/7, 32nd Cong., 1st sess. (1851), 436.

56. Georgia Exporting Company, memorial, August 26, 1850, HR 31A-G12.7, HCNA. Perhaps a factor in the failure of Congress to act on this line was that, as W. Stephen Belko observes, Green's long-standing political influence dramatically diminished around 1850, though Belko does not pursue Green's activities in detail after 1848 (*The Invincible Duff Green: Whig of the West* [Columbia: University of Missouri Press, 2006], 439).

57. William Caldwell Templeton to Nathan K. Hall, January 8, 1851, HR 32A-G15.3, HCPO. The notion that certain trade, particularly that of the West Indies, Mexico, South America, and the Far East, naturally belonged to the United States was a common refrain. See "Mail Communication with the West Coast of South America," HR 34A-G14.2, HCPO.

58. Ambrose W. Thompson, memorial, September 17, 1850, Sen 31A-H12.1, SCNA; Pennsylvania State Senate and House of Representatives, resolutions, February 15, 1851, HR 31A-G12.7, HCNA.

59. See the range of memorials on file in HR 32A-G13.1, HCNA.

60. *House Journal*, July 13, 1852, 889–90. American senators and British MPs alike would have been horrified to learn that despite their patriotic rhetoric and their committing themselves to serve as naval auxiliaries in time of war in order to prevent ruinous competition, Collins and Cunard secretly agreed in 1850 to a rate-fixing and revenue-sharing agreement brokered by two major investors in their firms, the Brown brothers, William and James. William also served as Collins's agent in Liverpool, and James performed that role in New York while serving as the steam line's president. The nature of this relationship only became clear beginning in the 1970s; see Edward W. Sloan in "Collins versus Cunard: The Realities of a North Atlantic Steamship Rivalry, 1850–1858," *International Journal of Maritime History* 4, no. 1 (1992): 83–100.

61. Gansevoort Melville to Maria Gansevoort Melville, August 18, 1845, in Parker, *Gansevoort Melville's 1846 London Journal*, 61.

62. John J. Currier, *"Ould Newbury": Historical and Biographical Sketches* (Boston: Damrell and Upham, 1896), 651–58.

63. William Wheelwright, *Statements and Documents Relative to the Establishment of Steam Navigation in the Pacific* (London: Whiting, Beaufort House, 1838), 7–8; see also Juan Bautista Alberdi, *The Life and Industrial Labors of William Wheelwright in South America* (Boston: A. Williams & Co., 1877) and Crosbie Smith, "'We Never Make Mistakes': Constructing the Empire of the Pacific Steam Navigation Company," in *The Victorian Empire and Britain's Maritime World, 1837–1901*, ed. Miles Taylor (New York: Palgrave Macmillan, 2013), 82–112, which discusses in great detail both Wheelwright's plans, challenges in coaling, and ultimate good luck.

64. Wheelwright, *Statements and Documents*, 13. Whether FitzRoy had personal knowledge of this coal is unclear; he spends two paragraphs in his account of the *Beagle* voyage citing other voyagers testifying to the presence and quality of coal along the western South American coast (*Narrative of the Surveying Voyages of His Majesty's Ships Adventure and Beagle between the Years 1826 and 1836*, 3 vols. [London: Henry Colburn, 1859], 2:423–24).

65. William Wheelwright, *To the Proprietors of the Pacific Steam Navigation Company* (n.p.: n.p., 1843), 4–5, William Wheelwright, *Mr. Wheelwright's Report on Steam Navigation* (n.p.: n.p., 1843), 2–3.

66. Wheelwright, *To the Proprietors*, 4; Wheelwright, *Mr. Wheelwright's Report*, 2–5, 34.

67. Ibid., 5, 17–21; Bollaert, "Observations on the Coal Formation in Chile, S. America" *Journal of the Royal Geographical Society of London* 25 (1855): 173.

68. Logbook of the steamer *Oregon*, 1849–1851, FAC 1614, HL.

69. Ibid.

70. William H. Aspinwall to Alfred Robinson, December 8, 1848, PMSC.

71. William H. Aspinwall to Alfred Robinson, March 19, 1849, William H. Aspinwall to Alfred Robinson, April 19, 1849, and Samuel W. Comstock to Alfred Robinson, February 16, 1850, PMSC.

72. William H. Aspinwall to Alfred Robinson, May 26, 1849, and Samuel W. Comstock to William C. Stout, October 15, 1849, PMSC. Still, as late as April, 1850, no Vancouver coal had yet reached the company (Samuel W. Comstock to Alfred Robinson, April 13, 1850, PMSC).

73. Samuel W. Comstock to Alfred Robinson, March 15, 1850, and William H. Aspinwall to Robinson, Bissell and Co., June 13, 1850, PMSC.

74. William H. Aspinwall to Robinson, Bissel and Co., October 12, 1850, PMSC.

75. William H. Aspinwall to Robinson, Bissell and Co., July 8, 1850 and August 13, 1850, PMSC.

76. Samuel W. Comstock to John Van Dewater, October 15, 1849, and Samuel W. Comstock to John Can Dewater, November 12, 1849, PMSC; Charles B. Stuart, *The Naval and Mail Steamers of the United States* (New York: Charles B. Norton, 1853), 144.

77. Marshall O. Roberts to Parradise, Saffarans and Co., December 6, 1848, Marshall O. Roberts to J. C. Burnham and Co., January 5, 1849, Marshall O. Roberts to Padelford, Fay and Co., January 6, 1849, Roberts to Newell, Sturtevant and Co., January 8, 1849, Marshall O. Roberts to Padelford, Fay and Co., January 27, 1849, Marshall O. Roberts to J. C. Burnham and Co., January 31, 1849, and Marshall O. Roberts for D. N. Carrington to Newell, Sturtevant and Co., February 5, 1849, vol. 1, Roberts Letter Book; Marshall O. Roberts to J. F. Schenck, February 13, 1850, vol. 3, Roberts Letter Book.

78. Davis, Brooks and Co. to William A. Graham, May 5, 1852, roll 272, M124.

79. S. Ex. Doc. No. 1/5, 31st Cong., 2nd sess. (1850), 196; S. Ex. Doc. No. 1/6, 31st Cong., 2nd sess. (1850), 236.

80. H.R. Rep. No. 87, 36th Cong., 2nd sess. (1861), 2–3; on the shipbuilding projects of the 1850s, see Kurt Hackemer, *The U.S. Navy and the Origins of the Military-Industrial Complex, 1847–1883* (Annapolis, MD: Naval Institute Press, 2001), 1–66.

81. H.R. Ex. Doc. No. 2/7, 28th Cong., 1st sess. (1843), 526; S. Ex. Doc. No. 1, 36th Cong., 2nd sess. (1860), 242.

82. H.R. Ex. Doc. No. 82, 35th Cong., 2nd sess. (1858), 2–3.

83. H.R. Rep. No. 184, 35th Cong., 2nd sess. (1859), 31–45.

84. Ibid, 33–37.

85. S. Ex. Doc. No. 50, 32nd Cong., 1st sess. (1852), 94.

86. S. Ex. Doc. No. 50, 32nd Cong., 1st sess. (1852), 114–18.

87. David Dixon Porter to Matthew Perry, August 7, 1850, enclosed in Matthew Perry to William Ballard Preston, September 2, 1850, Roll 347, Captains' Letters.

88. Among Porter and the *Panama*'s first passengers were Sam Ward, the future master lobbyist; William McKendree Gwin, soon to be elected senator in California; Major Joseph Hooker, who was assuming a post as assistant adjutant general for the U.S. Army in the Pacific; Thomas Butler King, who was leaving the House as a Georgia representative to become collector of the port of San Francisco; and Jessie Benton Frémont, who was traveling to California to reunite with her husband John. See William H. Aspinwall to Alfred Robinson, February 3, 1849, and William H. Aspinwall to Alfred Robinson, March 14, 1849, PMSC, "For California," *Trenton (NJ) State Gazette*, February 16, 1849, 2, "California Items," *Trenton (NJ) State Gazette*, February 17, 1849, 1, Cotton [pseud.], "Twenty Years Ago—The Steamer 'Panama' and Her Passengers," *San Francisco Daily Evening Bulletin*, June 4, 1869, 3, Kathryn Allamong Jacob, *King of the Lobby: The Life and Times of Sam Ward, Man-About-Washington in the Gilded Age.* (Baltimore, MD: Johns Hopkins University Press, 2010), 44–45, and H. Edwin Tremain, *In Memoriam: Major-General Joseph Hooker* (Cincinnati: Robert Clarke, 1881), 4.

89. Lewis R. Hamersly, *The Records of Living Officers of the U.S. Navy and Marine Corps* (Philadelphia: Lippincott, 1870); Lewis R. Hamersly, *The Records of Living Officers of the U.S. Navy & Marine Corps* 3rd. ed. (Philadelphia: Lippincott, 1878); Matthew Perry to William A. Graham, May 1, 1851, roll 348, Captains' Letters.

90. In a typical example, steam mail promoter Christian Hansen insisted in a letter dated January 18, 1858, to William H. English, the chairman of the Committee on Post Offices and Post Roads, that "the mercantile community should be able to reckon with certainty on the punctual departure of the mails at the appointed times, and to calculate with precision the times of their arrival." "Private enterprise," he added, "cannot be depended upon for providing such communication." Even at the end of the 1850s, private ocean mail steamers remained too expensive, too risky, and too disadvantaged when competing with steamers subsidized by other nations (HR 35A-G16.3, HCPO).

Chapter 2 · Engineering Economy

1. Daniel Webster, "The Boston Mechanics' Institution," in *The Writings and Speeches of Daniel Webster*, vol. 2 (Boston: Little, Brown, 1903), 36.

2. Around the 1930s, according to Mitchell, economists led by Keynes began speaking of "the economy"—a thing—that had not previously existed conceptually (*Carbon Democ-*

racy: Political Power in the Age of Oil [London: Verso, 2011], 124). According to J. Adam Tooze, the concept emerged several decades earlier in Germany; see his "Imagining National Economies: National and International Economic Statistics, 1900–1950," in *Imagining Nations*, ed. Geoffrey Cubitt (Manchester, UK: Manchester University Press, 1998), 212–4. I thank Quinn Slobodian for bringing this essay to my attention.

3. Lydia Maria Child, *The American Frugal Housewife, Dedicated to Those Who Are Not Ashamed of Economy*, 27th ed. (New York: Samuel S. and William Wood, 1841), 7. Cf. the influential political economist Henry C. Carey: "The relation of husband and wife, and that of parent and child, are both essential to the development of all that is good and kind, gentle and thoughtful. The desire to provide for the wife and the child prompts the husband to labour, for the purpose of acquiring the means of present support, and to economy as a means of preparation for the future. The desire to provide for the husband and the children prompts the wife to exertions that would otherwise have been deemed impossible, and to sacrifices that none but a wife or a mother could make" (*The Harmony of Interests, Agricultural, Manufacturing, and Commercial* [Philadelphia: J. S. Skinner, 1851], 201).

4. Noah Webster, *An American Dictionary of the English Language*, rev. ed. (New York: White and Sheffield, 1841), s.v. "efficiency" and "efficacy."

5. Aristotle, *The Metaphysics*, trans. Hugh Tredennick (Cambridge, MA: Harvard University Press, 1933), 211.

6. Gilbert mathematically defined efficiency as the product of the force applied to a machine and the distance over which the force operated; with the development of thermodynamics midcentury, this quantity would be reinterpreted as the work done on a system; see his "On the Expediency of Assigning Specific Names to All Such Functions of Simple Elements as Represent Definite Physical Properties," *Philosophical Transactions of the Royal Society* 117 (1827): 27.

7. William Whewell, *The First Principles of Mechanics* (Cambridge: Cambridge University Press, 1832), 52–56.

8. "Work," in this context, meant what Rankine defined as "moving against resistance," as in calculated "by the product of the resistance into the distance through which its point of application is moved" (*A Manual of Applied Mechanics* [London: Richard Griffin, 1858], 477). Elsewhere, Rankine notes that the total work performed by a machine, useful as well as lost, is equal to the total energy imparted to the machine (*A Manual of Applied Mechanics*, 610, 625). For Rankine's earlier accounts of efficiency from 1854 and 1855, see his *Miscellaneous Scientific Papers*, 2 vols. (London: Charles Griffin, 1881), 2:226, 364–85.

9. Frederick Winslow Taylor, *The Principles of Scientific Management* (New York: Harper and Brothers, 1911), 11; on the evolution of the term "efficiency," see Jennifer Karns Alexander, *The Mantra of Efficiency: From Waterwheel to Social Control* (Baltimore, MD: Johns Hopkins University Press, 2008).

10. Robertson Buchanan, *A Treatise on the Economy of Fuel, and Management of Heat* (Glasgow: printed for the author, 1815), vi.

11. N. L. S. Carnot, *Reflections on the Motive Power of Heat*, ed. R. H. Thurston (New York: John Wiley and Sons, 1897), 126. As he helped develop the modern science of thermodynamics, William Thomson (Lord Kelvin) understood Carnot's writing in a manner very different from Carnot himself. As suggested by Carnot's reference to "the man called to direct," Carnot thought very much in a prethermodynamic way. His original essay was published in 1824 as *Réflexions sur la puissance motrice du feu*.

12. Walter R. Johnson, "On the Economy of Fuel with Reference to Its Domestic Applications," *American Journal of Science and Arts* 23, no. 2 (1833): 319.

13. James Renwick, *Treatise on the Steam Engine* (New York: Carvill, 1830), 291.

14. Thomas Parke Hughes, *Networks of Power: Electrification in Western Society, 1880–1930* (Baltimore, MD: Johns Hopkins University Press, 1983).

15. Thomas Jefferson, *Notes on the State of Virginia* (London: John Stockdale, 1787), 42–43; Sean Patrick Adams, *Old Dominion, Industrial Commonwealth: Coal, Politics, and Economy in Antebellum America* (Baltimore, MD: Johns Hopkins University Press, 2004), 27, 51–52. Laurel Mountain lies along the western edge of the Alleghenies, now in West Virginia.

16. William B. Rogers, *Report of the Progress of the Geological Survey of the State of Virginia for the Year 1839* (Richmond, VA: Samuel Shepherd, 1840), 88; Henry D. Rogers, *First Annual Report of the State Geologist* (Harrisburg, PA: Samuel D. Patterson, 1836), 17. In the early years of Henry Rogers's survey work, much to his frustration, the already powerful collieries of Pennsylvania did not appreciate his impolitic skepticism with respect to the future prospects of the state's anthracite fields; likewise, William Rogers was thwarted by agrarian political interests in the eastern part of Virginia, who did not share his belief that there were potentially vast coalfields in the west. On the search for and coal in the early to mid-nineteenth century, see Paul Lucier, *Scientists and Swindlers: Consulting on Coal and Oil in America, 1820–1890* (Baltimore, MD: Johns Hopkins University Press, 2008), esp. 79–94, "Sketch of Denison Olmsted," *Popular Science Monthly*, January 1895), 404, Walter B. Hendrickson, "Nineteenth-Century State Geological Surveys: Early Government Support of Science," *Isis* 52, no. 3 (1961): 357–71, Francis P. Boscoe, "'The Insanities of an Exalted Imagination: The Troubled First Geological Survey of Pennsylvania," *Pennsylvania Magazine of History and Biography* 127 (July, 2003): 291–308, Sean Patrick Adams, "Partners in Geology, Brothers in Frustration: The Antebellum Geological Surveys of Virginia and Pennsylvania," *Virginia Magazine of History and Biography* 106, no. 1 (1998): 5–34, and Benjamin R. Cohen, "Surveying Nature: Environmental Dimensions of Virginia's First Scientific Survey, 1835–1842," *Environmental History* 11, no. 1 (2006): 37–69. On the extent of Maryland coal, see, for example, "Mineral Wealth of Allegany County, and Considerations on the Best Means of Developing It," in *Annual Report of the Geologist of Maryland* (n.p.: n.p., 1840): 27–45; for Ohio, see "Report of C. Briggs, Jr., Fourth Assistant Geologist," in *First Annual Report of the Geological Survey of the State of Ohio* (Columbus, OH: Samuel Medary, 1838), 83–87; for Indiana, see D. D. Owen, *Continuation of Report of a Geological Reconnoissance of the State of Indiana Made in the Year 1837* (Indianapolis, IN: John C. Walker, 1859), 8, 10–12, 17, 19, 31–42, 44–45, 52.

17. "Government Contract for Conveying the Mail by Steam Coaches," *Morning Post* (London), October 8, 1829, 2. This article was republished with commentary as "From the U.S. Telegraph," *Providence (RI) Patriot*, January 6, 1830, 1, and "The Mechanical Age," *Niles' Weekly Register*, December 5, 1829, 236–37. The original also appeared in the *Richmond (VA) Enquirer*, January 14, 1830, 4, among other papers.

18. The concerns appear well warranted at the time; the captain of the *Sirius* blamed poor American coal for the slow voyage back to Plymouth ("Plymouth, July 16," *Richmond (VA) Enquirer*, August 10, 1838, 2; "The Coal Mines of Cumberland," *Albion* 6, no. 24 [1838]: 191; Stephen Fox, *Transatlantic: Samuel Cunard, Isambard Brunel, and the Great Atlantic Steamships* [New York: Perennial, 2003], 76–80).

19. "Steam Ship Clarion and Anthracite Coals," *New York Herald*, October 11, 1841, 2.

20. Renwick, *Treatise on the Steam Engine*, 49.

21. R. C. Taylor, *Statistics of Coal*, 2nd ed. (Philadelphia: J. W. Moore, 1855), 338, 424, 566. The contemporary understanding of combustion played an important role in the selection of coal. Since burning hydrogen produced more heat than burning carbon, then, for equal weights of fuel, coal containing more hydrogen should generate more heat. In the 1820s, the operators of steam engines indeed preferred the soft, hydrogen-rich pinewood to hardwoods and, similarly, hydrogen-rich bituminous coal to anthracite. But the advantages of hydrogen-rich fuels could only be realized when they could be burned in their entirety. In practice, hydrogen gas, liberated from its wood or coal, often escaped without combusting, carrying with it an additional thick smoke of carbon particulates (Renwick, *Treatise on the Steam Engine*, 51–52). On wheat and hog commodification, see William Cronon, *Nature's Metropolis: Chicago and the Great West* (New York: Norton, 1991).

22. Antoine Lavoisier, "Expériences sur l'effet compare de différents combustibles," in *Oeuvres de Lavoisier,* 6 vols. (Paris: Imprimerie impériale, 1862): 2:377–90. Lavoisier's observation about the interdependence of fuels, markets, and politics would become a constant theme in the growth of fossil fuels.

23. Count Rumford, "Observations Relative to the Means of Increasing the Quantities of Heat Obtained in the Combustion of Fuel" (1801), "Inquiries Relative to the Structure of Wood" (1812), "Chimney Fireplaces, with Proposals for Improving them to Save Fuel" (1796), and "On the Management of Fire and the Economy of Fuel" (1797), in Count Rumford, *The Collected Works of Count Rumford*, 5 vols., ed. Sanborn C. Brown (Cambridge, MA: Belknap Press of Harvard University Press, 1968–70), 2:155–61, 171–219, 221–295, 309–477, quote on 310.

24. Bull's research was widely disseminated and his findings on wood combustion would remain authoritative at least through the 1840s, when Walter Johnson ensured their reproduction in the American edition of an influential German chemistry manual; see Friedrich Ludwig Knapp, *Chemical Technology; or, Chemistry, Applied to the Arts and to Manufactures*, vol. 1., trans. and ed. Edmund Ronalds and Thomas Richardson, with notes and additions by Walter R. Johnson (Philadelphia: Lea and Blanchard, 1848), vi, 24, and Marcus Bull, "Experiments to Determine the Comparative Quantities of Heat Evolved in the Combustion of the Principal Varieties of Wood and Coal Used in the United States, for Fuel," *Transactions of the American Philosophical Society*, n.s., 3, no. 1 (1830): 2, 21–22.

25. Benjamin Henry Latrobe Jr. to John B. Jervis, May 28, 1845, BHL.

26. Walter Johnson, "Use of Anthracite Coal in Locomotives," *Journal of the Franklin Institute,* ser. 3, 14, no. 2 (1847): 110–16.

27. Ibid., 111.

28. George W. Whistler Jr., *Report upon the Use of Anthracite Coal in Locomotive Engines on the Reading Rail Road* (Baltimore, MD: John D. Toy, 1849), 12ff.

29. The line's wood budget included the costs of cutting and transporting it. The Reading's Transportation Department had a total budget of $651,876.29 (*Report of the President and Managers of the Philadelphia & Reading Rail Road Co. to the Stockholders, January 12, 1847* [Philadelphia: Isaac M. Moss, 1847], 25–26).

30. *Report of the President and Managers of the Philadelphia and Reading Rail Road Company to the Stockholders, January 12, 1857* (Philadelphia: Moss and Brother, 1855), 31; see also

John H. White Jr., "Railroads: Wood to Burn," in *Material Culture of the Wooden Age*, ed. Brooke Hindle (Tarrytown, NY: Sleepy Hollow Press, 1981), 199–203.

31. Higgins, *The Fourth Annual Report of James Higgins, MD, State Agricultural Chemist, to the House of Delegates of the States of Maryland* (Baltimore, MD: Sherwood, 1854), 72–73; on the chemistry of coal combustion, see also Benjamin Franklin Isherwood, "Notes on the U.S. Steamer 'Princeton,'" *Appleton's Mechanics' Magazine and Engineers' Journal* 1, no. 10 (1851): 621.

32. Higgins, *The Fourth Annual Report of James Higgins*, 72–73.

33. Walter R. Johnson, *The Coal Trade of British America, with Researches on the Characters and Practical Values of American and Foreign Coals* (Washington, DC: Taylor and Maury, 1850), 6. Not all of Johnson's experiences with government science were positive; in 1836, Commodore Thomas ap Catesby Jones appointed him to study electricity, magnetism, and astronomy on the proposed United States South Seas Exploring Expedition, but when Jones was replaced by the imperious Lieutenant John Wilkes, Johnson was cut from the voyage; see "Johnson, Walter Rogers," in *Appleton's Cyclopaedia of American Biography* (New York: Appleton, 1898), 3:451, H.R. Ex. Doc. No. 396, 25th Cong., 2nd sess. (1838), Walter R. Johnson, "The Advantages and Means of a National Foundry," *National Gazette*, July 18, 1839, 2, S. Ex. Doc. No. 229, 26th Cong., 1st sess. (1840), S. Ex. Doc. No. 167, 28th Cong., 1st sess. (1844), *House Journal*, December 20, 1838, 25th Cong., 3rd sess., 110, John Quincy Adams, *Memoirs of John Quincy Adams, Comprising Portions of His Diary from 1795 to 1848*, 12 vols., ed. Charles Francis Adams (Philadelphia: Lippincott, 1874–77), 10:86, William Stanton, *The Great United States Exploring Expedition of 1838–1842* (Berkeley: University of California Press, 1975), 48, 55, 61, 66, and Bruce Sinclair, *Philadelphia's Philosopher Mechanics: A History of the Franklin Institute, 1824–1865* (Baltimore, MD: Johns Hopkins University Press, 1974), 126, 135, 152–55, 184–85, 252.

34. S. Ex. Doc. No. 386, 28th Cong., 1st sess. (1843), vi, xi–xii, 5, Mason quote on 1; Walter R. Johnson to the Board of Navy Commissioners, June 26, 1841, box 707, XF 1841–51, subject file, U.S. Navy, RG 45, NARA-1. On public-private governance, see Brian Balogh, *A Government Out of Sight: The Mystery of National Authority in Nineteenth-Century America* (New York: Cambridge, 2009).

35. S. Ex. Doc. No. 386, 209.

36. On two other important characteristics for a steaming coal—its propensity to form clinker and quickly evaporate water—Cumberland varieties also performed well. By the clinker measure, some anthracites tended to come out ahead, and on rapid evaporation, Virginia bituminous varieties showed well, but taking all measures into account, Johnson's results for Cumberland coal helped characterize the fuel as ideal for steaming (S. Ex. Doc. No. 386, 592–99).

37. "Experiments on Coal," *Baltimore Sun*, January 2, 1844, 1.

38. Proposals for Cumberland coal for specific vessels found in H.R. Ex. Doc. No. 1/8, 30th Cong., 2nd sess. (1848), 700–701, H.R. Ex. Doc. No. 5/8, 31st Cong., 1st sess. (1849), 515, 546, S. Ex. Doc. No. 1/8, 31st Cong., 1st sess. (1849), 515, and S. Ex. Doc. No. 1/6, 31st Cong., 2nd sess. (1850), 302. The department still sometimes ordered or accepted bids for bituminous coal from western Pennsylvania; see, for example, H.R. Ex. Doc. No. 8/7, 30th Cong., 1st sess. (1847), 1045, *Abstract of Professor Johnson's Report to the Secretary of the Navy, of the United States, Respecting Forest Improvement Coal* (New York: George F. Nesbitt, 1845), and S. Misc. No. 117, 31st Cong., 1st sess. (1850).

39. Anthrax, "The Cumberland Coal Field," *Philadelphia Public Ledger*, May 16, 1845, 1.

40. Estimates of fuel consumption in the mid-nineteenth century are always approximate, but in 1850 the American coal industry probably produced a little over 8 million tons, roughly split between bituminous and anthracite. A year later, navy records indicate the purchase of fewer than 5,000 tons of bituminous and anthracite coal (though this figure does not include coal purchased on foreign stations, which probably no more than doubled this figure at the most). See Sam H. Schurr et al., *Energy in the American Economy: An Economic Study of Its History and Prospects* (Baltimore, MD: Johns Hopkins University Press, 1960), 491.

41. Taylor, *Statistics of Coal*, 37–38.

42. Henry de la Beche and Lyon Playfair, *First Report on the Coals Suited to the Navy* (London: William Clowes, 1848), 12; Henry de la Beche and Lyon Playfair, *Second Report on the Coals Suited to the Navy* (London: William Clowes, 1849), 3–6ff.; Henry de la Beche and Lyon Playfair, *Third Report on the Coals Suited to the Navy* (London: William Clowes, 1851). Due to small samples, some of the foreign coals were only tested chemically and not in actual boilers. Also, it was not only Americans who were interested in this research; a German translation appeared in Vienna almost immediately; de la Beche and Playfair, *Erster Bericht über die zer Dampfschiffahrt geeigneten Steinkohlen Englands*, trans. A. Schrötter (Vienna: Kaiserlich-königlichen Hof- und Staats-Druckerei, 1849).

43. de la Beche and Playfair, *Third Report*, 3.

44. de la Beche and Playfair, *First Report*, 17ff.

45. On the development of thermodynamics and its connection to steam engineering, see Crosbie Smith, *The Science of Energy: A Cultural History of Energy Physics in Victorian Britain* (Chicago: University of Chicago Press, 1998), and Crosbie Smith and M. Norton Wise, *Energy and Empire: A Biographical Study of Lord Kelvin* (New York: Cambridge University Press, 1989), 282–347.

46. Charles W. Morgan to William Ballard Preston, May 15, 1849, roll 345, Captains' Letters.

47. H.R. Rep. No. 184, 35th Cong., 2nd sess. (1859), 133; "Report to the Coal Mining Association," *Hazard's Register of Pennsylvania*, 13, No. 20 (1834): 310–12.

48. H.R. Rep. No. 184, 129; An Act Making Appropriations for the Naval Service, September 28, 1850, ch. 80, 9 *Stat.* 515; *Cong. Globe*, September 28, 1850, 2060. Two failed amendments a week before the passage of this provision, sponsored by Representative Robert Schenck of Ohio and Representative Joseph Chandler of Pennsylvania, explicitly called for appointing government agents to inspect coal at mines or ports, both to protect the government interest (especially for coal purchased for overseas service) as well as that of mine operators, whom Schenck claimed "themselves not only assented to the proposition, but . . . earnestly desired that it should be adopted as a matter of protection to their own interests" in preventing an unusually poor shipment from ruining their reputations (*Cong. Globe*, September 21, 1850, 1902–3).

49. B., "On the Comparative Value of Anthracite and Bituminous Coals for the Purpose of Generating Steam," *Journal of the Franklin Institute*, 3rd ser. 23, no. 6 (1852): 418–19; Benjamin H. Springer to James Cooper, February 18, 1851; Benjamin H. Springer to Merrick and Son, January 21, 1851, box 707, XF 1841–1851, subject file, U.S. Navy, RG 45, NARA-1.

50. S. Rep. No. 356, 32nd Cong., 1st sess. (1852), 2; "Resolution Requesting Our Representatives in Congress to Obtain from the Secretary of the Navy, a Copy of Instructions and

the Report Upon the Subject of Anthracite and Semi-bituminous Coals," in *Laws Made and Passed by the General Assembly of the State of Maryland at a Session Begun and Held at Annapolis, on Wednesday, the 7th Day of January, 1852, and Ended on Monday, the 31st of May, 1852* (Annapolis: B. H. Richardson and Cloud's, 1852), 464–65; "Anthracite vs. Cumberland Coal," *Baltimore Sun*, April 12, 1851, 1; "Anthracite vs. Cumberland Coal," *Baltimore Sun*, June 19, 1851, 4; "Reports from Congressional Committees," *Baltimore Sun*, May 26, 1852, 4.

51. S. Ex. Doc. No. 74, 32nd Cong., 1st sess. (1852), 8–11; Charles Stuart, *The Naval and Mail Steamers of the United States* (New York: Charles B. Norton, 1853), 95.

52. Charles Stuart, "A Report of the Engineer in Chief of the Navy, on the Comparative Value of Anthracite and Bituminous Coals," *Journal of the Franklin Institute*, 3rd ser., 24, no. 4 (1852): 217–22, quote on 217.

53. Charles Miner to William A. Graham, n.d., roll 270, M124, probably originally enclosed in W. A. Bradley to William A. Graham, April 5, 1852, roll 271, M124; Charles Miner to Samuel J. Packer, November 17, 1833, in *Report of the Committee of the Senate of Pennsylvania, upon the Subject of the Coal Trade* (Harrisburg, PA: Henry Welsh, 1834), 93–96; Charles Francis Richardson and Elizabeth Miner Richardson, *Charles Miner: A Pennsylvania Pioneer* (Wilkes-Barre, PA: n.p., 1916), 66–67.

54. Benjamin Franklin Isherwood, *Engineering Precedents for Steam Machinery*, 2 vols. (New York: Baillière Brothers, 1859), 2:1–38; "What the Rebels Propose to Do with Our Coal Mines," *Scientific American* 9, no. 3 (18, 1863): 35; David D. Porter, *The Naval History of the Civil War* (New York: Sherman, 1886), 369.

55. Ben Marsden, *Watt's Perfect Engine: Steam and the Age of Invention* (New York: Columbia University Press, 2002); Richard L. Hills, *Power from Steam: A History of the Stationary Steam Engine* (New York: Cambridge University Press, 1989).

56. H.R. Rep. No. 184; Sherman, *Recollections of Forty Years in the House, Senate, and Cabinet: An Autobiography*, 2 vols. (Chicago: Werner Company, 1895), 1:158–61. Sherman was familiar with the intersection between the state and engineering; at fourteen he had served as a junior rodman for the Muskingum River Improvement, an Ohio project opening the river to traffic. During the time he worked on the project, he quickly rose to supervise lock and dam assembly. The dismissal of a superintendent by a new state Democratic administration in 1838 and subsequent firing of Sherman himself forcefully revealed to him the inextricable relationship between large-scale engineering projects and politics. Afterward, he gave up a prospective surveying career for one as a Whig and later Republican politician; see Theodore Burton, *John Sherman* (Boston: Houghton and Mifflin Company, 1906), 7–8.

57. H.R. Rep. No. 184, 93–97, 116, 144, 153.

58. Langdon Winner, "Do Artifacts Have Politics?," *Daedalus* 109, no. 1 (1980): 121–36; for further debate over the question, see Bernward Joerges, "Do Politics Have Artefacts?" *Social Studies of Science* 29, no. 3 (1999): 411–31, Steve Woolgar and Geoff Cooper, "Do Artefacts Have Ambivalence: Moses' Bridges, Winner's Bridges and Other Urban Legends in S&TS?," *Social Studies of Science* 29, no. 3 (1999): 433–57, and Bernward Joerges, "Scams Cannot Be Busted: Reply to Woolgar and Cooper," *Social Studies of Science* 29, no. 3 (1999): 450–57.

59. On antebellum naval steam engineering, see Kurt Hackemer, *The U.S. Navy and the Origins of the Military-Industrial Complex, 1847–1883* (Annapolis, MD: Naval Institute Press, 2001), Brendan Foley, "Fighting Engineers: The U.S. Navy and Mechanical Engineering, 1840–1905" (PhD diss., Massachusetts Institute of Technology, 2003), and Monte A. Calvert,

The Mechanical Engineer in America, 1830–1910: Professional Cultures in Conflict (Baltimore, MD: Johns Hopkins University Press, 1967), 245–61; see also Robert Greenhalgh Albion, *Makers of Naval Policy, 1798–1947* (Annapolis, MD: Naval Institute Press, 1980), 178–96.

60. Josiah Parkes, "On Steam-Boilers and Steam-Engines," pt. 2, *Transactions of the Institution of Civil Engineers* 3, no. 2 (1840): 50. This research won the prestigious Telford Metal in Gold in 1840; see "Telford Premiums," *Minutes of Proceedings of the Institution of Civil Engineers* 68, no. 1 (1841): 19.

61. H.R. Ex. Doc. No. 20, 31st Cong., 1st sess. (1850), 601.

62. Thomas Ewbank, petition, SEN 34A-H13.1, SCNA.

63. Thomas Ewbank, "On the Paddles of Steamers—their Figure, Dip, Thickness, Material, Number, &c.," *Journal of the Franklin Institute*, 3rd series, 17, no. 1 (1849): 42.

64. Ewbank, "On the Propulsion of Steamers," 595. For more on Ewbank, see William A. Bate, "Thomas Ewbank: Commissioner of Patents, 1849–1852," *Records of the Columbia Historical Society, Washington, D.C.*, 49 (1973–74): 111–24.

65. Charles B. Stuart to Charles W. Skinner, July 24, 1851, SEN 34A-H13.1, SCNA.

66. *Cong. Globe*, January 17, 1853, 312.

67. George T. Parry, memorial, January 13, 1857, SEN 34A-H13.1, SCNA; for a description of the device, see George T. Parry, "Anti Friction Roller," U.S. Patent No. 9,912, issued August 2, 1853; Stuart, *The Naval and Mail Steamers*, 120; S. Rep. No. 447, 34th Cong., 3rd sess. (1857), 3–8.

68. S. Rep. No. 447, 1–3; S. Rep. No. 317, 35th Cong., 1st sess. (1858), 2.

69. "Naval Economy," *Washington (DC) Republic*, February 27, 1851, 4; "Fresh Water for the Boilers of Sea Steamers," *New York Herald* article, quoted in *Additional and Fresh Evidence of the Practical Working of Pirsson's Steam Condenser* (Washington: Gideon and Co., 1851), 21; *Cong. Globe*, August 30, 1852, 2446–48; see also "Pirsson's Patent Surface Condensers," *Mechanics Magazine*, November 13, 1858, 457–60. Though Secretary William Graham requested Congress's aid in January, he left the resolution to his successor, John P. Kennedy, who took over as secretary on July 26, 1852.

70. *Cong. Globe*, August 30, 1852, 2448. Discussing a retrofit of the steamers *Princeton* and *Allegheny*, Pirsson himself suggested that his condenser's fuel savings for a single vessel would likely amount to 5 or 6 tons per day. In the Atlantic, where coal was cheaper, that would mean about $1,000 a month; in the more expensive Pacific, it would amount to $4,000 a month (Joseph P. Pirsson to William A. Graham, March 13 and March 17, 1852, roll 270, M124).

71. *Cong. Globe*, August 30, 1852, 2449–50.

72. "Professor Page's Electro-magnetic power," *National Intelligencer*, August 29, 1850.

73. Ibid.

74. *Cong. Globe*, Friday, August 9, 1850, 1554; Robert C. Post, "The Page Locomotive: Federal Sponsorship of Invention in Mid-19th-Century America," *Technology and Culture* 13, no. 2 (1972): 151–52.

75. Post, "The Page Locomotive," 156–65.

76. George H. Babcock, "Substitutes for Steam," *Transactions of the American Society of Mechanical Engineers* 7 (1885–86): 680–741. For more on these steam substitutes and Ericsson's early efforts, see Ben Marsden, "Blowing Hot and Cold: Reports and Retorts on the Status of the Air-Engine as Success or Failure," *History of Science* 36, no. 4 (1998): 373–420,

and Ben Marsden, "Superseding Steam: The Napier and Rankine Hot-air Engine," *Transactions of the Newcomen Society* 76, no. 1 (2006): 1–22. Late nineteenth-century engineers were exasperated that these fuel-saving engines kept reappearing as supposedly revolutionary devices despite the proof of their limitations provided by thermodynamics; see, for example, W. P. Trowbridge, "The Bisulphide of Carbon Engine," *School of Mines Quarterly* 7 (1885–86): 212–18.

77. John O. Sargent, *A Lecture on the Late Improvements in Steam Navigation and the Arts of Naval Warfare* (New York: Wiley and Putnam, 1844), 55; William Conant Church, *The Life of John Ericsson*, 2 vols. (New York: Charles Scribner's Sons, 1891), 1:36–37, 67–83, quote on 71; John Ericsson, *Contributions to the Centennial Exhibition* (New York: Nation Press, 1876), 425–38.

78. An explosion of one of the ship's twelve-inch guns, the "Peacemaker," during a demonstration in 1844 killed the secretaries of state and navy, along with four others.

79. Church, *The Life of John Ericsson*, 1:185, 188–89; for descriptions of the ship and the operation of its engines, see Eugene S. Ferguson, "John Ericsson and the Age of Caloric," *United States National Museum Bulletin* 228 (1963): 44–45, and Edward E. Daub, "The Regenerator Principle in the Stirling and Ericsson Hot Air Engines" *British Journal for the History of Science* 7, no. 3 (1974): 259–77. Ericsson's chief supporter was New York merchant John B. Kitching, who also invested in early telegraphy and the Atlantic cable ("John B. Kitching's Funeral," *New York Times*, July 23, 1887).

80. John Ericsson to John Sargent, August 31, 1852, roll 1, JEP.

81. Privately, Joseph Henry of the Smithsonian and Alexander Bache of the Coast Survey expressed their doubts about caloric power to navy secretary John P. Kennedy, who suggested the two form a commission to examine the ship and its engine. Henry demurred, in Kennedy's account, for concern over "rather too much responsibility." Kennedy ultimately declined to appoint the commission out of concern its work would delay a trial visit of the ship from New York to Washington for an inspection by Congress. With the second session of the thirty-second Congress set to conclude in just a few weeks, Kennedy evidently hoped to impress official Washington with the ship's merits and have the government commission Ericsson for new caloric vessels before the session ended. See John Pendleton Kennedy Journal 7g, February 27, 1853, roll 2, JPKP, John B. Kitching, *Ericsson's Caloric Engine* (New York: Slater and Riley, 1860), 8, 17–19, 21, 24–26, 28, 47 (on breathing) and 23, 28, 35, 42 (for newspaper quotes from the *Express*, *Tribune*, and *Daily Times*), and "The Hot Air Ship Ericsson," *Scientific American* 8, no. 19 (1853): 149.

82. *New Orleans Times-Picayune*, February 23, 1853, 2.

83. "Annual Festival of the Maryland Historical Society," *Baltimore Sun*, February 19, 1853, 1.

84. "Caloric Ships for the Navy," *National Intelligencer*, March 3, 1853; "Distinguished Visitors to the Ericsson," *Baltimore Sun*, February 25, 1853, 4; "The President and President Elect, Etc.," *New York Weekly Herald*, February 26, 1853, 1; Washington Irving to Sarah Irving, February 25, 1853, in Washington Irving, *Letters, 1846–1859*, ed. Ralph M. Aderman, Herbert L. Kleinfield, and Jenifer S. Banks (Boston: Twayne, 1982), 376–77; Henry T. Tuckerman, *The Life of John Pendleton Kennedy* (New York: G. P. Putnam and Sons, 1871), 223–24; John Pendleton Kennedy Journal 7g, February 27, 1853, roll 2, JPKP.

85. In "John Ericsson and the Age of Caloric," Ferguson (46) writes that the House Naval Affairs Committee ignored Kennedy's ostensibly misguided request. In fact, the

committee pursued the matter to the floor of the House; see *Cong. Globe*, February 25, 1853, 854, 861.

86. Church, *The Life of John Ericsson*, 195–98; John Ericsson to John Sargent, July 28, 1854, and John Ericsson to John P. Kennedy, January 15, 1854, roll 14, JPKP.

87. Church, *The Life of John Ericsson*, 202–3; Kitching, *Ericsson's Caloric Engine.*

88. H.R. Rep. No. 184, 53; *Cong. Globe*, June 13, 1860, 2950–51.

89. H.R. Rep. No. 184, 183.

90. Benjamin Franklin Isherwood, "Water-Tube and Fire-Tube Boilers," *Journal of the Franklin Institute* 107, no. 1 (1879): 14.

91. H.R. Rep. No. 184, 85–86.

92. Edward William Sloan III, *Benjamin Franklin Isherwood, Naval Engineer: The Years as Engineer in Chief, 1861–1869* (Annapolis, MD: Naval Institute Press, 1965), 105–18.

93. H.R. Rep. No. 184, 18–19.

Chapter 3 · The Economy of Time and Space

1. *Cong. Globe*, January 25, 1849, 291–92, 359; H.R. Rep. No. 275, 30th Cong., 1st sess. (1848).

2. David Henshaw to John Percival January 22, 1844, in K. Jack Bauer, ed., *Diplomatic Activities*, vol. 3 of *The New American State Papers: Naval Affairs* (Wilmington, DE: Scholarly Resources, 1981), 375–76. For background on the voyage, see Tyrone G. Martin, *A Most Fortunate Ship: A Narrative History of Old Ironsides*, rev. ed. (Annapolis, MD: Naval Institute Press, 1997), 266–68; David Foster Long, *"Mad Jack": The Biography of Captain John Percival, USN, 1779–1862* (Westport, CT: Greenwood Press, 1993), 132–40, and James H. Ellis, *Mad Jack Percival: Legend of the Old Navy* (Annapolis, MD: Naval Institute Press, 2002), 152–86.

3. Timothy Jenkins to John Percival, March 16, 1844, enclosed to Thomas Butler King, roll 2, TBKP; a Citizen of Boston, "To the Editor of the Transcript," *Boston Evening Transcript*, February 13, 1844, 2; "Naval," *Boston Daily Atlas*, February 17, 1844, 2; Citizen, "East India Trade," *Salem (VA) Register*, February 19, 1844, 2; "The New Minister to Brazil," *Charleston (SC) Southern Patriot*, February 19, 1844, 2.

4. At least, that was what Bright was reported as saying in the *Congressional Globe*. In a discussion several days later on changing the contract for reporting floor debates, he noted that while he had been quoted as taunting about "possum catching," he had in fact said "fossil hunting" (*Cong. Globe*, January 18, 1849, 291, and January 30, 1849, 395).

5. *Cong. Globe*, January 25, 1849, 359–60.

6. *Cong. Globe*, January 25, 1849, 360–61; H.R. Rep. No. 275. Support for the relief of Percival was bipartisan but unequally divided: nearly two-thirds of the votes for the original appropriation came from Whigs, while of the thirteen votes against, all but one came from Democrats.

7. Wolfgang Schivelbusch, *The Railway Journey: The Industrialization of Time and Space in the 19th Century* (Berkeley: University of California Press, 1986), 33–44; William Cronon, *Nature's Metropolis: Chicago and the Great West* (New York: Norton, 1991), ch. 5; Richard R. John, *Spreading the News: The American Postal System from Franklin to Morse* (Cambridge, MA: Harvard University Press, 1995), 10. For a more critical view of the transformation of the perception of space framed around costs, see Richard White, *Railroaded: The Transcontinentals and the Making of Modern America* (New York: Norton, 2011), ch. 4.

8. Francis Lieber, ed., *Encyclopaedia Americana* (Philadelphia: Lea and Blanchard, 1845), s.v. "machinery."

9. "Annual Report of the Secretary of the Treasury," *Baltimore Sun*, December 12, 1848, 1. A glance at the Google Books Ngram Viewer also shows the phrases "economy of time" and "economy of space" were much more often found in print than the "annihilation" of either. For examples of positive references to the economy of time and the economy of space, see "Steam Navigation to India," *Athenaeum*, July 13, 1839, 516, "Memorial and Remonstrance of the Directors of the Boston and Worcester Railroad," *Boston Evening Transcript*, March 19, 1847, 4, "Economy of Time," *Maine Cultivator and Hallowell Gazette*, April 5, 1845, 1, "Pacific Railroad," *Boston Courier*, November 1, 1849, 1, "Passenger Railroads in Philadelphia," *Philadelphia Public Ledger*, September 16, 1853, 2, "The Atlantic and Pacific Oceans—Interesting and Important Statistics," *San Francisco Daily Placer Times and Transcript*, November 21, 1855, 1, and "Suez Canal," *Boston Press and Post*, November 13, 1856, 1.

10. Bob Reece, "Two Missionaries in Brunei in 1837: George Tradescant Lay and the Revd J. T. Dickinson," *Sarawak Museum Journal*, n.s., 57, no. 78 (2002): 179–204; William Jackson Hooker and G. A. Walker Arnott, *The Botany of Captain Beechey's Voyage* (London: Henry G. Bohn, 1841); Frederick William Beechey et al., *The Zoology of Captain Beechey's Voyage* (London: Henry G. Bohn, 1839); Frederick William Beechey and John Tuckerman Tower, *Narrative of a Voyage to the Pacific and Beering's Strait* (London: Colburn and Bentley, 1831).

11. George Tradescant Lay, *Notes Made During the Voyage of the "Himmaleh,"* vol. 2 of *The Claims of Japan and Malaysia upon Christendom, Exhibited in Notes of Voyages Made in 1837, from Canton, in the Ship "Morrison" and Brig "Himmaleh"* (New York: French, 1839), 138–39. The modern spelling is "Kianngeh."

12. John Niven, ed., *The Salmon P. Chase Papers*, vol. 1 (Kent: Kent State University Press, 1993), 149; see also Richard C. Taylor, "On the Anthracite and Bituminous Coal Fields in China," *Journal of the Franklin Institute*, 3rd ser., 10, no. 1 (1845): 56.

13. "Borneo Coal and Mineral Resources of India," *Singapore Free Press*, September 15, 1842.

14. Rodney Mundy and James Brooke, *Narrative of Events in Borneo and Celebes, Down to the Occupation of Labuan*, 2 vols., 2nd ed. (London: John Murray, 1848), 2:21, 27–28, 342–46. See also Nicholas Tarling, *Britain, the Brookes, and Brunei* (New York: Oxford University Press, 1971).

15. Mundy and Brooke, *Narrative of Events in Borneo and Celebes*, 1:338–39.

16. James Brooke to John C. Templer, December 31, 1844, in James Brooke, *The Private Letters of Sir James Brooke, KCB, Rajah of Sarawak, Narrating the Events of His Life, from 1838 to the Present Time*, 3 vols., ed. John C. Templer (London: Richard Bentley, 1853), 2:43.

17. Mundy and Brooke, *Narrative of Events in Borneo*, 2:342, 349, 46–68. Playfair did note, however, that this sample might not represent the genuine quality of Labuan coal, since it had been removed from the exposed surface (rather than mined from underground) and then transported eight thousand miles to England under harsh conditions, including "upon the back of a camel from Suez across the desert." Subsequent tests would indeed find Labuan coal with even higher carbon and lower ash content than the original experiments (Wemyss Reid and Lyon Playfair, *Memoirs and Correspondence of Lyon Playfair* [New York: Harper and Brothers, 1899], 95).

18. James Brooke to John C. Templer, April 4, 1845, in Brooke, *The Private Letters of Sir James Brooke*, 2:57–58.

19. Frederick E. Forbes, *Five Years in China, from 1842 to 1847* (London: Richard Bentley, 1848), 310–11; Mundy and Brooke, *Narrative of Events in Borneo*, 2:349. Following a subsequent discovery of coal on the Borneo mainland, Mundy anticipated that the diligent efforts of Brooke and his naval associates would reduce the cost of coal at Singapore by at least three-quarters and open "this rich and magnificent country . . . to the commercial enterprise of Great Britain" (Mundy and Brooke, *Narrative of Events in Borneo*, 2:176, 335).

20. "A New British Settlement," *Singapore Free Press*, October 10, 1845. The article was originally published in *The Atlas (for India, China and the Colonies)* on August 2. A gap in this chain existed between Galle and Singapore; the authors of this article anticipated filling it with a station on either the Andaman or Nicobar islands—preferably the latter because a Danish expedition there had discovered coal. See also "The Island of Lobuan," *Singapore Free Press*, August 13, 1846.

21. David Henshaw to Percival January 22, 1844, in *Diplomatic Activities*, 375; Long, *"Mad Jack,"* 136.

22. Henry George Thomas and Alan B. Flanders, *Around the World in Old Ironsides: The Voyage of USS Constitution, 1844–1846* (Lively, VA: Brandylane Publishers, 1993), 80–81.

23. Benjamin F. Stevens, "A Cruise on the Constitution," *United Service*, 3rd ser., 5, no. 5 (1904): 546; William C. Chaplin and John Percival to George Bancroft, April 9, 1845, roll 321, Captains' Letters. The microfilm is difficult to read, and "deposited" represents my best guess of the word used in this sentence.

24. William C. Chaplin and John Percival to George Bancroft, April 9, 1845, roll 321, Captains' Letters. Though he was correct in believing that Omar Ali was unprepared to negotiate with him, Chaplin was, in fact, mistaken in concluding that Captain Bethune had actually completed a commercial treaty with the sultan. Britain and Brunei would not sign a commercial treaty until May 27, 1847, and only then formalize Brooke's hoped-for claim on Labuan (Edward Hertslet, comp., *General Index Arranged in Order of Countries and Subjects, to Hertslet's Commercial Treaties* [London: Butterworth's, 1885], 58, 249; *British and Foreign State Papers, 1846–1847*, [London: Harrison and Sons, 1860], 14–17).

25. J. C. Reinhardt, "Report of J. C. Reinhardt, Naturalist," *Fourth Bulletin of the National Institute for the Promotion of Science* (Washington, DC: William Q. Force, 1846), 534, 549–50.

26. Mundy and Brooke, *Narrative of Events in Borneo*, 2:33–34; James Brooke to J. E. B Stuart, July 4, 1845, in Brooke, *The Private Letters of Sir James Brooke*, 2:77.

27. Mundy and Brooke, *Narrative of Events in Borneo*, 2:22; James Brooke to John C. Templer, May 22, 1845, in Brooke, *The Private Letters of Sir James Brooke*, 2:65–67.

28. Mundy and Brooke, *Narrative of Events in Borneo*, 2:295–96, 335.

29. Forbes, *Five Years in China*, 322–23; Horace Stebbing Roscoe St. John, *The Indian Archipelago: Its History and Present State*, 2 vols. (London: Longman Brown Green and Longmans, 1853), 2:349–50; Hugh Low, *Sarawak: Its Inhabitants and Productions: Being Notes During a Residence in That Country with H. H. The Rajah Brooke* (London: Richard Bentley, 1848), 13–4, 16; Robert Fisher, "The Coal-Fields of Labuan, Borneo," *Transactions of the Federated Institution of Mining Engineers* 7 (1893–94): 590.

30. Lay, *Notes Made During the Voyage of the "Himmaleh,"* vii–xi; Stevens, "A Cruise on the Constitution," 543–54; for Balestier's early work as a commissions agent in Singapore, see Joseph Balestier, circular letter, December 16, 1833, MHS.

31. John Clayton to Joseph Balestier August 16, 1849, in roll 152, DIDS.

32. Entries for December 23, 1849, and February 21, May 26, and June 6, 1850, George P. Welsh, "Journal Aboard the U.S.S. Plymouth, 1848–1851," box 3, GPWP.

33. Joseph Balestier to John M. Clayton June 24, 1850, roll 9, DSADS.

34. Kang Chao, "The Chinese-American Cotton-Textile Trade, 1830–1930," in *America's China Trade in Historical Perspective: The Chinese and American Performance*, ed. Ernest R. May and John K. Fairbank (Cambridge, MA: Committee on American-East Asian Relations of the Department of History in collaboration with the Council on East Asian Studies Harvard University, 1986), 105.

35. H.R. Ex. Doc. No. 13, 29th Cong., 1st sess. (1845); H.R. Ex. Doc. No. 147, 34th Cong., 1st sess. (1855); S. Ex. Doc. No. 20, 36th Cong., 2nd sess. (1860). For cotton's role in the world economy, see Sven Beckert, "Emancipation and Empire: Reconstructing the Worldwide Web of Cotton Production in the Age of the American Civil War," *American Historical Review* 109, no. 5 (2004): 1405–38.

36. Webster, evidently dissatisfied with reports of Balestier's atrocious conduct in Siam, expressed himself in the circumspect but admonishing language of diplomacy: "It is, by no means, necessary for you to return to the United States, merely to be the bearer of the unimportant Convention, which you announce as having been concluded by yourself with the Sultan of Brunei" (Daniel Webster to Joseph Balestier February 15, 1851, roll 152, DIDS).

37. H.R. Rep. No. 34, 31st Cong., 2nd sess. (1851), 33. The literature on the United States and the opening of Japan is vast. Essential works include Peter Booth Wiley, *Yankees in the Land of the Gods: Commodore Perry and the Opening of Japan* (New York: Penguin, 1990), William McOmie, *The Opening of Japan, 1853–1855* (Folkestone, UK: Global Oriental, 2006), William L. Neumann, "Religion, Morality, and Freedom: The Ideological Background of the Perry Expedition," *Pacific Historical Review* 23, no. 2 (1954): 247–57, and the primary sources collected in *The Perry Mission to Japan*, 8 vols. (Surrey, UK: Japan Library, 2002).

38. H.R. Rep. No. 596, 30th Cong., 1st sess. (1848).

39. Ambrose W. Thompson, memorial, in SEN 31A-H12.1, SCNA; Pennsylvania State Senate and House of Representatives, resolutions, HR 31A-G12.7, HCNA; H.R. Rep. No. 34, 6; Ambrose W. Thompson, *Letter to the Hon. Fred. P. Stanton* (Philadelphia: B. Franklin Jackson, 1852).

40. Webster was sensitive to the long-standing reclusiveness of Japan and was merely seeking its minimal help to facilitate American steamers reaching China rather than a more substantial "opening" of the country itself to the wider world of commerce. Accordingly, in his early instructions to Perry's predecessor, John Aulick, Webster allowed that if the Japanese preferred that American ships not land, it would suffice if the Japanese government would send its own ships to deposit coal on a nearby island and for Americans to refuel there, "avoiding thus the necessity of an intercourse with any large number of the people of the country" (Daniel Webster to John H. Aulick, June 10, 1851, in Charles M. Wiltse, ed., *Papers of Daniel Webster, Diplomatic Papers*, vol. 2 [Hanover, NH: University Press of New England, 1988], 290–91).

41. John H. Schroeder, *Matthew Calbraith Perry: Antebellum Sailor and Diplomat* (Annapolis, MD: Naval Institute Press, 2001), 86, 123–53; Samuel Eliot Morison, *"Old Bruin," Commodore Matthew C. Perry, 1794–1858* (Boston: Little, Brown, 1967), 179–251.

42. William Skiddy to William Ballard Preston, September 29, 1849, Matthew Perry, William D. Salter, and Charles W. Copeland to William Ballard Preston, November 22,

1849, and Matthew Perry to William Ballard Preston, December 18, 1849, roll 345, Captains' Letters; Matthew Perry to William Ballard Preston, July 11, 1849, roll 346, Captains' Letters; Matthew Perry to Willard Ballard Preston, April 2, 1850, roll 347, Captains' Letters. The invention is described in William H. Lindsay, "Fluid-Meter," U.S. Patent 6,130, issued February 20, 1849. On the perils of the *Teredo navalis* shipworm and the benefits of copper sheathing, see Robert Martello, *Midnight Ride, Industrial Dawn: Paul Revere and the Growth of American Enterprise* (Baltimore, MD: Johns Hopkins University Press, 2010), 219–24. The New York agent for the line explained to the secretary of the navy, William Ballard Preston, that George Law deliberately chose to postpone coppering until after running these ships once or twice up the Mississippi River to allow the fresh water to remove barnacles; see Marshall O. Roberts to William Ballard Preston, September 17, 1849, vol. 1, Roberts Letter Book.

43. Matthew Perry to William A. Graham, January 27, 1851, in J. G. de Roulhac Hamilton, ed., *Papers of William A. Graham*, 8 vols. (Raleigh, NC: State Department of Archives and History, 1961), 4:16–22; *Cong. Globe*, April 1, 1852, 943.

44. Beverley Tucker to William A. Graham, April 18, 1852, roll 271, M124; Thomas M. Blount to William A. Graham, November 24, 1852, in Hamilton, *Papers of William A. Graham*, 4:434–36; S. Ex. Doc. No. 2, 33rd Cong., special sess. (1853).

45. *Cong. Globe*, May 17, 1852, 1373. As an added indignity, the regular anthracite agent, Benjamin Springer, found that after rejecting several coal shipments from Howland and Aspinwall, the navy began paying him only 5 percent of the value of the coal itself, rather than the gross cost (which included the high costs of shipment, insurance, and so on). The difference was between the 5 percent of $34,448.85 that he was paid and the 5 percent of $131,121.64 that he was not. After several months of protests, he was fired from the position altogether (Thomas M. Blount to William A. Graham, November 24, 1852; S. Misc. Doc. No. 144, 35th Cong., 1st sess. [1858]).

46. "Establishing Relations with Japan," *Trenton (NJ) State Gazette*, February 3, 1852, 2.

47. "The Invasion of Japan," *Cleveland (OH) Plain Dealer*, March 5, 1852, 2.

48. "The Invasion of Japan," *Pennsylvania Freeman* (Philadelphia), March 18, 1852, 45.

49. "Intervention for Robery [*sic*] and Plunder," *Anti-Slavery Bugle*, March 13, 1852.

50. "The Japanese Expedition," *North American and United States Gazette*, February 7, 1852, 2; "The Japanese Expedition," *Boston Evening Transcript*, February 11, 1852, 2. Concern over the consequences of a successful mission did not immediately fade. In June 1852, when reports surfaced of the discovery of a gigantic mine in Kentucky whose coal burned fast and hot, a Michigan newspaper wryly noted that "its quality is supposed to be far superior to the undiscovered coal mines of Japan, in search of which the Administration has sent a large naval force" (*Northern Islander* [St. James, MI], June 17, 1852, 3).

51. "The Events of the Week," *New York Herald*, April 3, 1852, 108.

52. "Island of Formosa," *Daily Alta California*, May 1, 1851, 2.

53. Among otherwise outstanding accounts of the opening of Japan, see, for example, Wiley, *Yankees in the Land of the Gods*, 92–93, and McOmie, *The Opening of Japan*, 57.

54. H.R. Rep. No. 34, 19–33.

55. Daniel Ammen to William Gwin, January 10, 1852, SEN 32A-E8, SCNA.

56. Ibid; see also John H. Schroeder, "Matthew Calbraith Perry: Antebellum Precursor of the Steam Navy," in *Quarterdeck and Bridge: Two Centuries of American Naval Leaders*, ed. James Bradford (Annapolis, MD: Naval Institute Press, 1997), 116.

57. Matthew Perry, *The Japan Expedition, 1852–1854: The Personal Journal of Commodore Matthew C. Perry* (Washington, DC: Smithsonian Institution Press, 1968), 3.

58. H.R. Rep. No. 184, 35th Cong., 2nd sess. (1859), 33–37.

59. Perry, *The Japan Expedition*, 20, 35.

60. Matthew Perry to William A. Graham, June 12, 1852, roll 350, Captains' Letters. Long-term arrangements with the P&O for coal were not a success; in Ceylon, the British maintained a large supply of coal, but the frequent arrival of steamships, some ten a month, led the P&O, which oversaw the supply, to forbid the sale of British coal to foreign naval ships. Perry could only scrounge a meager supply from the local Bengali government (Perry, *The Japan Expedition*, 38).

61. S. Ex Doc. No. 79, 33rd Cong., 2nd sess. (1856), 1:129.

62. Perry, *The Japan Expedition*, 51.

63. Arthur Walworth, *Black Ships Off Japan: The Story of Commodore Perry's Expedition* (New York: Knopf, 1946), 127–28.

64. S. Ex. Doc. No. 34, 33rd Cong, 2nd sess. (1855), 142.

65. *New York Tribune*, May 2, 1854, 4.

66. S. Ex. Doc. No. 79, 2:59.

67. Ibid., 2:53–54.

68. Ibid., 2:137–38, 42.

69. Ibid., 2:153. What Jones perceived as deception and misdirection was likely a consequence of fear and precaution, as was documented elsewhere in Perry's expedition. The account of a grocer and local government head from Hakodate, Japan, about Americans landing there in 1854 described official edicts warning villagers of the danger Americans posed, providing instructions to hide food, women, and children (Matajiro Kojima, *Commodore Perry's Expedition to Hakodate*, in *The Perry Mission to Japan*, 8:3–11).

70. S. Ex Doc. No. 79, 2:156–57.

71. Ibid. The official whom Jones called the "hip-toy" was likely the "hip t'oi," or 協臺, which translates to a Chinese deputy lieutenant general; see Ernest John Eitel, *A Chinese Dictionary in the Cantonese Dialect* (London: Trübner, 1877), 158. I thank Jia-Chen Fu and Lane Harris for this reference.

72. S. Ex. Doc. No. 79, 2:157–59.

73. Breese, quoted in John Glendy Sproston, *A Private Journal of John Glendy Sproston, USN*, ed. Shiho Sakanishi (Tokyo: Sophia University, 1968), 70.

74. Ibid, 70–72; S. Ex Doc. No. 79, 2:159, 1:178–82.

75. Matthew Perry to John P. Kennedy, October 18, 1853, roll 16, JPKP.

76. S. Ex. Doc. No. 34, 112–13, 153–55.

77. Matthew Perry, *A Paper by Commodore M. C. Perry, USN, Read before the American Geographical and Statistical Society, at a Meeting Held March 6th, 1856* (New York: Appleton, 1856), 7–8.

78. Ibid., 12.

Chapter 4 · The Slavery Solution

1. "Address on Colonization to a Deputation of Negroes," in *Collected Works of Abraham Lincoln*, 9 vols., ed. Roy P. Basler (New Brunswick, NJ: Rutgers University Press, 1953–55), 5:370–75; Kate Masur, "The African American Delegation to Abraham Lincoln: A Reappraisal," *Civil War History* 56, no. 2 (2010): 117–44; Eric Foner, "Abraham Lincoln, Coloni-

zation, and the Rights of Black Americans," in Walter Johnson, Eric Foner, and Richard Follett, *Slavery's Ghost: The Problem of Freedom in the Age of Emancipation* (Baltimore, MD: Johns Hopkins University Press, 2011), 43.

2. "Address on Colonization to a Deputation of Negroes," 373–74; Samuel C. Pomeroy to Abraham Lincoln, April 16, 1863, folder 2, CIC-AL.

3. Warren A. Beck, "Lincoln and Negro Colonization in Central America," *Abraham Lincoln Quarterly* 6, no. 3 (1950): 162–83; Gabor S. Boritt, "The Voyage to the Colony of Linconia: The Sixteenth President, Black Colonization, and the Defense Mechanism of Avoidance," *Historian* 37, no. 4 (1975): 619–32; Michael Vorenberg, "Abraham Lincoln and the Politics of Black Colonization," *Journal of the Abraham Lincoln Association* 14, no. 2 (1993): 22–45. See also William W. Freehling, " 'Absurd' Issues and the Causes of the Civil War: Colonization as a Test Case," in his *The Reintegration of American History: Slavery and the Civil War* (New York: Oxford, 1994), 138–57, Paul J. Scheips, "Lincoln and the Chiriqui Colonization Project," *Journal of Negro History* 37, no. 4 (1952): 418–53, Paul J. Scheips, "Ambrose W. Thompson: A Neglected Isthmian Promoter" (masters thesis, University of Chicago, 1949), and Sebastian N. Page, "Lincoln and Chiriquí Colonization Revisited," *American Nineteenth Century History* 12, no. 3 (2011): 289–325. On the other contemporary colonization project in Haiti, see Willis D. Boyd, "James Redpath and American Negro Colonization in Haiti, 1860–1862," *The Americas* 12, no. 2 (1955): 169–82, Willis D. Boyd, "The Île a Vache Colonization Venture, 1862–1864" *The Americas* 16, no. 1 (1959): 45–62, and James D. Lockett, "Abraham Lincoln and Colonization: An Episode That Ends in Tragedy at L'Ile à Vache, Haiti, 1863–1864," *Journal of Black Studies* 21, no. 4 (1991): 428–44.

4. Jacob Collamer to Ambrose W. Thompson, April 6, 1849, 1849–51 folder, box 4, RWT; Ambrose W. Thompson, memorials, SEN 31A-H12.1, SCNA; Pennsylvania General Assembly, resolution (P.L. 735, No. 4); Ambrose W. Thompson obituary, *Philadelphia Inquirer*, May 30, 1882, 2; Ambrose W. Thompson, "Propeller," U.S. Patent 7,907, issued January 21, 1851; citizens of Albany, NY, memorial, June 9, 1852, HR 32A-G15.3, HCPO; Ambrose W. Thompson, petition, January 12, 1852, HR 32A-G13.3, HCNA.

5. Jorge Enrique Illueca, "A Socioeconomic History of the Panamanian Province of Chiriquí: A Case Study of Development Problems in the Humid Neotropics" (PhD diss., UCLA, 1983), 16, 88–89, 439; "Report by Robert McDowall, M.D., residing at David, Province of Chiriqui," *Mining Magazine* 8, no. 3 (1857): 255; Chiriquí expedition notes (undated, untitled, and unsigned but probably written by Newton S. Manross ca. 1860), folder 10, box 3, RWT.

6. Aims McGuiness, *Path of Empire: Panama and the California Gold Rush* (Ithaca, NY: Cornell University Press, 2008), 29–31; John Haskell Kemble, *The Panama Route, 1848–1869* (Columbia: University of South Carolina Press, 1990), ix–x, 1–57; Paul J. Scheips, "Gabriel Lafond and Ambrose W. Thompson: Neglected Isthmian Promoters," *Hispanic American Historical Review* 36, no. 2 (1956), 214–16.

7. "Private Notes for Dr Clark=being a retrospective history of the Chiriqui matters," Descriptions of Chiriqui folder, box 3, RWT; J. Eugene Flandin to Ambrose W. Thompson, October 19, 1857, 1857 folder, box 4, RWT.

8. Ibid; Scheips, "Gabriel Lafond and Ambrose W. Thompson," 211–19; Thomas D. Schoonover, *The French in Central America: Culture and Commerce, 1820–1930* (Wilmington, DE: Scholarly Resources, 2000), 30–31; Gabriel Lafond de Lurcy, *Notice sur le Golfo Dulce, dans l'état de Costa Rica* (Paris: Chez Fontaine, 1856), 45–46; "The Secretary of Foreign

Relations of Costa Rica to the Governor of the Province of Chiriquí, Republic of New Granada," in *Documents Annexed to the Argument of Costa Rica before the Arbitrator Hon. Edward Douglass White, Chief Justice of the United States*, 4 vols. (Rosslyn, VA: Commonwealth Company, 1913), 2:101–2. On the challenges of development and inculcating expertise in Colombia, see Frank Safford, *The Ideal of the Practical: Colombia's Struggle to Form a Technical Elite* (Austin: University of Texas Press, 1976).

9. La Barrière to Victor Herrán, September 13, 1850, in James Silk Buckingham, ed., *Colonisation of Costa Rica* (London: Effingham Wilson, 1852), 27–28; H.R. Rep. No. 568, 36th Cong., 1st sess. (1860), 75, 78–79.

10. After William Walker's filibustering expedition to Nicaragua a few years later, the legislature of Chiriquí attempted to induce Cornelius Vanderbilt to abandon his investments there and ally instead further south with Thompson's Chiriqui Improvement Company. "That the peaceable and industrious inhabitants of 'Chiriqui' instead of being in any manner opposed to the characteristic spirit of civilization and enterprise of the nearby Sons of 'Washington,' have a vehement desire to see themselves regenerated by them," they wrote; see James D. Bowie to Cornelius Vanderbilt, September 27, 1856, 1856 folder, box 4, RWT, and John Whiting to Ambrose W. Thompson, November 30, 1852, November–December 1852 folder, box 4, RWT.

11. Robert MacDowall to Ambrose W. Thompson, May 2, 1852, John Whiting to Ambrose W. Thompson, September 15, 1852, and Ambrose W. Thompson to the legislature of the province of Chiriqui," January–September 1852 folder, box 4, RWT; "Private Notes for Dr Clark=being a retrospective history of the Chiriqui matters," Descriptions of Chiriqui folder, box 3, RWT; "Chiriqui Grants: Letter, to the President of the United States," Chiriqui Improvement Company folder, pamphlets, box 3, RWT.

12. John Whiting to Ambrose W. Thompson, September 15, 1852, January–September 1852 folder, box 4, RWT.

13. Scheips, "Ambrose W. Thompson," 27–30; H.R. Rep. No. 568, 8–32; "The Chiriqui Improvement Company and Ambrose W. Thompson: Abstract of Titles," 1–6, box 45, AWT. Though the Spanish spelling of "Chiriquí" includes the acute diacritic on the final letter, I have followed the spelling of sources by leaving it off in direct quotations and in the name of Thompson's company.

14. John Whiting to Ambrose W. Thompson, September 15, 1852, January–September 1852 folder, box 4, RWT; Ambrose W. Thompson to Stephen R. Crawford, October 16, 1852, October 1852 folder, box 4, RWT; unsigned (Ambrose W. Thompson?) to James C. Dobbin, September 21, 1853, and J. Eugene Flandin to Esteban F. Cordero, November 4, 1853, June–December 1853, folder, box 4, RWT; J. Eugene Flandin to Justo Arosemena, January 27, 1854, 1854 folder, box 4, RWT.

15. James Cook, report (copy), February 1, 1853, January–May 1853 folder, box 4, RWT; *Chiriqui Improvement Company: Geological Report of Professor Manross, with Accompanying Papers, Maps, &c.* (New York: George F. Nesbitt, 1856), 3–18, 26; J. A. Moore to Joseph Foulke, October 6, 1856, 1856 folder, box 4, RWT; "C. S. Richardson, Chiriqui Improvement Company—Report," *Mining Magazine* 8, no. 4 (1857): 348–55; "Chiriqui Improvement Company," *Panama Star and Herald* (Panama City), June 26, 1855, 2; H.R. Rep. No. 568, 36.

16. J. W. King to J. D. B. Curtis, September 9, 1864, Consul of London's Paper Pertaining to Panama folder, box 3, RWT; Timothy Hough to Aaron V. Brown, August 7, 1857,

and J. A. Morel to Joseph Foulke, November 27 and December 6, 1857, 1857 folder, box 4, RWT; *Report of Capt. Almy, U.S.N.* (Washington, DC: n.p., 1858), 4.

17. Ambrose W. Thompson to Aaron V. Brown, December 20, 1857, 1857 folder, box 4, RWT; Ambrose W. Thompson to Aaron V. Brown, June 18, 1858, 1858 folder, box 4, RWT; "The Belly Intrigues in Central America—Impudent Declaration of Presidents Mora and Martinez," *New York Herald*, June 18, 1858, 4; *Memoir Relating to the Isthmus Crossing at the Chiriqui Lagoon Addressed to the President of the United States, June 9, 1858* (Washington, DC: Polkinhorn's Steam Printing Office, 1858), 11.

18. Timothy Hough to Aaron V. Brown, August 7, 1857, Ambrose W. Thompson to Aaron V. Brown, December 20, 1857, Ambrose W. Thompson to William A. Harris, February 5, 1858, J. Eugene Flandin to Ambrose W. Thompson, March 11, 1858, and Ambrose W. Thompson to Lewis Cass, June 24, 1858, 1858 folder, box 4, RWT; Ambrose W. Thompson to Lewis Cass, July 1, 1858, and Richard W. Thompson to Ambrose W. Thompson, July 1, 1858, box 6, AWT; *Memoir Relating to the Isthmus Crossing at the Chiriqui Lagoon*, 12; *Second Memoir Relating to the Isthmus Crossing at the Chiriqui Lagoon Addressed to the President of the United States, June 9, 1858* (Washington, DC: Polkinhorn's Steam Printing Office, 1858); "Quasi Ratification of the Cass-Herran Treaty," *New York Times*, July 28, 1858, 4. The Conservative government of Mariano Ospina did, in fact, investigate the prospects of an American annexation, but American racial politics made at least some officials wary of the consequences of annexation given the diverse population of the country; see McGuiness, *Path of Empire*, 179–80.

19. Ambrose W. Thompson to William A. Harris, October 8 and October 19, 1857, 1857 folder, box 4, RWT; William A. Harris to Aaron V. Brown, July 2, 1858, 1858 folder, box 4, RWT.

20. At least as of 1862, Thompson was to receive 20 percent of the profits of the venture; see agreement with Richard W. Thompson, September 12, 1862, September 1862 folder, box 4, RWT, and Ambrose W. Thompson to Richard W. Thompson, September 21, 1858, and Ambrose W. Thompson to Reverdy Johnson, September 21, 1858, 1858 folder, box 4, RWT.

21. Francisco Párraga to Ambrose W. Thompson, November 28, 1859, August–December 1859 folder, box 4, RWT; Ambrose W. Thompson to Amalia Herrán, March 24, 1859, and Ambrose W. Thompson to Pedro Alcántra Herrán, April 4, 1859, January–July 1859 folder, box 4, RWT; H.R. Rep. No. 568, 32–35. This and subsequent quotations from Francisco Párraga are from English translations prepared for Thompson from Párraga's Spanish originals; for most letters, both copies can be found in the R. W. Thompson collection, and the translations appear accurate.

22. Ambrose W. Thompson to Isaac Toucey, November 4, 1858, Ambrose W. Thompson to Aaron V. Brown, November 13, 1858, and Ambrose W. Thompson to Aaron V. Brown, November 20, 1858, 1858 folder, box 4, RWT; secretary of navy, draft letter to the Senate Committee on Finance, Jeremiah S. Black to Isaac Toucey, March 14 and May 11, 1859, and James S. Green to Isaac Toucey, May 12, 1859, January–July 1859 folder, box 4, RWT. A copy of the contract may be found in H.R. Misc. Doc. No. 239, 51st Cong., 1st sess. (1891), 19:51–53.

23. "Our Navy: Annual Report of the Secretary of the Navy for the Year 1859," *New York Herald*, December 28, 1859, 3–4; see also "Report of the Secretary of the Navy," *Philadelphia Inquirer*, December 28, 1859, 4, "Reports from the Departments," *Boston Evening Transcript*,

December 28, 1859, 4, S. Ex. Doc. No. 2/15, 36th Cong., 1st sess. (1859), 1151–52, and Francisco Párraga to Ambrose W. Thompson, March 12, 1860, January–March 1860 folder, box 4, RWT.

24. Jones expressed his displeasure by venting to Thompson about the shortsightedness of the legislature; Párraga took a different tack, confronting Nuñez for spreading lies about the Chiriquí project and its objectives. "I went in search of him," Párraga wrote Thompson, "provided with a good, strong horse whip: met him in the street, and having told him all I thought of him, I had the gratification to cut him three times across the face with my whip. I do not know what he will do; if he challenges me, I will whip him again." See Francisco Párraga to Ambrose W. Thompson, November 28 and December 12, 1859, August–December 1859 folder, box 4, RWT, Francisco Párraga to Ambrose W. Thompson, February 12, 1860, George W. Jones to Ambrose W. Thompson, February 25, 1860, and Francisco Párraga to Ambrose W. Thompson, February 28, 1860, January–March 1860 folder, box 4, RWT, and George W. Jones to Lewis Cass, January 28, 1860, George W. Jones to J. A. Pardo, January 22, 1859, and George W. Jones to Lewis Cass, February 12, February 28, March 12, and March 26, 1860, roll 14, T33. It is plausible that Jones, too, was a "friend" of Thompson, but the nature of their relationship is ambiguous from their correspondence; see George W. Jones to Ambrose W. Thompson, February 11, 1860, January–March 1860 folder, box 4, RWT. On "friends" and corruption in the United States, see Richard White, *Railroaded: The Transcontinentals and the Making of Modern America* (New York: Norton, 2011), 93–133, and Mark W. Summers, *The Plundering Generation: Corruption and the Crisis of the Union, 1849–1861* (New York: Oxford, 1987).

25. Francisco Párraga to Ambrose W. Thompson, March 12 and March 26, 1860, January–March 1860 folder, box 4, RWT; George W. Jones to Lewis Cass, March 12, 1860, roll 14, T33.

26. Thomas Francis Meagher to Ambrose W. Thompson, January 16 and February 2, 1860, and Jesús Jiménez to Thomas Francis Meagher, February 24, 1860, January–March 1860 folder, box 4, RWT; "The Latest News: Received by Magnetic Telegraph," *New York Tribune*, August 6, 1860, 8.

27. Luís Molina Bedoya to Ambrose W. Thompson, January 29, 1861, and Francisco Párraga to Ambrose W. Thompson, January 30, 1861, 1861 folder, box 4, RWT.

28. Isaac Toucey to James H. Hammond, June 12, 1860, April–June 1860 folder, box 4, RWT; *Senate Journal*, 36th Cong., 1st sess., June 18, 1860, 695, and June 21, 1860, 726, 729; *House Journal*, 36th Cong., 1st sess., June 21, 1860, 1182–83.

29. "The Chiriqui Surveying Expedition," *New Orleans Picayune*, October 18, 1860, 4; "Letter from Panama: The U.S. Expedition to Chiriqui," *New Orleans Picayune*, September 16, 1860, 1; "The Chiriqui Expedition," *New Orleans Picayune*, December 7, 1860, 8; John Evans, report, September 30, 1860, September–October 1860 folder, box 4, RWT; H.R. Ex. Doc. 41, 36th Cong., 2nd sess. (1861), 45–55; "Chiriqui Survey," *Baltimore Sun*, July 19 1860, 1; "By Telegraph for the Baltimore Sun," *Baltimore Sun*, July 26 1860, 1; "The American Navy: Annual Report of the Secretary of the Navy," *Philadelphia Inquirer*, December 6 1860, 2; Paul J. Scheips, "Buchanan and the Chiriqui Naval Station Sites," *Military Affairs* 18, no. 2 (1954): 75–79; Richard X. Evans, "Dr. John Evans, U.S. Geologist, 1851–1861," *Washington Historical Quarterly* 26, no. 2 (1935): 83–89.

30. "The Chiriqui Expedition," *Philadelphia Inquirer*, November 27, 1860, 3; "Result of the Chiriqui Commission," *Philadelphia Inquirer*, November 29, 1860, 4; "The Chiriqui Commission—Arrival of the Brooklyn," *Baltimore Sun*, November 29, 1860, 2; "Our Wash-

ington Letter," *Philadelphia Inquirer*, December 1, 1860, 4; "The Chiriqui Expedition," *Philadelphia Inquirer*, December 3, 1860, 4.

31. Ambrose W. Thompson to James Buchanan, October 18, 1860, September–October 1860 folder, box 4, RWT; *Senate Journal*, 36th Cong., 2nd sess., January 17, 1861, 109. Notably, future Lincoln cabinet members Simon Cameron and William Seward both voted for the measure.

32. *Cong. Globe*, January 31, 1861, 671–67; *Cong. Globe*, February 1, 1861, 692–65; *Cong. Globe*, February 2, 1861, 715–78, 729–35; *House Journal*, 36th Cong., 2nd sess., February 5, 1861, 270–71.

33. Francisco Párraga to Ambrose W. Thompson, February 28 and April 4, 1861, 1861 folder, box 4, RWT.

34. Francisco Párraga to Ambrose W. Thompson, February 28, 1861, 1861 folder, box 4, RWT.

35. Stephen C. Rowan to Gideon Welles, September 3, 1861, and Stephen C. Rowan to Silas H. Stringham, September 21, 1861, ORUCN 1, 6:160–61, 245–46.

36. H.R Rep. No. 184, 35th Cong., 2nd sess. (1859), 1–5, 57–59; "From Washington," *Ohio State Journal* (Columbus), June 2, 1858, 2; "From Philadelphia," *New York Tribune*, June 15, 1858, 6; "Navy Department Corruption" and "The Climax of Corruption," *New York Tribune*, February 25, 1859, 4, 6; "Naval Intelligence," *New Orleans Times-Picayune*, May 5, 1859, 1; "From Washington," *New York Tribune*, July 6, 1859, 5.

37. David D. Porter, *The Naval History of the Civil War* (New York: Sherman, 1886), 369; "Anthracite Coal for the Navy," *National Republican* (Washington, DC), July 10, 1861, 3. In Hatteras Inlet, Stephen Rowan complained that a bituminous fire could "be seen from one end of the sound to the other" (Stephen C. Rowan to Silas H. Stringham, September 13, 1861, ORUCN 1, 6:207; see also William G. Temple to Gideon Welles, December 21, 1861, ORUCN 1, 1:255, Samuel F. Du Pont to A. K. Hughes, March 1, 1863, ORUCN 1, 13:710, and G. H. Scott to Gideon Welles, August 21, 1861, ORUCN 1, 1:69–70). The risks of burning bituminous coal were not hypothetical; the commander of the CSS *Florida* in 1863, for example attributed his ship's detection by union vessels to its consumption of soft coal ("Extracts from the Journal of Lieutenant John N. Maffitt," ORUCN 1, 2:667).

38. Ambrose W. Thompson to Abraham Lincoln, April 11, 1861, ALP.

39. Francisco Párraga to Ambrose W. Thompson, February 13, 1862, January–May 1862 folder, box 4, RWT; Bushnell, *The Making of Modern Colombia: A Nation in Spite of Itself* (Berkeley: University of California Press, 1993), 119–20; Francis Adams to the directors of the Chiriqui Road Company, May 3, 1853, January–May 1853 folder, box 4, RWT; "Independencia del Estado: Acta de Chiriquí," *Panama Star and Herald* (Panama City), April 18, 1861, 2–3.

40. H.R. Rep. Doc. 568, 36th Cong. 1st sess. (1860), 10–13; Charles B. Sedgwick to Gideon Welles, August 7, 1861, box 7, AWT.

41. Ambrose W. Thompson to Gideon Welles, August 8, 1861, ALP; "Coal Memoranda," 1860 folder (misfiled), box 4, RWT; draft indenture between Ambrose W. Thompson and Gideon Welles, August, 1861, 1861 folder, box 4, RWT.

42. Ninian W. Edwards to Abraham Lincoln, August 9 and August 10, 1861, ALP.

43. Lincoln repeated these words during his first debate with Stephen Douglas in 1858 to counter Douglas's claim that he "was engaged at that time in selling out and Abolitionizing the old Whig party." See "Speech at Peoria, Illinois," in *Collected Works of Abraham*

Lincoln, 2:255–56, Harold Holzer, ed., *The Lincoln-Douglas Debates: The First Complete, Un-expurgated Text* (New York: Harper Collins, 1993), 61, "Annual Meeting of the American Colonization Society," *African Repository* 32, no. 2 (1856): 48, Foner, "Abraham Lincoln, Colonization, and the Rights of Black Americans," 39–40, Foner, *Free Soil, Free Labor, Free Men: The Ideology of the Republican Party Before the Civil War* (New York: Oxford University Press, 1970), 268–74, and Frank P. Blair Jr., "Colonization and Commerce: An Address Before the Young Men's Mercantile Library Association of Cincinnati, Ohio, Nov. 29, 1859," *National Era*, 13, no. 678 (1859): 208. On Bates, who heartily supported colonization but not colonization tied to coal, see Howard K. Beale, ed. *The Diary of Edward Bates, 1859–1866.* (Washington, DC: Government Printing Office, 1933), 113, 192, 262–64.

44. Schoonover, "Misconstrued Mission: Expansionism and Black Colonization in Mexico and Central America during the Civil War," *Pacific Historical Review* 49, no. 4 (1980): 611–13; Gideon Welles, "The History of Emancipation," *Galaxy* 14, no. 6 (1872): 849, Caleb B. Smith to Abraham Lincoln, November 4, 1861, box 7, AWT; Abraham Lincoln to Caleb B. Smith, October 23, 1861, ALP.

45. Caleb B. Smith to Abraham Lincoln, November 4, 1861, box 7, AWT; Gideon Welles, *Diary of Gideon Welles: Secretary of the Navy under Lincoln and Johnson*, 3 vols. (Boston: Houghton Mifflin, 1911), 1:151.

46. Francis P. Blair Sr. to Abraham Lincoln, November 16, 1861 with memoranda enclosed; Ambrose W. Thompson to Francis P. Blair Sr., November 17 and November 18, 1861, and Francis P. Blair Sr. to Simon Cameron, draft, December, 1861, ALP. Blow had found financial success manufacturing paints and processing white lead in St. Louis; before the war he branched into lead mining and smelting (C. R. Barnes, ed., *The Commonwealth of Missouri: A Centennial Record* [St. Louis: Bryan, Brand, 1877], 738–40).

47. An Act for the Release of Certain Persons held to Service or Labor in the District of Columbia, 12 *Stat.* 378; memorandum marked in pencil with "R.W.T.," January–May 1862 folder, box 4, RWT.

48. Caleb B. Smith to Ambrose W. Thompson, April 26, 1862, box 7, AWT.

49. Caleb B. Smith to Abraham Lincoln, May 9, 1862, box 7, AWT; Ambrose W. Thompson to Caleb B. Smith, May 14 and May 16, 1862, January–May 1862 folder, box 4, RWT.

50. Draft articles of agreement, June, 1862, June–August 1862 folder, box 4, RWT; John P. Usher to Abraham Lincoln, August 2, 1862, ALP. On Usher's involvement, see Page, "Lincoln and Chiriquí Colonization Revisited," 299–300, and John P. Usher to Richard W. Thompson, July 25, 1862, RWT 1, in which Usher strategizes with Thompson about persuading Lincoln to support the project, emphasizing the political advantages in Indiana of encouraging the removal of blacks and also the appeal of campaigning on using "the rebels money" to pay for it.

51. Samuel Whiting to Gideon Welles, December 16, 1861, ORUCN 1, 1:246; William Grenville Temple to Charles Rogers Nesbitt, December 17, 1861, ORUCN 1, 1:253; A. J. Thompson to William Grenville Temple, December 18, 1861, ORUCN 1, 1:254. Though Temple tried to track a blockade runner nearby, he was forced to abandon the islands altogether when his depleted coal bunkers compelled him to steam to Key West to refuel (William Grenville Temple to Gideon Welles, December 21, 1861, January 1, 1862, and January 7, 1862 [letter misdated 1861], ORUCN 1, 1:255, 1:268–70, 1:273). See also Kenneth J. Blume, "Coal and Diplomacy in the British Caribbean During the Civil War," *Civil War History* 41, no. 2 (1995): 116–41.

52. David D. Porter to Gideon Welles, August 23, 1861, ORUCN 1, 1:71–72; John Decamp to Gustavus Vasa Fox, January 8, 1862, ORUCN 1, 1:273–74.

53. Grace Palladino, *Another Civil War: Labor, Capital, and the State in the Anthracite Regions of Pennsylvania, 1840–68* (Urbana: University of Illinois Press, 1990), 95–120, quote on 99.

54. Joseph Henry to John Peter Lesley, May 28, 1862, and Joseph Henry to Alexander Dallas Bache, August 21, 1862, in Marc Rothenberg, Kathleen W. Dorman, and Frank R. Millikan, eds., *Papers of Joseph Henry*, 12 vols. (Washington, DC: Smithsonian Institution Press, 1972), 10:268–9, 278–81. On Henry's early life and career, see Thomas Coulson, *Joseph Henry* (Princeton: Princeton University Press, 1950), and Albert E. Moyer, *Joseph Henry: The Rise of an American Scientist* (Washington, DC: Smithsonian Institution Press, 1997).

55. Joseph Henry to Alexander Dallas Bache, August 21, 1862.

56. Joseph Henry to Frederick W. Seward, September 5, 1862; John Peter Lesley (unsigned) to Joseph Henry, September 5, 1862, ALP.

57. "From Washington," *New York Tribune*, August 26, 1862, 4–5; "The New Negro Colony in Central America," *Boston Evening Transcript*, August 26, 1862, 1; "Senator Pomeroy," *Hartford (CT) Daily Courant*, August 27, 1862, 2; "The Negro Colonization Scheme," *Farmer's Cabinet*, September 18, 1862, 2.

58. Welles, in fact, speculated that Pomeroy held a financial stake in the colonization plan, an idea tempered only by the continued support of Lincoln, postmaster general Montgomery Blair, his father, the power broker Francis P. Blair Sr., as well as "one or two men of integrity and character" (*Diary of Gideon Welles*, 1:123).

59. Schoonover, "Misconstrued Mission," 611–15; H.R. Ex. Doc. No. 1/2, 37th Cong., 3rd sess. (1862), 887–88, 892, 898, 903–4.

60. H.R. Ex. Doc. No. 1/2, 899–900; "The Colonization Scheme," *New York Tribune*, September 20, 1862, 7; "Northern News," *Macon (GA) Telegraph*, October 6, 1862, 4.

61. Francisco Párraga to Ambrose W. Thompson, November 17, 1861, 1861 folder, box 4, RWT; Francisco Párraga to Ambrose W. Thompson, January 4, 1862, January–May 1862 folder, box 4, RWT; Pedro Alcantára Herrán to Abraham Lincoln, June 14, 1862, Francisco Párraga to Samuel C. Pomeroy, August 26, 1862, and Ambrose W. Thompson to Francisco Párraga, August 30, 1862, June–August 1862 folder, box 4, RWT; Francisco Párraga to Ambrose W. Thompson, September 12, 1862, September 1862 folder, box 4, RWT.

62. S. Ex. Doc. No. 55, 39th Cong., 1st sess. (1866), 13–16.

63. Masur, "The African American Delegation to Abraham Lincoln"; "Letter from St. Louis," *San Francisco Evening Bulletin*, December 23, 1862, 1.

64. William H. Seward to Caleb B. Smith, October 13, 1862, C. S. Dyer to Ambrose W. Thompson, October 14, "Conflicting Claims of New Granada & Costa Rica to Lands in Chiriqui," and James Mitchell to Caleb B. Smith, October 14, 1862, October–December 1862 folder, box 4, RWT; H.R. Ex. Doc. No. 1/2, 202–4; S. Ex. Doc. No. 55, 55.

65. Page, "Lincoln and Chiriquí Colonization Revisited," 302; Ambrose W. Thompson to John P. Usher, April 10, 1863, 1863 folder, box 4, RWT; Boyd, "The Île a Vache Colonization Venture," 45–62.

66. James Mason Hoppin, *Life of Andrew Hull Foote, Rear-Admiral United States Navy* (New York: Harper and Brothers, 1874), 366; H.R. Ex. Doc. No. 1/15, 38th Cong., 1st sess. (1863), 761; Andrew Hull Foote to Garrett J. Pendergrast, October 29, 1862, and Andrew Hull Foote to H. Roland, November 10, 1862, vol. 1, Fair Copies of Letters Sent to Commandants

of Navy Yards and Stations, 1862–65, RG 24, NARA I; Andrew Hull Foote to David Glasgow Farragut, November 29, 1862, Andrew Hull Foote to Henry W. Morris, December 3, 1862, Andrew Hull Foote to James A. Doyle, February 17, 1863, and Andrew Hull Foote to David Glasgow Farragut, February 24 and March 19, 1863, vol. 1, Fair Copies of Letters Sent to Officers, RG 24, NARA I; Cornelius K. Stribling to Andrew Hull Foote, November 28, 1862, vol. 1, Letters Received from Commanders of Navy Yards, RG 24, NARA I. The army, too, needed to fuel its troop and supply carriers, and in fact it consumed even more coal than the navy. By 1865, the army owned or chartered 113 ocean steamers, 209 river and bay steamers, and 71 steam tugs (along with over 350 additional nonsteam vessels). Between the summers of 1863 and 1864 alone, the army shipped over 255,000 tons of anthracite from its own depot in Philadelphia—nearly double the navy's amount. The following year, the army's quartermaster purchased nearly 500,000 tons; see H.R. Ex. Doc. No. 83, 38th Cong., 2nd sess. (1865), 134–35, 143, and H.R. Ex. Doc. No. 1/20, 39th Cong., 1st sess. (1865), 296.

67. Darius N. Couch to Edward D. Townsend, November 13, 1863, ORUCN 1, 29.2.451; Palladino, *Another Civil War*, 140–65.

68. Darius N. Couch to Edwin Stanton, June 15, 1863, ORUCN 1, 27.3.131.

69. Douglas Southall Freeman, *R. E. Lee*, 4 vols. (New York: Charles Scribner's Sons, 1935), 3:57–58; James M. McPherson, *Battle Cry of Freedom: The Civil War Era* (New York: Oxford, 1988), 648–53.

70. The *Whig* remained uncertain as to what was more valuable—seizing control of the anthracite fields or simply depriving the Union of them. "Of one thing we may be sure," the paper wrote, "that whatever is best to be done will be done by Gen. Lee, and if he thinks fit to destroy the Pennsylvania mines they will certainly be destroyed. Should he leave them untouched it will be for the best of reasons. But it is impossible not to indulge the hope that he will avail himself of the tremendous power which the possession of the coal fields, even temporarily, would confer." Whether Lee indeed had designs on the anthracite fields remains unknown. It is certain, at least, that he sought to cultivate sympathetic feeling in the state, and given the history of draft conflict in mining counties, that prospect alone might have, to Lee, justified a march there ("The Coal Fields of Pennsylvania," *Richmond Whig*, July 2, 1863). On strategy, see Robert E. Lee to Jefferson Davis, June 10, 1863, ORUCN 1, 27.3.880–82; for representative northern responses, see "What the Rebels Propose to Do with Our Coal Mines," *New York Herald*, July 7, 1863, 7, and "The Rebels Propose to Set Fire to Our Coal Mines," *New Haven (CT) Daily Palladium*, July 8, 1863, 1; on sentiment in Richmond, see Thomas Cooper De Leon, *Four Years in Rebel Capitals: An Inside View of Life in the Southern Confederacy, from Birth to Death* (Mobile, AL: Gossip Printing Company, 1892), 253–54, "An Important Rumor From Pennsylvania," *Richmond Whig*, July 1, 1863, "The Reported Capture of Harrisburg," *Richmond Whig*, July 2, 1863, and "From Gettysburg," *Richmond Whig*, July 8, 1863. Reality only set in a week after the battle; see "From Gen. Lee's Army," *Richmond Whig*, July 9, 1863.

71. Andrew Hull Foote to Cornelius K. Stribling, March 9, 1863, and Albert N. Smith to Cornelius K. Stribling, August 14, October 8, and November 6, 1863, vol. 1, Fair Copies of Letters Sent to Commandants of Navy Yards and Stations, 1862–65, RG 24, NARA I; Hiram Paulding to Andrew Hull Foote, January 2, 1863 and enclosure, and Andrew A. Harwood to Andrew Hull Foote, January 9, 1863, vol. 1, Letters Received from Commanders of Navy Yards, RG 24, NARA I; Albert N. Smith to Samuel P. Lee, April 27, 1864, vol. 2, Fair Copies of Letters Sent to Officers, RG 24, NARA I.

72. John K. Mitchell to Stephen R. Mallory, November 16, 1863, and Stephen R. Mallory to Jefferson Davis, November 30, 1863, ORUCN 2, 2:534, 543–46; William N. Still, *Confederate Shipbuilding* (Columbia: University of South Carolina Press, 1987), 57.

73. William P. Browne to Colin J. McRae, January 25, 1862 and January 29, 1862, CJMP; William P. Browne to John K. Mitchell, February 26, 1863, WPBFP.

74. William Quinn to Stephen R. Mallory, June 29, 1863, Mss2Q4495a1, Virginia Historical Society, Richmond.

75. Invoices, Confederate States Navy, 1861–65, subject file, rolls 47–48, M1091, RG 45, NARA 1; secretary of the navy, reports, 1861, 344, 346; 1862, 704; 1863, 764; 1864, 911; Spencer Tucker, *Blue & Gray Navies: The Civil War Afloat* (Annapolis, MD: Naval Institute Press, 2006), 1, 9; William N. Still Jr., ed., *The Confederate Navy: The Ships, Men and Organization, 1861–65* (Annapolis, MD: Naval Institute Press, 1997), 39, 68. It is unfortunately impossible to figure the precise quantities of coal and wood purchased by the Confederate navy, since its invoices include a variety of measures (like cords, loads, and barrels), the conversions between which are impossible to make with certainty.

76. Ambrose W. Thompson to William H. Seward, March 30, 1863, roll 76, WHSP.

77. John P. Usher to Ambrose W. Thompson, March 18, 1864, box 43, AWT.

Chapter 5 · The Debate over Coaling Stations

1. "Foreign Coaling Stations," *New York Times*, May 6, 1891, 4.

2. Gideon Welles, *Diary of Gideon Welles: Secretary of the Navy under Lincoln and Johnson*, 3 vols. (Boston: Houghton Mifflin, 1911), 2:283; Olive Risley Seward, "A Diplomatic Episode," *Scribner's Magazine* 2, no. 5 (1887): 588; Frederick W. Seward, *Reminiscences of a War-Time Statesman and Diplomat, 1830–1915* (New York: G. P. Putnam's Sons, 1916), 263–343; Ex. Doc. No. K1, 40th Cong., 3rd sess. (1869), 2; Halvdan Koht, "The Origin of Seward's Plan to Purchase the Danish West Indies," *American Historical Review* 50, no. 4 (1945): 762–77.

3. Frederick W. Seward, *Seward at Washington, as Senator and Secretary of State: A Memoir of His Life, With Selections from His Letters, 1861–1872* (New York: Derby and Miller, 1891), 307–8; see also S. Ex. Doc. No. 231, pt. 8, 56th Cong., 2nd sess. (1901), 202. "In the event of a foreign war," wrote one American citizen visiting St. Thomas in 1866, "the possession of it as a coaling station and general *entrepot* for our own ships would be of incalculable value to the United States" ("The Cruise of the 'Monadnock,' No. 1," *Overland Monthly and Out West Magazine* 3, no. 1 [1869]: 18).

4. Sumner Welles, *Naboth's Vineyard: The Dominican Republic, 1844–1924*, 2 vols. (New York: Payson and Clark, 1928), 1:315–33, 346–58; Seward, *Reminiscences of a War-Time Statesman and Diplomat*, 344–56.

5. "Our New Possessions," *New York Times*, November 5, 1867, 4; Welles, *Diary of Gideon Welles*, 2:393, 406, 466–67; on needing coaling stations, see also, for example, "The Acquisition of St. Thomas," *Daily National Intelligencer* (Washington, DC), January 25, 1868, 2, "The Key to the Gulf of Mexico," *New York Herald*, December 22, 1869, 3, and untitled, *Baton Rouge Daily Advocate*, December 29, 1869, 2.

6. Welles had trouble understanding why Seward was so intent on securing a Caribbean naval station. "It is a scheme, personal and political, on the part of Seward," Welles had concluded by January. "A tub thrown to assure Thad Stevens and Fessenden." Welles speculated (wrongly) that the entire effort was part of Seward's continued presidential ambitions,

which in reality he had already abandoned (Welles, *Diary of Gideon Welles*, 2:631, 643; 3:40, 124–25; Walter Stahr, *Seward: Lincoln's Indispensable Man* [New York: Simon and Schuster, 2012], 454).

7. Roy A. Watlington and Shirley H. Lincoln, eds., *Disaster and Disruption in 1867: Hurricane, Earthquake, and Tsunami in the Danish West Indies* (St. Thomas, VI: Eastern Caribbean Center, University of the Virgin Islands, 1997), 5–10, 37. On the Monongahela, see "The Late Earthquake at St. Thomas," *Harper's Weekly* 12, no. 578 (1868), reprinted in Watlington and Lincoln, eds., *Disaster and Disruption in 1867*, 31, Confidential Ex. Doc. No. W, 40th Cong., 2nd sess. (1868), 1, and Confidential Ex. Doc. No. AA, 40th Cong., 2nd sess. (1868), 1–21.

8. "Admiral David D. Porter's Statement: The Island of Santo Domingo," SEN 41B-B7, SCFR; Charles Callan Tansill, *The United States and Santo Domingo, 1798–1873: A Chapter in Caribbean Diplomacy* (Baltimore, MD: Johns Hopkins Press, 1938), 123–29; G. Pope Atkins and Larman C. Wilson, *The Dominican Republic and the United States: From Imperialism to Transnationalism* (Athens: University of Georgia Press, 1998), 15–16.

9. Orville E. Babcock, statement, and Rufus Ingalls, statement, SEN 41B-B7, SCFR. Probably alluding to St. Thomas, Babcock added that judging from centuries-old buildings, Santo Domingo was also likely free from earthquakes.

10. Allan Nevins, *Hamilton Fish: The Inner History of the Grant Administration*, 2 vols. (New York: Dodd, Mead, 1937), 1:310.

11. Samuel L. M. Barlow to W. M. Gabb, January 25, 1870, Samuel L. M. Barlow to Colonel Talcott, January 25, 1870, Samuel L. M. Barlow to William L. Cazneau, January 25, 1870, and Samuel L. M. Barlow to Thomas F. Bayard, January 25, February 1, and March 28, 1870, roll 15, Barlow Letterbooks, SLMB. Cf. the account in Tansill, *The United States and Santo Domingo*, 389–90.

12. Nevins, *Hamilton Fish*, 1:316–22; David Herbert Donald, *Charles Sumner* (1960; repr., New York: Da Capo, 1996), 434–44; Eric T. L. Love, *Race Over Empire: Racism and U.S. Imperialism, 1865–1900* (Chapel Hill: University of North Carolina Press, 2004), 27–72.

13. *Cong. Globe*, April 7, 1871, 525. The Democratic *New York Sun* was particularly (if predictably) scathing; see "Samana" (February 22, 1871, 2), and "The Santo Domingo Job: Samana Bay Good for Nothing, and the Island Useless" (February 22, 1871, 3).

14. Barry Rigby, "The Origins of American Expansion in Hawaii and Samoa, 1865–1900," *International History Review* 10, no. 2 (1988): 228–33; H.R. Ex. Doc. No. 1, 42nd Cong., 3rd sess. (1872), 3:13.

15. Frederic B. Vinton, "Samoa and the Samoans," *United Service* 13, no. 4 (1885): 439, 446–47; "Samoa to Be Defended," *Philadelphia Inquirer*, January 24, 1889, 1; H.R. Ex. Doc. No. 238, 50th Cong., 1st sess. (1888), 32.

16. George H. Bates, "Some Aspects of the Samoan Question," *Century* 37, no. 6 (1889): 947; "No Coal for Pango Pango: Commodore Schley Says He Knows of None Being Purchased," *New York Herald*, February 10, 1889, 13; "Coal for Samoa: Secretary Whitney Makes Inquiries as to Purchase and Transportation," *New York Herald*, February 8, 1889, 6; "Our Navy's Disaster," *New York Herald*, March 31, 1889, 13; Paul Kennedy, *The Samoan Tangle: A Study in Anglo-German-American Relations, 1878–1900* (Dublin: Irish University Press, 1974), 86.

17. Benjamin Harrison, "Inaugural Address," in James D. Richardson, ed., *A Compilation of the Messages and Papers of the Presidents, 1789–1897* (New York: Bureau of National

Literature, 1897), 12:5440; James C. Cresap, "The Samoan Question," *Independent* 41, no. 2099 (1889): 6.

18. H.R. Ex. Doc. No. 1, pt. 1, 53rd Cong., 3rd sess. (1895), 17; Rigby, "The Origins of American Expansion in Hawaii and Samoa," 223–24; Ralph S. Kuykendall, *The Hawaiian Kingdom, 1854–1874: Twenty Critical Years* (Honolulu: University of Hawaii Press, 1953), 229.

19. Barton S. Alexander to A. A. Humphreys, March 11, 1872, and John M. Schofield to A. H. Humphreys, March 12, 1872, folder 1, Hawaii, box 78, JMSP; John M. Schofield, *Forty-six Years in the Army* (New York: Century Company, 1897), 431–34.

20. A. H. Humphreys to John M. Schofield, May 20, 1872, William W. Belknap to John M. Schofield, June 24, 1872, and William T. Sherman to John M. Schofield, copy of telegram, December 27, 1872, folder 1, Hawaii, box 78, JMSP.

21. John M. Schofield to William T. Sherman, February 15, 1873 in John Y. Simon, ed., *The Papers of Ulysses S. Grant*, 31 vols. to date (Carbondale: Southern Illinois University Press, 1967–), 24:72; Alfred S. Hartwell to John M. Schofield, March 12, 1873, folder 1, Hawaii, box 78, JMSP.

22. John M. Schofield to William T. Sherman, February 15, 1873, folder 1, Hawaii, box 78, JMSP; S. Ex. Doc. No. 77, 52nd Cong., 2nd sess. (1893), 150–54. A further investigation of the coral by Rear Admiral Alexander Pennock, who had brought Schofield and Alexander to the islands, found the scale of the needed excavation to be about two and a half times greater than what the generals had initially estimated; see John M. Schofield and Barton S. Alexander to William W. Belknap, September 30, 1873, folder 1, Hawaii, Box 78, JMSP.

23. Merze Tate, *Hawaii: Reciprocity or Annexation* (East Lansing: Michigan State University Press, 1968), 82–134; David D. Porter, "Hawaiian Islds, 1875," box 28, DDPFP.

24. Tate, *Hawaii*, 121, 183–214; H.R. Ex. Doc. No. 1, pt. 1, 53rd Cong., 3rd sess. (1895), 171, 337, 351; Edward P. Crapol, *James G. Blaine: Architect of Empire* (Wilmington, DE: Scholarly Resources, 2000), 124–29.

25. H.R. Ex. Doc. No. 46, 47th Cong., 1st sess. (1882), 6, 33–44; David M. Pletcher, *The Awkward Years: American Foreign Relations under Garfield and Arthur* (Columbia: University of Missouri Press, 1962), 28–29. Among the *Sun*'s blistering coverage of the proposal were "A Startling Scheme: Is Mr. Evarts Imitating Lord Beaconsfield's Policy?—Possible Trouble Ahead" (February 17, 1880, 1), "A Foothold on the Isthmus: The U.S. Steamer Adams Establishing a Coaling Station at the Golfa [*sic*] Dulce" (April 13, 1880, 1), and "Mr. Rogers as a Lobbyist: Mr. Hayes's Secretary Accused of Influencing Legislation" (February 26, 1881, 3). Richard Thompson, the Chiriqui Company's longtime lawyer and Hayes's navy secretary, claimed that he had nothing to do with the revived scheme, though the company's new lawyer reported meeting with him several times, and after Ambrose Thompson's death in 1882, he refused to surrender the papers of the old company, insisting he was still owed $50,000—money he would likely never see unless the plan was funded by Congress (Richard W. Thompson, undated statements to the president [David William Henry?], box 4, RWT).

26. [William McKendree Gwin?] to William E. Chandler, undated draft ca. 1883, E. H. Carmick to William McKendree Gwin, November 9, 1883, E. H. Carmick to William McKendree Gwin, November 26, 1883, William Sharon to Chester A. Arthur, copy ca. 1884, and William Sharon to William McKendree Gwin, January 18, 1884, box 1, WMGP.

27. E. H. Carmick to William McKendree Gwin, December 7, 1883, William McKendree Gwin to Chester A. Arthur, copy? ca. 1884, and E. H. Carmick to William McKendree Gwin, December 7, 1884, and February 11, 1885, box 1, WMGP.

28. H.R. Ex. Doc. No. 1, pt. 3, 48th Cong., 1st sess. (1883), 1:32, 40–41.

29. H.R. Ex. Doc. No. 1, pt. 3, 46th Cong., 3rd sess. (1880), 24–27; Frederick C. Drake, *The Empire of the Seas: A Biography of Rear Admiral Robert Wilson Shufeldt, USN* (Honolulu: University of Hawaii Press, 1984), 153–71, 176–256; David Foster Long, *Gold Braid and Foreign Relations: Diplomatic Activities of the U.S. Naval Officers, 1798–1883* (Annapolis, MD: Naval Institute Press, 1988), 396. Fernando Pó is today known as Bioko; it is an island off the coast of modern-day Cameroon in the Gulf of Guinea where at the time the British Admiralty maintained a coaling station ("One Who Knows the Facts," *The Truth About the Navy and Its Coaling Stations* [London: Pall Mall Gazette, 1884], 26).

30. A. L. C. Portman to William H. Seward, June 10, 1865, in *Executive Documents Printed by Order of the House of Representatives, During the First Session of the Thirty-Ninth Congress, 1865–'66*, vol. 1, pt. 3 (Washington, DC: Government Printing Office, 1866), 250–51; Seward W. Livermore, "American Naval-Base Policy in the Far East, 1850–1914," *Pacific Historical Review* 13, no. 2 (1944): 114.

31. The Peterhoff, 72 U.S. 28 (1866); William Beach Lawrence, *Lawrence's Wheaton: Elements of International Law*, 2nd annot. ed. (Boston: Little, Brown, 1863), 767–96.

32. S. Ex. Doc. No. 2, 36th Cong., 1st sess. (1860), 32–35.

33. U.S. Department of State, *Papers Relating to the Treaty of Washington*, 6 vols. (Washington, DC: Government Printing Office, 1872), 1:89–190; 2:433–38, 513–19; Adrian Cook, *The Alabama Claims: American Politics and Anglo-American Relations, 1865–1872* (Ithaca, NY: Cornell University Press, 1975), 207–40; Nevins, *Hamilton Fish*, 1:400–401; Simon, ed., *Papers of Ulysses S. Grant*, 20:302–5.

34. U.S. Department of State, *Papers Relating to the Treaty of Washington*, 4:148–50.

35. U.S. Department of State, *Papers Relating to the Treaty of Washington*, 4:11–12, 50–51, 74–75, 418–46, 497–99; Frank Warren Hackett, *Reminiscences of the Geneva Tribunal of Arbitration, 1872: The Alabama Claims* (Boston: Houghton Mifflin, 1911), 348–50.

36. Charles H. Stockton, "The Reconstruction of the United States Navy," *Overland Monthly and Out West Magazine* 16, no. 94 (1890): 381–86; Godfrey Lushington, *A Manual of Naval Prize Law* (London: Butterworths, 1866), 36; Freeman Snow, *International Law: Lectures Delivered at the Naval War College*, ed. Charles H. Stockton (Washington, DC: Government Printing Office, 1895), 138–39. Then there was David Porter, who (wrongly) claimed in 1883 that the Paris Declaration of 1856 had made coal entirely contraband, by which he meant that in the case of war, "all the coaling stations of the world would be closed against us." Porter's solution was a return to sail (H.R. Ex. Doc. No. 1, pt. 3, 48th Cong., 1st sess. [1883], 1:388; H. A. Smalley, "A Defenseless Sea-Board," *North American Review* 138, no. 328 [1884]: 237). On Butler, see "General Butler on Treaties: How They Are Made-Specimens of American Diplomacy," *New York Sun*, January 21, 1885, 2.

37. "Washington's Farewell Address," in *Annals of Congress: Appendix*, 4th Cong., 2879. According to Alfred Mahan, the best protection from Great Britain as an adversary is "a cordial understanding with that country," even as a "formal alliance between the two is out of the question" ("The United States Looking Outward," *Atlantic* 66, no. 398 [1890]: 823–24).

38. J. Ross Browne, *A Sketch of the Settlement and Exploration of Lower California* (San Francisco: H. H. Bancroft, 1869), 177; H.R. Ex. Doc. No. 1, pt. 3, 47th Cong., 1st sess. (1881), 179; H.R Ex. Doc. No. 1, pt. 3, 47th Cong., 2nd sess. (1882), 102; H.R. Ex. Doc. No. 1, pt. 3, 48th Cong., 1st sess. (1883), 237.

39. Livermore, "American Strategy Diplomacy in the South Pacific," 33–37; Stephen A. Hurlbut to James G. Blaine, October 5, 1881 and James G. Blaine to Stephen A. Hurlbut, November 22 and December 3, 1881, in *Index to the Executive Documents of the House of Representatives for the First Session of the Forty-Seventh Congress, 1881–'82*, vol. 1, *Foreign Relations* (Washington, DC: Government Printing Office, 1882), 938–40, 948–51, 955–57; Benjamin Harrison to Blaine, December 31, 1891, in Albert T. Volwiler, ed., *Correspondence Between Benjamin Harrison and James G. Blaine, 1882–1893* (Philadelphia: American Philosophical Society, 1940), 223–24; "War Ships for Chili," *New York Herald*, August 31, 1892, 7. Blaine also admonished Hurlbut for his blatant attempt to insert himself as trustee for the railroad before anticipating turning it over to an American company.

40. Albert T. Volwiler, "Harrison, Blaine, and American Foreign Policy, 1889–1893," in *The Shaping of American Diplomacy: Readings and Documents in American Foreign Relations, 1750–1955*, ed. William Appleman Williams (Chicago: Rand McNally, 1956), 356–58; Benjamin Harrison to James G. Blaine, August 3, 1891, and James G. Blaine to Benjamin Harrison, August 10, 1891, in Volwiler, *Correspondence Between Benjamin Harrison and James G. Blaine*, 169–70, 173–74.

41. George Washington Littlehales, *The Development of Great Circle Sailing*, 2nd ed. (Washington, DC: Government Printing Office, 1899), 9.

42. This simplification, of course, assumes a spherical globe. The earth is more approximately, though still not perfectly, an oblate spheroid, the figure traced by an ellipse rotated about its shorter axis. The earth's radius at the equator is 21.39 kilometers greater than its radius at the poles. See C. M. R. Fowler, *The Solid Earth: An Introduction to Global Geophysics* (New York: Cambridge University Press, 1990), 163–66, 452. For more detail on great circle routes, see S. T. S. Lecky, *"Wrinkles" in Practical Navigation*, rev. and enlarged ed. (London: George Philip and Son, 1903), 407–13, Nathaniel Bowditch, *The American Practical Navigator: Being an Epitome of Navigation and Nautical Astronomy*, rev. ed. (Washington, DC: Government Printing Office, 1888), 22, 53–60, and Littlehales, *The Development of Great Circle Sailing*.

43. Charles H. Cotter, "The Early History of Great Circle Sailing," *Journal of Navigation* 29, no. 3 (1976): 254–62; Littlehales, *The Development of Great Circle Sailing*, 9; John Davis, *The Seamans Secrets* (London: Thomas Dawson, 1599), 43. The other two sailings are horizontal, by which ships trace courses parallel with the equator and other lines of latitude, and Mercator, also known as paradoxal or rhumb sailing, by which ships' courses trace constant angles with meridians and thus appear as straight lines on Mercator charts (Henry Phillippes, *The Geometrical Sea-Man; or, the Art of Navigation Performed by Geometry*, 2nd ed. [London: Robert and William Leybourn, 1657], 48). On Bowditch, compare the seventeenth edition of *The New American Practical Navigator* (1847) with the eighteenth edition (1848; see ii, 452–58). On the evolution of the work, see John F. Campbell, *History and Bibliography of the New American Practical Navigator and the American Coast Pilot* (Salem, MA: Peabody Museum, 1964).

44. Ferdinand Labrosse, *The Navigation of the Pacific Ocean, China Seas, Etc.*, trans. by Jacob W. Miller (Washington, DC: Government Printing Office, 1875), 57–59, 117, 219–21.

45. Richard A. Proctor, "Chart for Great Circle Sailing," *Scientific American Supplement* 20, no. 501 (1885): 7991–92; John Thomas Towson, *Tables to Facilitate the Practice of Great Circle Sailing, and the Determination of Azimuths*, 6th ed. (London: Hydrographic Office, Admiralty, 1861), 48; Littlehales, *The Development of Great Circle Sailing*, 10. See also

"Scientific News in Washington: Recent Developments in Great Circle Sailing," *Science* 12, no. 291 (1888): 105–6. Even with successive improvements in great circle calculation techniques in the latter half of the nineteenth century, dissatisfaction with their use persisted well into the twentieth. In 1919, a U.S. navy commander griped that of the two major approaches then available, one required "a rather long preliminary study of the method" and the other "require[d] the navigator to burden his memory with a rarely used formula," both approaches still tending toward error (H. G. S. Wallace, "Great Circle Sailing—a Few 'Wrinkles' to Save Time," *United States Naval Institute Proceedings* 45, no. 7 (1919): 1197–99.

46. S. Rep. Com. No. 116, 39th Cong., 1st sess. (1866), 1–7; H.R. Ex. Doc. No. 1/19, 39th Cong., 2nd sess. (1866), 8–9; H.R. Ex. No. Doc. 275, 40th Cong., 2nd sess. (1868), 206; *Cong. Globe*, July 2 and 17, 1866, 3530–36, 3858–66; H.R. Misc. Doc. No. 113, 42nd Cong, 3rd sess. (1874), 252; untitled, *Boston Daily Advertiser*, December 14, 1866, 2; "Trans-Pacific Mail Service," *Richmond Whig*, March 12, 1867, 1; *Cong. Globe*, February 11, 1867, 1140; H.R. Ex. Doc. No. 12, 40th Cong., 2nd sess. (1867).

47. Dan O'Donnell, "The Pacific Guano Islands: The Stirring of American Empire in the Pacific Ocean," *Pacific Studies* 16, no. 1 (1993): 56–57; Christina Duffy Burnett, "The Edges of Empire and the Limits of Sovereignty: American Guano Islands," *American Quarterly* 57, no. 3 (2005): 779, 781–82, 801; S. Ex. Doc. No. 79, 40th Cong., 2nd sess. (1868), 3–13; S. Rep. Com. No. 194, 40th Cong., 3rd sess. (1869), 18.

48. G. B. Airy, "On a Method of Very Approximately Representing the Projection of a Great Circle upon Mercator's Chart," *Monthly Notices of the Royal Astronomical Society* 18, no. 5 (1858): 150–55; J. H. C. Coffin, "Memoir of William Chauvenet, 1820–1870," *National Academy of Sciences: Biographical Memoirs* 1, no. 8 (1877): 240; "Professor Chauvenet's 'Great Circle Protractor,'" *New York Times*, March 21, 1855, 4; Littlehales, *The Development of Great Circle Sailing*.

49. H.R. Ex. Doc. No. 1, 39th Cong., 1st sess. (1865), 161; H.R Ex. Doc. No. 1, pt. 3, 49th Cong., 1st sess. (1885), 130; Charles D. Sigsbee, "Graphical Method for Navigators," *Proceedings of the United States Naval Institute* 11, no. 2 (1885): 241–63; H.R. Ex. Doc. No. 1, pt. 3, 53rd Cong., 3rd sess. (1894), 213. As a captain nine years later in 1898, Sigsbee would be aboard the USS *Maine* when it sunk in Havana.

50. Robert Greenhalgh Albion, "Adolph E. Borie," in *American Secretaries of the Navy*, vol. 1, ed. Paolo Coletta (Annapolis, MD: Naval Institute Press, 1980), 363–66; Magnus S. Thompson, ed., *General Orders and Circulars Issued by the Navy Department from 1863 to 1887* (Washington, DC: Government Printing Office, 1887), 68–84; Robert Greenhalgh Albion, *Makers of Naval Policy* (Annapolis, MD: Naval Institute Press, 1980), 200–201; Stephen Howarth, *To Shining Sea: A History of the United States Navy, 1775–1991* (New York: Random House, 1991), 218. Lance Buhl, who notes a range of other historians criticizing Porter, views the admiral's move and others like it through the lens of competition for professional status between line officers and naval engineers during a time of technological ambiguity ("Mariners and Machines: Resistance to Technological Change in the American Navy, 1865–1869," *Journal of American History* 61, no. 3 [1974]: 703–27).

51. D. B. Harmony, "Navy Department Circular No. 36," November 2, 1887.; B. F. Tracy, "Navy Department General Order No. 390," May 22, 1891; Truman H. Newberry, "Navy Department Special Order No. 8," February 12, 1906. Unless otherwise noted, these and other circulars and orders are cataloged in volume 5 of the *CIS Index to U.S. Executive Branch Documents, 1789–1909: Guide to Documents Listed in Checklist of U.S. Public Documents,*

1789–1909, Not Printed in the U.S. Serial Set (Bethesda, MD: Congressional Information Service, 1990). Porter himself noted that restricting coal use had other consequences, too, observing in 1873 the results of replacing the positions of firemen and coal heavers with seamen-firemen and seamen-coal heavers. "The work is unpopular because, as a rule, steam is seldom used, and the extra pay allowed for these occasions will not even compensate for the clothing worn out." As a consequence, Porter noted, "most of the seamen-firemen and seamen coal-heavers desert" (H.R. Ex. Doc. No. 1, pt. 3, 43rd Cong., 1st sess. [1873], 280). On the slow development of the New Navy, see Timothy S. Wolters, "Recapitalizing the Fleet: A Material Analysis of Late-Nineteenth-Century U.S. Naval Power," *Technology and Culture* 52, no. 1 (2011): 103–26.

52. David D. Porter to Gideon Welles, August 23, 1861, ORUCN 1, 1:71–72; David D. Porter to W. A. Parker, David D. Porter to J. M. Frailey, and David D. Porter to W. A. Parker, telegram, November 30, 1864, ORUCN 1, 11:111; David D. Porter to H. A. Adams, December 20, 1864, ORUCN 1, 11:202; David D. Porter to Emanuel Mellach, ORUCN 1, 11: 392; David D. Porter to J. M. Berrien, December 30, 1864, ORUCN 1, 11: 393; David D. Porter to Gideon Welles, January 6, 1865, ORUCN 1, 11:413; David D. Porter to senior naval officer, January 9, 1865, ORUCN 1, 11:416; David D. Porter to W. C. West, January 14, 1865, ORUCN 1, 11:597; David G. Farragut to Gideon Welles, June 16, 1862, ORUCN 1, 18:561; Ulysses S. Grant to David G. Farragut, March 22, 1863, ORUCN 1, 20:7; David G. Farragut to Gideon Welles, March 27, 1863, ORUCN 1, 20:35; Farragut to James. S. Palmer, May 6, 1863, ORUCN 1, 20: 77; David D. Porter to Henry Walke, October 26, 1862, ORUCN 1, 23:450; David D. Porter to Henry Walke, November 2, 1862, ORUCN 1, 23:463; David D. Porter to Thomas O. Selfridge, November 15, 1862, ORUCN 1, 23:485; David D. Porter to Henry Walke, November 19, 1862, ORUCN 1, 23:495; David D. Porter to W. Brenton Boggs, November 23, 1862, ORUCN 1, 23:500; David D. Porter to Henry Walke, November 28, 1862, ORUCN 1, 23:512; David D. Porter to Thomas O'Reilly, December 12, 1862, ORUCN 1, 23:542; David D. Porter to E. K. Owen, December 21, 1862, ORUCN 1, 23:560–61; David D. Porter to W. T. Sherman, and David D. Porter to A. M. Pennock, December 21, 1862, ORUCN 1, 23:644.

53. "Present Condition of the U.S. Navy," *International Review* 6, no. 3 (1879): 273; Clark, N. B. "Deflecting Armor," *Proceedings of the United States Naval Institute* 7, no. 15 (1881): 22; Stephen B. Luce, "Naval Training," *Proceedings of the United States Naval Institute* 16, No. 3 (1890): 380–85; J. C. Wilson, "Comments on S. B. Luce, 'Naval Training,'" *Proceedings of the United States Naval Institute* 16, No. 3 (1890): 429.

54. Benjamin Franklin Isherwood, *Experimental Researches in Steam Engineering* (Philadelphia: William Hamilton, 1863); H.R. Ex. Doc. No. 1/18, 38th Cong., 2nd sess. (1864), 1096.

55. Ruth D'Arcy Thompson, *The Remarkable Gamgees: A Story of Achievement* (Edinburgh: Ramsay Head Press, 1974), 157–72, 66; John Gamgee, "Thermo Dynamic Engine," U.S. Patent 240,400, filed February 26, 1881, and issued April 19, 1881; Benjamin Franklin Isherwood, "The Gamgee Perpetual Motion," *Scientific American* 44, no. 21 (1881): 324. For more on Isherwood's contentious career in the navy, see Brendan Foley, "Fighting Engineers: The U.S. Navy and Mechanical Engineering, 1840–1905" (PhD diss., Massachusetts Institute of Technology, 2003), 134–45, and Edward William Sloan III, *Benjamin Franklin Isherwood, Naval Engineer: The Years as Engineer in Chief, 1861–1869* (Annapolis, MD: Naval Institute Press, 1965).

56. J. S. C., "The Zero-Motor: Steam Possibly to be Superseded by Ammonia and Water," *Cleveland Leader*, April 25, 1881, 2; "An Enthusiastic Inventor," *New York Tribune*, August 4, 1881, 8; J. J. Brown, "Prof. Gamgee," *Northern Christian Advocate* September 1, 1881, 3; J. J. Brown, "The Zero Motor Again," *Northern Christian Advocate*, May 19, 1881, 3; "No New Motive Power in the Zero Motor," *New York Herald*, April 30, 1881, 3; "The Zero Motor," *Vicksburg (MS) Daily Commercial*, May 4, 1881, 2; John Gamgee, "Electricity Boxed," *New York Herald*, May 25, 1881, 11; "The Gamgee Perpetual Motion," *Scientific American* 44, no. 20 (1881): 305; "The Gamgee Zeromotor," *Scientific American*, 44, no. 21 (1881): 321; "The Gamgee Perpetual Motion," *Scientific American* 44, no. 22 (1881): 337; "Perpetual Motion Delusions," *Scientific American*, 44, no. 23 (1881): 352.

57. Samuel R. Franklin, *Memories of a Rear-Admiral* (New York: Harper and Brothers, 1898), 178; Charles Edwin Taylor, *St. Thomas, as a Naval and Coaling Station* (St. Thomas, DWI: J. N. Lightbourn, 1891), 12–13. Elsewhere, Taylor uses the example of the coaling women to disprove common stereotypes of Caribbean people. Meaningful labor produced purpose in life: "All this movement may seem incredible to those who have been accustomed to associate life in the tropics with laziness and a disinclination to exertion," he writes, "especially when the negroes are concerned; give them work, and pay them properly for it, and they will do it quite as promptly, and far more good-naturedly, than their white brother in a like station of life, who, the slave perhaps of some trade union, is far worse off to-day than the negro ever was at the time of slavery" (*An Island of the Sea: Descriptive of the Past and Present of St. Thomas, Danish West Indies* [St. Thomas, DWI: published by the author, 1895], 34). St. Thomas visitor Maturin M. Ballou also observed the coaling women. "A hundred women and girls, wearing one scant garment reaching to the knees, are in line, and commence at once to trot on board in single file, each one bearing a bushel basket of coal upon her head, weighing, say sixty pounds. Another gang fill empty baskets where the coal is stored, so that there is a continuous line of negresses trotting into the ship at one port and, after dumping their loads into the coal bunkers, out at the other, hastening back to the source of supply for more." Ballou also noted their song, and witnessed the women engaged in "a firefly dance" performed to the light of burning coal after completing their night's labor (*Equatorial America: Descriptive of a Visit to St. Thomas, Martinique, Barbadoes, and the Principal Capitals of South America* [Boston: Houghton, Mifflin, 1892], 29–31). See also Seward, *Reminiscences of a War-Time Statesman and Diplomat*, 288.

58. Isherwood, "The Gamgee Perpetual Motion"; Charles D. Sigsbee, "Comments on W. F. Fullam, 'The System of Naval Training and Discipline Required to Promote Efficiency and Attract Americans,'" *Proceedings of the United States Naval Institute* 16, no. 4 (1890): 532; "Why Spend Money for Coaling Stations? A Plan for Taking Coal on Board a Cruiser at Sea," *New York Herald*, March 8, 1891, 29; "Coaling Cruisers at Sea," *Scientific American* 69, no. 19 (1893): 290–91. Although coaling at sea remained impractical, the navy did build vessels such as the battleship *Indiana* in 1895, which was smaller than its British counterparts but capable nevertheless of firing larger and more devastating rounds. The *Indiana* was built to hew close to the U.S. coast, which mean it required less fuel and ammunition ("The Battle Ship Indiana," *Scientific American* 73, no. 17 [1895]: 258–59).

59. George Frederick Zimmer, *The Mechanical Handling of Material* (London: Crosby Lockwood and Son, 1905), 222–33; "Coaling at Sea," *Scientific American* 57, no. 19 (1887): 288; "Coaling Warships at Sea," *Scientific American* 76, no. 17 (1897): 257–58; "Coaling Vessels at Sea," *Journal of the American Society of Naval Engineers* 11, no. 4 (1899): 1049–50; "Coal-

ing Ships at Sea," *Journal of the American Society of Naval Engineers* 12, no. 1 (1900): 215–19; Spencer Miller, "The Coaling of the U.S.S. Massachusetts at Sea," *Transactions of the Society of Naval Architects and Marine Engineers* 8 (1900): 155–65; Spencer Miller, "Coaling Warships at Sea—Recent Developments," *Transactions of the Society of Naval Architects and Marine Engineers* 12 (1904): 177–200; "Coaling Warships at Sea," *International Marine Engineering* 15, no. 9 (1910): 370–2; Spencer Miller to George Dewey, January 31, February 11, and 15, 1901, and George Dewey to Spencer Miller, February 14, 1901, box 12, GDP; Spencer Miller to George Dewey, September 17, 1902, box 13, GDP. David Snyder finds evidence of navy disappointment with the Miller apparatus, though by 1911, six naval colliers contained 110 winches between them based on Miller's patents (these, however, were used to load steamers anchored alongside the collier rather than while steaming at sea, which was Miller's more radical idea). Miller's naval work led to his appointment by Josephus Daniels in 1915 as one of the original members of the Naval Consulting Board; see David Allan Snyder, "Petroleum and Power: Naval Fuel Technology and the Anglo-American Struggle for Core Hegemony, 1889–1922" (PhD diss., Texas A&M University, 2001), 93–98, Lidgerwood Manufacturing Company advertisement, *International Marine Engineering* 16, no. 3 (1911): 41, and "Spencer Miller," *Journal of the American Society of Mechanical Engineers* 40, no. 1 (1918): 74.

60. H.R. Ex. Doc. No. 1, pt. 3, 48th Cong., 2nd sess. (1884), 1:229; "Report of the Policy Board," *Proceedings of the United States Naval Institute* 16, no. 2 (1890): 205–19, 232, 248; T. F. Jewell, "Comments on S. B. Luce, 'Naval Training,'" *Proceedings of the United States Naval Institute* 16, no. 3 (1890): 410.

61. "Our Large War Ship: The Ironclad New York to be Launched This Month," *Cleveland Plain Dealer*, November 9, 1891, 4; "The Big Cruiser, New York, Beautifully Launched at Philadelphia Yesterday," *Wheeling (WV) Register*, December 3, 1891, 1; "Satisfactory Trial of the War Ship New York," *Scientific American* 68, no. 22 (1893): 338; Albert Franklin Matthews, "Trial Trip of the Cruiser 'New York,'" *Chautauquan* 17, no. 5 (1893): 548–50; H.R. Ex. Doc. No. 1, pt. 3, 51st Cong., 2nd sess. (1890), 10.

62. H.R. Ex. Doc. No. 1, pt. 3, 51st Cong., 2nd sess. (1890), 14–15; H.R. Ex. Doc. No. 1, pt. 3, 52nd Cong., 1st sess. (1891), 3; "Launch of the Columbia," *Scientific American* 67, no. 6 (1892): 84; William S. Aldrich, "Speed in American War-Ships," *North American Review* 162, no. 470 (1896): 53; Ira N. Hollis, "Coal Endurance and Machinery of the New Cruisers," *Journal of the American Society of Naval Engineers* 4, no. 4 (1892): 637–83. At the same time, British engineers complained that the Royal Navy's twenty-four new second-class cruisers could carry only between 400 and 550 tons of coal, not even enough to steam sixteen hundred miles ("Our New Cruisers and Their Coal Supply," *Journal of the American Society of Naval Engineers* 4, no. 4 [1892]: 796–98).

63. H.R. Ex. Doc. No. 1, pt. 3, 51st Cong., 2nd sess. (1890), 13; "The Speed Trial of the United States Battleship Massachusetts," *Scientific American* 74, no. 19 (1896): 296–97; "Battleships Nos. 5 and 6," *Scientific American* 73, no. 24 (1895): 376–77; "The Three New Battleships," *Scientific American* 75, no. 5 (1896): 119; C. H. Stockton, "Response to A. P. Cooke, 'Our Naval Reserve and the Necessity for its Organization,'" *Proceedings of the United States Naval Institute* 14, no. 1 (1888): 220; Stephen B. Luce, "Naval Administration," *Proceedings of the United States Naval Institute* 14, no. 4 (1888): 736.

64. British Naval Maneuvers," *Scientific American* 57, no. 11 (1887): 160–61; Theodore Ayrault Dodge, "The Needs of Our Army and Navy," *Forum* 12 (October 1891): 254–55.

65. George P. Scriven, "The Nicaragua Canal in Its Military Aspects," *Journal of the Military Service Institution of the United States* 15, no. 67 (1894): 11.

66. Robert Seager II, "Alfred Thayer Mahan: Christian Expansionist, Navalist, and Historian," in *Admirals of the New Steel Navy: Makers of the American Naval Tradition, 1880–1930*, ed. James C. Bradford (Annapolis, MD: Naval Institute Press, 1990): 41–42.

67. Robert Seager II, *Alfred Thayer Mahan: The Man and His Letters* (Annapolis, MD: Naval Institute Press, 1977), 154; Alfred Thayer Mahan to William C. Whitney, April 6, 1885, in Robert Seager II and Doris D. Maguire, eds., *Letters and Papers of Alfred Thayer Mahan*, 3 vols. (Annapolis, MD: Naval Institute Press, 1975), 1:598–99.

68. Alfred Thayer Mahan, *The Influence of Sea Power Upon History, 1660–1783* (Boston: Little, Brown, 1890), 31, 83, 329–30, 540–41; Alfred Thayer Mahan, "The United States Looking Outward," *Atlantic* 66, no. 398 (1890): 823.

69. Alfred Thayer Mahan, "Contingency Plan of Operations in Case of War with Great Britain, New York, December 1890," in Seager and Maguire, *Letters and Papers of Alfred Thayer Mahan*, 3:562–63, 570, 574–75.

70. Alfred Thayer Mahan to John M. Brown, December 13, 1897, in Seager and Maguire, *Letters and Papers of Alfred Thayer Mahan*, 2:532; Alfred Thayer Mahan, "Needed as a Barrier: To Protect the World from an Invasion of Chinese Barbarism," *New York Times*, February 1, 1893, 5.

71. Alfred Thayer Mahan, "Hawaii and our Future Sea-Power," *Forum* 15 (March 1893): 8; Alfred Thayer Mahan to James H. Kyle, February 4, 1898, in Seager and Maguire, *Letters and Papers of Alfred Thayer Mahan*, 2:538–39.

72. Irving M. Scott, "Naval Needs of the Pacific," *Overland Monthly* 24, no. 142 (1894): 367. Scott was a San Francisco shipbuilder who had his own reasons for supporting naval expansion; see "Far Ahead of America: Irving M. Scott's Opinion of Some Foreign Shipyards," *New York Times*, August 30, 1891, 2.

73. Marsden Manson, "Naval Control of the Pacific Ocean," *Overland Monthly* 25, no. 145 (1895): 56–61; see also John A. Harman, "The Political Importance of Hawaii," *North American Review* 160, no. 460 (1895): 374–77, S. Ex. Doc. No. 62, 55th Cong., 2nd sess. (1898), 1, S. Rep. No. 681, 55th Cong., 2nd sess. (1898), 51–52, 98, and John R. Proctor, "Hawaii and the Changing Front of the World," *Forum* 24 (September 1897): 41–42.

74. Francis X. Hezel, *The First Taint of Civilization: A History of the Caroline and Marshall Islands in Pre-Colonial Days, 1521–1885* (Honolulu: University of Hawaii Press, 1983), 307–14; Colin Walter Newbury, "The Administration of French Oceania, 1842–1906" (PhD diss., Australian National University, 1956), 239–40; S. Ex. Doc. No. 188, 55th Cong., 2nd sess. (1898), 7–10, 14–17.

75. S. Ex. Doc. No. 188, 15, 17; S. Ex. Doc. No. 315, 55th Cong., 2nd sess. (1898), 1–2. The author of the passage Melville quoted, Captain Charles F. Winter, the governor general of Canada's foot guards, went on to add in an unquoted portion of his essay that "there is always this consolation, that whoever is supreme in sea power in the Pacific can at any time possess the islands therein, but in the meantime we must do with less favourable coaling depôts, etc., which for the protection of trade and commerce appears to be the principal consideration" ("The Protection of Commerce During War," *Journal of the Royal United Service Institution* 42, no. 243 [1898]: 523).

76. For example, see South Dakota's first senator Richard F. Pettigrew in the Senate, *Cong. Record Appendix*, 53rd Cong., 2nd sess., July 2, 1894, 163, and Pettigrew's 1898 Senate

speeches against Hawaiian annexation, collected as "The Strategic Value of Hawaii" in his *The Course of Empire* (New York: Boni and Liveright, 1920), 137–62. In the House, see remarks by Kentucky's James McCreary, *Cong. Record*, February 20, 1895, 2471.

77. S. Ex. Doc. No. 82, 55th Cong., 2nd sess. (1898), 14; *Cong. Record*, July 2, 1894, 7060.

78. *Cong. Record*, February 8, 1895, 1943.

79. Alfred Thayer Mahan, "A Distinction Between Colonies and Dependencies: Remarks to the New York State Chapter of the Colonial Order, New York, November 30, 1898," in Seager and Maguire, *Letters and Papers of Alfred Thayer Mahan*, 2:581–91; Alfred Thayer Mahan to John D. Long, [August 15–20], 1898, in Seager and Maguire, *Letters and Papers of Alfred Thayer Mahan*, 3:596; W. D. McCrackan, "Our Foreign Policy," *Arena*, no. 44 (1893): 147.

Chapter 6 · Inventing Logistics

1. H.R Ex. Doc. No. 3, 55th Cong., 3rd sess. (1898), 22–23; F. V. McNair to George Dewey, December 31, 1897, 5, box 47, GDP.

2. Commander in chief to secretary of navy, telegrams, February 27, March 1, March 11, March 12, and April 2, 1898, commander in chief to U.S. minister in Tokyo, telegram, April 3, 1898, U.S. minister in Tokyo to commander in chief, telegram, April 4, 1898, commander in chief to the chief officer of the *Monocacy* [Commander Oscar. W. Farenholt], telegrams, April 4 and 5, 1898, and the chief officer of the *Monocacy* to commander in chief, telegrams, April 5 and 6, 1898, box 53, GDP.

3. Secretary of navy to commander in chief, telegram, April 5, 1898, commander in chief to secretary of navy, telegram, April 5, 1898, chief officer of the *Monocacy* to commander in chief, telegram, April 6, 1898, commander in chief to secretary of navy, telegram, April 6, 1898, secretary of navy to commander in chief, telegram, April 6, 1898, commander in chief to secretary of navy, telegram, April 8, 1898, secretary of navy to commander in chief, telegram, April 9, 1898, and commander in chief to secretary of navy, telegrams, April 9 and April 19, 1898, box 53, GDP. In his memoirs, Dewey made a point of recounting how he ignored Long's orders to "man and arm" the colliers, as that would have made them subject to the same neutrality restrictions with respect to coal that other naval steamers were subject to in wartime. In his telegraphic correspondence on April 9, Dewey did, however, tell Long he would "arm equip and man vessel immediately," and it does not appear that he decided to keep them as merchant ships (which were cleared for the Bonin Islands and Guam and were expected to be fueled at Shanghai) until about a week later, a decision he promptly explained to Washington. Moreover, it is unclear why Dewey believed merchant ships would not be subject to contraband laws, unless he believed he could keep the destinations of these vessels a permanent secret. For the memoir, see Dewey, *Autobiography of George Dewey: Admiral of the Navy* (New York: Charles Scribner's Sons, 1913), 190–92. See also Caspar F. Goodrich, "Dewey at Manila," 1901, 15–16, folder 22, box 2, subseries 2, CFGP.

4. Commander in chief to secretary of navy, telegram, May 4 and May 25, 1898, Bureau of Equipment [signed by Royal B. Bradford] to commander in chief, telegram, June 25, 1898, commander in chief to secretary of navy, telegram, July 17, 1898, and Bureau of Equipment to commander in chief, telegram, August 10, 1898, box 53, GDP; J. Warren Coulston to Dewey, May 25, 1901, box 12, GDP; George Dewey to secretary of navy, January 12, 1899, box 46, GDP; Wesley Merritt to adjutant general [Henry C. Corbin], June 23 and June 24,

1898, in *Correspondence Relating to the War with Spain and Conditions Growing out of the Same*, 2 vols. (Washington, DC: Government Printing Office, 1902), 2:711–12.

5. *Alleged Structural Defects in Battle Ships: Hearings before the Committee on Naval Affairs, United States Senate*, 60th Cong., 1st sess., March 9, 1908, 250; "Dewey Has a Good Supply of Coal," *Cleveland Plain Dealer*, May 3, 1898, 2; "Dewey Has Plenty of Coal: Ten Thousand Tons Captured by Him at Manila," *New York Tribune*, May 10, 1898, 1; "How Dewey Got His Coal: Good Work Done by Naval Equipment Bureau Before the War Had Begun," *Grand Forks (ND) Daily Herald*, October 23, 1898, 3; "Coal in Philippines: Admiral Dewey Reports Plenty of It There–Quality Good," *Wilkes-Barre (PA) Times*, October 29, 1898, 1.

6. "A Reading of the Hands of Admiral Dewey, by CHEIRO," box 51, GDP.

7. Undated memorandum," box 43, GDP; George Dewey, memorandum, June 3, 1901, box 51, GDP. On the broader development of naval thought in this period, see Dirk Bönker, *Militarism in a Global Age: Naval Ambitions in Germany and the United States before World War I* (Ithaca, NY: Cornell University Press, 2012).

8. The pioneering work of this sort is Martin L. van Creveld, *Supplying War: Logistics from Wallenstein to Patton* (New York: Cambridge University Press, 1977); see also Martin L. van Creveld, "World War I and the Revolution in Logistics," in *Great War, Total War: Combat and Mobilization on the Western Front, 1914–1918*, ed. Roger Chickering and Stig Förster (New York: Cambridge University Press, 2000), 57–72, and John A. Lynn, ed., *Feeding Mars: Logistics in Western Warfare from the Middle Ages to the Present* (Boulder, CO: Westview Press, 1993). In his important study of Revolutionary War military supply, historian James A. Huston observes that the word "logistics" was unused in eighteenth-century America without examining the significance of this fact to the relationship between ideas and actions in warfare (*Logistics of Liberty: American Services of Supply in the Revolutionary War and After* [Newark: University of Delaware Press, 1991], 9).

9. Carl T. Vogelgesang, "Logistics—Its Bearing Upon the Art of War," 14–16, folder 26, box 2, RG 14, USNWC.

10. Naval officers' unfamiliarity with the subject might have been for responsible for Luce's defining the term in a footnote (as "the branch of the military art which has to do with the details of moving and supplying armies or fleets")—and this to an audience of professional naval officers. His one notable example of the importance of logistics was coal, a substance "of the very first importance" and which "under certain conditions coal may rank above ammunition in the scale of military values" ("Naval Training," *Proceedings of the United States Naval Institute* 16, no. 3 [1890]: 383). Similarly, Mahan's example of the most important logistic work involved coaling the fleet ("The Naval War College," *North American Review* 196, no. 680 [1912]: 72–75). For usages of the word, see the *United States Naval Institute Proceedings* between 1890 and 1913.

11. Stephen B. Luce, "War Schools," *Proceedings of the United States Naval Institute* 9, no. 5 (1883): 633–57; T. J. Cowie to Josephus Daniels, March 15, 1915, "Outline History of the Naval War College," July 1915, enclosures to Austin Knight to Josephus Daniels, August 16, 1916 (a June 1913 synopsis and an untitled history of the Naval War College), and Josephus Daniels to T. J. Cowie, March 19, 1915, roll 100, navy subject file, JDP; Ronald H. Spector, *Professors of War: The Naval War College and the Development of the Naval Profession* (Newport, RI: Naval War College Press, 1977); John B. Hattendorf, B. Mitchell Simpson III, and John R. Wadleigh, *Sailors and Scholars: The Centennial History of the U.S. Naval War College* (Newport, RI: Naval War College Press, 1984).

12. Theodore Roosevelt to Caspar Frederick Goodrich, May 28 and June 16, 1897, in Elting E. Morison, ed., *Letters of Theodore Roosevelt*, 8 vols. (Cambridge: Harvard University Press, 1951–54), 1:617–68, 626; C. F. Goodrich to assistant secretary of the navy [Theodore Roosevelt], June 23, 1897, and attached coal supply memorandum, 1, 7, folder 6, box 104, RG 8, USNWC.

13. Goodrich's classification of war theory derived from the late nineteenth-century West Point engineer Junius B. Wheeler, who in turn drew from the Swiss theorist Antoine-Henri Jomini. Following Wheeler, Goodrich had first divided the study of the art of war into four areas, strategy, tactics, engineering, and logistics; in his second set of notes, in which he focused specifically on naval warfare, he replaced engineering with coast defense and logistics with convoy and supply. The switch reflected the uncertainty of applying a technical military term to the naval context (two drafts of "Definitions," box 2, CFGP; Junius Brutus Wheeler, *A Course of Instruction in the Elements of the Art and Science of War* [New York: D. Van Nostrand, 1878], 7–10; Junius Brutus Wheeler, "Logistics," circa 1901, 1–9, box 3, CFGP).

14. Antoine Henri Jomini, *The Art of War* (Westport, CT: Greenwood Press, 1971), 252; Edward S. Farrow, *Farrow's Military Encyclopedia: A Dictionary of Military Knowledge* (New York: published by the author, 1885), 2:230; Rudolf von Caemmerer, *The Development of Strategical Science During the 19th Century*, trans. Karl von Donat (London: Hugh Rees, 1905), 44. Von Caemmerer does not mention in this text the word itself, which is "λογιστικό." See also "Modern War in Theory and Practice: Its Foreign Lessons and Domestic Teachings," *United States Service Magazine* 1, no. 1 (1864): 60, and H. Wager Halleck, *Elements of Military Art and Science; or, Course of Instruction in Strategy, Fortification, Tactics of Battles, &c*, 3rd ed. (New York: Appleton, 1862), 88.

15. Bradley A. Fiske, *The Navy as a Fighting Machine* (New York: Charles Scribner's Sons, 1916), 278.

16. Carl T. Vogelgesang, "Logistics—Its Bearing Upon the Art of War," 4, folder 26, box 2, RG 14, USNWC.

17. T. J. Cowie, "Logistics," March 5, 1915, 10, folder 5, box 89, RG 8, USNWC.

18. "Outline of Organization of the Naval War College," roll 100, JDP; George Cyrus Thorpe, *Pure Logistics: The Science of War Preparation* (Kansas City, MO: Franklin Hudson, 1917), 6.

19. H. P. Huse, "Logistics—Its Influence Upon the Conduct of War and Its Bearing Upon the Formulation of War Plans," September 15, 1916, folder 313, box 8, RG 4, USNWC. This War College logistics thesis soon appeared in print as "Logistics—Its Influence Upon the Conduct of War and Its Bearing Upon the Formulation of War Plans" (*United States Naval Institute Proceedings* 43, no. 2 [1917]: 245–53). Some twenty-five years earlier, Huse had served as Rear Admiral Bancroft Gherardi's aid and translator on his mission to secure Môle-Saint-Nicolas in Haiti as a coaling station. His career thus bridges the navy's ad hoc and scientific approaches to coaling.

20. T. J. Cowie, "Logistics," March 5, 1915, 32, folder 5, box 89, RG 8, USNWC. On the Napoleonic logistics to which Cowie refers, see Martin L. van Creveld, *Supplying War: Logistics from Wallenstein to Patton*, 2nd ed. (New York: Cambridge University Press, 2004), especially 40–74. Compare with War College instructor Carl T. Vogelgesang, who also emphasized the reciprocity between logistics and strategy: "The strategic conception may be that of a genius, but if it be not based on a solid foundation of logistic facts, it can have no

force and will be of no effect; unless, indeed, it leads directly to disaster" ("Logistics—Its Bearing Upon the Art of War," 1, 5–7, folder 26, box 2, RG 14, USNWC).

21. Daniel Joseph Costello, "Planning for War: A History of the General Board of the Navy, 1900–1914" (PhD diss., Fletcher School of Law and Diplomacy, Tufts University, 1968), esp. 45–49, also 23–105, 173–225, more generally; Richard D. Challener, *Admirals, Generals, and American Foreign Policy, 1898–1914* (Princeton, NJ: Princeton University Press, 1973), 81–110, 179–98, 323–32; Arent S. Crowninshield to George Dewey, July 11, 1901, box 43, GDP. Bradford defended his territory over coaling stations by withholding requested documents detailing the present and anticipated future capacities of naval stations and the quantity of ships capable of coaling, as well as by refusing to answer correspondence from the board unless it came through the proper channels of the Bureau of Navigation (Royal B. Bradford, "Coaling-Stations for the Navy," *Forum* 26 [February 1899]: 732–47). Ultimately, the conflict was resolved by placing Bradford on the General Board itself. On the General Board and coaling, see H. C. Taylor to George Dewey, September 10, 1902, F. E. Chadwick to chief of the Bureau of Equipment, August 20, 1902, William H. Moody to the president of the General Board, July 18, 1902, and General Board minutes for August, September, and October, 1902, box 43, GDP, GB no. 28, April 17, 1900, 5, M1493, and John H. Maurer, "Fuel and the Battle Fleet: Coal, Oil, and American Naval Strategy, 1898–1925," *Naval War College Review* 34, no. 6 (1981): 60–77. As for developing coal supplies in the new American colony of the Philippines itself, the U.S. government did survey and reopen some old Spanish coal mines; see "The United States Government Mining Its Own Coal," *American Monthly Review of Reviews* 33 (February 1906): 218–19, Oscar Halvorsen Reinholt, "United States Enterprise in the Coal Trade of the Philippines," *Engineering Magazine* 30, no. 4 (1906): 491–517, Warren Du Pre Smith, *The Coal Deposits of Batan Island: With Notes on the General and Economic Geology of the Adjacent Region* (Manila: Bureau of Printing, 1905), and H. G. Wigmore, *Report of Examination of Coal Deposits on the Batan Military Reservation, Batan Island, P.I.* (Washington, DC: Government Printing Office, 1905). Still, American fuel in the Philippines continued to be imported from the United States.

22. T. J. Cowie, "Logistics," March 5, 1915, 11, folder 5, box 89, RG 8, USNWC.

23. J. S. McKean, *Naval Logistics: Lecture Delivered by Commander J. S. McKean, United States Navy, at the Naval War College Extension, Washington, D.C., March 10, 1913* (Washington, DC: Government Printing Office, 1915), 3; T. J. Cowie, "Logistics," March 5, 1915, 5, folder 5, box 89, RG 8, USNWC.

24. R. E. Bakenhus, "Lecture as to the Course in Logistics," December 1, 1926, 2–3, folder 1204, box 29, RG 4, USNWC.

25. "Strategic 49," July 1915, 1–2, folder 478, box 12, RG 4, USNWC. Other colors commonly used included black for Germany, red for Great Britain, and green for Mexico; more exotic ones included citron for Brazil and indigo for Iceland. On the color system, see "Appendix I: The Colors of the Rainbow," in Michael Vlahos, *The Blue Sword: Naval War College and the American Mission, 1919–1941* (Newport, RI: Naval War College Press, 1980), 163. For the development of U.S. war plans against Japan, see Edward S. Miller, *War Plan Orange: The U.S. Strategy to Defeat Japan, 1897–1945* (Annapolis, MD: Naval Institute Press, 1991), and Steven T. Ross, *American War Plans, 1890–1939* (London: Frank Cass, 2002).

26. See, for example, the letters and charts reporting fuel consumption for naval vessels in 1919 in folders 2, 3, and 5, box 38, RG 8, and Paul Foley, "Notes on the Preparation of the Logistic Sheet," 1911, folder 2, box 45, RG 8, USNWC.

27. T. J. Cowie, "Logistics," March 5, 1915, 9, folder 5, box 89, RG 8, USNWC.

28. T. J. Cowie, "Logistic Data on Production and Industry of the United States," March 1, 1917, 1, folder 86, box 2, RG 4, USNWC.

29. Carl T. Vogelgesang, "Logistics—Its Bearing Upon the Art of War," 9–10, folder 26, box 2, RG 14, USNWC.

30. Richard Wainwright, memorandum, March 4, 1910, and "Fuel Supply for the Fleet," circa July, 1910, box 39, GBSF.

31. T. J. Cowie, "Logistics," March 5, 1915, 8, 34, folder 5, box 89, RG 8, USNWC.

32. H. T. B. Harris to Samuel McGowan, October 31, 1906, H. T. B. Harris, "Report on the Fitness of Officers," G. A. Converse to Samuel McGowan, June 25, 1906, and February 11, March 9, May 11, and July 17, 1907, C. S. Sperry to J. E. Pillsbury, August 20, 1908, V. H. Metcalf to Samuel McGowan, May 26, 1908, Beekman Winthrop to Samuel McGowan, June 1, 1909, and Samuel McGowan to Philip Andrews, June 18, 1910, box 1, SMP.

33. Bulletin No. 533, June 25, 1914, Josephus Daniels to Samuel McGowan, July 1, 1914, W. S. Benson to Josephus Daniels, August 30, 1916, and Josephus Daniels to B. R. Tillman, January 29, 1917, box 1, SMP; Thomas R. Shipp, "Rear Admiral Samuel McGowan," *World's Work* 36, no. 1 (1918): 36–38.

34. Samuel McGowan, statement, January 1915, box 4, SMP; "Efficiency in War Work Illustrated," *Christian Science Monitor*, December 5, 1917, 9; Josephus Daniels to Harry Garfield, October 7, 1918, roll 36, JDP; Josephus Daniels, statement, box 1, FDRL; T. J. Cowie to Archibald McNeill and Sons Co., May 23, 1914, [Franklin D. Roosevelt], memorandum, July 13, 1914, Samuel McGowan, memorandum, September 23, 1914, and Samuel McGowan, confidential memorandum, box 5, FDRL; *Hearing Before the House Committee on Naval Affairs*, 63rd Cong., 3rd sess., December 16, 1914, 957–76.

35. Albert W. Fox, "Dilemma for Daniels," *Washington Post*, September 26, 1917, 1; "Rally for M'Gowan," *Washington Post*, September 27, 1917, 3; "To Oppose Change in Navy's System," *Washington Evening Star*, September 26, 1917, 17; "Navy Its Own Buyer: McGowan Is Victor in Struggle with War Industries Board," *Washington Post*, October 6, 1917, 2. For its part, the *Washington Post* was baffled at the disparity in organization between the army and navy; see "Army and Navy Supply Systems," *Washington Post*, December 22, 1917, 6, and Albert W. Fox, "What Navy Has Done: Everything Army Couldn't, So Congressional Report Says," *Washington Post*, January 14, 1918, 1. See also Martin J. Gillen, "The Mobilization and Supply of Army Material," *Journal of the Military Service Institution of the United States* 57, no. 198 (1915): 379–86. The best works on American mobilization remain Robert D. Cuff, *The War Industries Board: Business-Government Relations during World War I* (Baltimore, MD: Johns Hopkins University Press, 1973), and Paul A. C. Koistinen, *Mobilizing for Modern War: The Political Economy of American Warfare, 1865–1919* (Lawrence: University Press of Kansas, 1997).

36. Paymaster general [Samuel McGowan] to H. M. Robinson, February 27, 1920; Samuel McGowan to Chesapeake and Ohio Coal and Coke Company, March 22, 1920; secretary of the navy [Josephus Daniels] to unnamed coal dealers, August 18, 1920, box 5, SMP; *Hearing before the House Committee on Naval Affairs, Subcommittee for Investigation of Conduct and Administration of Naval Affairs*, 65th Cong., 2nd sess., December 19, 1917, 48–50). The figures for navy orders in 1917 run through December 10 (Samuel McGowan to Josephus Daniels, June 19, 1916, roll 56, JDP). Final prices could be determined by the contractor

accepting the navy's determination, fixture by the Federal Trade Commission, or on the authority of the president with the advice of the War Industries Board.

37. Conference of Coal Producers, minutes May 10, 1917, and Conference of Special Committee of Coal Producers, minutes, May 10, 1917, 9, box 12, FDRL.

38. Conference of Coal Producers, minutes, May 10, 1917, Conference of Special Committee of Coal Producers, minutes, May 10, 1917, Conference of Coal Producers, minutes, May 29, 1917, 2, 42, Conference of Coal Producers, minutes, June 5, 1917, 21, box 12, FDRL; Josephus Daniels to Harry Garfield, October 7, 1918, roll 36, JDP.

39. Harry Garfield to Josephus Daniels, March 3, 1921, roll 36, JDP; "City Struggles to Obey," *New York Times*, January 18, 1918, 1; John G. Clark, *Energy and the Federal Government: Fossil Fuel Policies, 1900–1946* (Urbana: University of Illinois Press, 1987), 49–80.

40. C. G. May, memorandum, December 21, 1920, box 2, SMP.

41. Samuel McGowan to Josephus Daniels, March 31, 1920, roll 57, JDP.

42. "How Navy Supply Bureau Saves Money: By a System of Charts a Force of Over 150 Men Estimates the Opportune Moments to Make Expenditures of Millions of Dollars Yearly," *Annalist*, October 22, 1917; "Efficiency in War Work Illustrated," *Christian Science Monitor*, December 5, 1917, 9; Frank G. Carpenter, "How the War Business of the United States Navy Is Being Managed," *Washington Evening Star*, May 12, 1918, 35; *Hearing before the House Committee on Naval Affairs, Subcommittee for Investigation of Conduct and Administration of Naval Affairs*, 65th Cong., 2nd sess., December 19, 1917, 50–51.

43. Fred C. Kelly, "How They Did It," *System* 33, no. 1 (1918): 48–50.

44. [W. G. Townes?] to Josephus Daniels, May 15, 1917, People's Power League of Illinois to Frank O. Lowden (copy), July 10, 1917, People's Power League, Committee on National Defense, report, July 15, 1917, roll 29, navy subject file, JDP.

45. Thomas H. Watkins to Josephus Daniels, September 18, 1917, roll 29, navy subject file, JDP; Thomas H. Watkins, "Causes of the Present Coal Situation," *Retail Coalman* 31, no. 4 (1917): 66a.

46. H. C. Davis, "Logistics: Its Influence Upon the Conduct of War and Its Bearing Upon the Preparation of War Plans," *United States Naval Institute Proceedings* 43, no. 5 (1917): 951–52.

47. Harrington Emerson, "The Coal Resources of the Pacific," *Engineering Magazine* 23, no. 2 (1902): 161, 164.

48. Ibid, 165.

49. Benton MacKaye, "Alaska—an Opportunity to Build a Nation," 1920, 3–4, 8, box 176, PMF; Larry Anderson, *Benton Mackaye: Conservationist, Planner, and Creator of the Appalachian Trail* (Baltimore, MD: Johns Hopkins University Press, 2002), 62–69, 76–77, 125, 302–3.

50. On the reimagining of geographic relationships, see Richard White, *Railroaded: The Transcontinentals and the Making of Modern America* (New York: Norton, 2011), 140–78, Anne Godlewska, "Map, Text and Image: The Mentality of Enlightened Conquerors: A New Look at the *Description de l'Egypte*," *Transactions of the Institute of British Geographers*, n.s., 20, no. 1 (1995): 5–28, D. Graham Burnett, *Masters of All They Surveyed: Exploration, Geography, and a British El Dorado* (Chicago: University of Chicago Press, 2000), Derek Gregory, "Between the Book and the Lamp: Imaginative Geographies of Egypt, 1849–50," *Transactions of the Institute of British Geographers*, n.s., 20, no. 1 (1995): 29–57, Susan Schulten, *The Geographical Imagination in America, 1880–1950* (Chicago: University of Chicago

Press, 2001), Susan Schulten, *Mapping the Nation: History and Cartography in Nineteenth-Century America* (Chicago: University of Chicago Press, 2012), Joan Schwartz, "The Geography Lesson: Photographs and the Construction of Imaginative Geographies," *Journal of Historical Geography* 22, no. 1 (1996): 16–45, Daniel Lord Smail, *Imaginary Cartographies: Possession and Identity in Late Medieval Marseille* (Ithaca, NY: Cornell University Press, 2000), Emma J. Teng, *Taiwan's Imagined Geography: Chinese Colonial Travel Writing and Pictures, 1683–1895* (Cambridge, MA: Harvard University Press, 2004), and of course Edward Said, *Orientalism* (New York: Pantheon, 1978), and Benedict Anderson, *Imagined Communities: Reflections on the Origin and Spread of Nationalism* (London: Verso, 1983).

51. "Dutch Harbor (Unalaska), Alaska; Summary of Correspondence," April 10, 1915, and senior member of General Board to secretary of the navy, August 30, 1915, box 37, GBSF; George Dewey for General Board to secretary of the navy, November 25, 1903, box 36, GBSF. See also General Board, minutes, December 19, 1900, and June 19, September 26, and September 27, 1902, roll 1, M1493. The Aleutians had been first extensively explored by Americans in the early 1870s, through a series of summer expeditions by the U.S. Coast Survey. Under William H. Dall, Coast Survey scientists measured currents and tides and recorded the meteorology of the chain. In 1873, they surveyed the islands of Attu, Kiska, the Davidoffs, Amchitka, Adakh, Atkha, Unalaska, and the Shumagin group. Kiska was then of particular importance; of all the islands surveyed, it alone provided a harbor adequate to build a relay station for a proposed telegraphic cable from North America to Japan. Dall described the chain as having a "mild and uniform" climate, "not so cold as that of Philadelphia," but buffeted by frequent fogs, rain, and "extreme fluctuations" in barometric pressure (William H. Dall, "Explorations in the Aleutian Islands and Their Vicinity," *Journal of the American Geographical Society of New York* 5 [1874]: 243–35).

52. George Dewey for General Board to secretary of the navy, November 25, 1903, and William H. Moody to commander in chief of the Pacific Squadron, June 13, 1903, box 36, GBSF. Although the great circle route from Puget Sound to Manila swung north of some Aleutian islands, ships following that path typically avoided steaming so high in order to avoid passing through the dangerous straights between individual islands. According to one General Board report of 1903, "The violent and irregular currents, frequent fogs, hidden rocks and imperfectly charted shores, have justly caused these passes to be regarded as difficult of navigation, and have deterred the trans-Pacific steamers from following the true great circle track to the northward of the islands" ("Dutch Harbor (Unalaska), Alaska; Summary of Correspondence," April 10, 1915, box 37, GBSF).

53. Henry Glass to H. C. Taylor, May 28 and May 5, 1903, and Henry Glass to secretary of the navy, August 13, 1903, box 36, GBSF; William H. Moody to commander in chief of the Pacific Squadron, June 13, 1903, box 36, GBSF.

54. Henry Glass to secretary of the navy, August 13, 1903, box 36, GBSF.

55. Royal B. Bradford, endorsement of Henry Glass's Aleutian report (dated August 13, 1903), September 26, box 36, GBSF.

56. George Dewey for General Board to secretary of the navy, November 25, 1903, George A. Converse to General Board, January 28, 1904, and Francis H. Sherman to chief of Bureau of Navigation, October 2, 1904, box 36, GBSF.

57. Francis H. Sherman to chief of Bureau of Navigation, October 2, 1904, secretary of the navy to commanding officer of the *Petrel* [Francis H. Sherman], March 29, 1904, George A. Converse, second endorsement of letter from George B. Cortelyou (dated

February 3, 1904), March 9, 1904, memorandum, June 18, 1904, and George A. Converse to General Board, January 28, 1904, box 36, GBSF. The Bureau of Equipment had at this time as little as $140,000 available in its appropriation for coal depots.

58. "Dutch Harbor (Unalaska), Alaska; Summary of Correspondence," April 10, 1915, and George W. Brown, report, October 11, 1915, box 37, GBSF; Theodore Roosevelt, executive order, June 13, 1902, box 36, GBSF. Squatters had already caused headaches near naval stations in the Caribbean in Culebra and Guantanamo (George Dewey for General Board to secretary of the navy, November 25, 1903, Theodore Roosevelt, executive order, December 9, 1903, and secretary of General Board to Theodore Roosevelt, December 8, 1903, box 36, GBSF).

59. George Dewey to secretary of the navy, March 29, 1905, and George Dewey, second endorsement of letter from Arthur R. Boyle (dated February 17, 1908), March 25, 1908, box 36, GBSF.

60. These stations remained unfortified as well; see "Dutch Harbor (Unalaska), Alaska; Summary of Correspondence," April 10, 1915, box 37, GBSF.

61. Ellicott had been a student at the War College in 1896 and served on its staff between 1900 and 1901 (*Register of Officers, 1884–1968* [(Newport, RI: Naval War College, 1968], Naval Historical Collection, USNWC). Within the navy, interest in assignment to the waters of Alaska was not widespread—Ellicott reported that the detail officer was happy to give him this command, explaining that "nobody ever asked for that district before" and adding that Ellicott would "find it the most harassing, hazardous, thankless job you ever undertook." Ellicott later confirmed this prediction ("Harbor Hunting in Alaska," *Proceedings of the United States Naval Institute* 63, no. 413 [1937]: 939).

62. Alfred H. Brooks, "Alaska Coal and Its Utilization," in *Mineral Resources of Alaska: Report on Progress of Investigations in 1909,* ed. Alfred H. Brooks (Washington, DC: Government Printing Office, 1910), 47–100; secretary of the navy to General Board, April 7, 1910, and "Extract from 'Geology and Mineral Resources of the Controller Bay Region, Alaska,' U.S. Geological Survey, 1908," box 36, GBSF.

63. George F. Kay, "The Bering River Coal Field, Alaska," *Popular Science Monthly* 79 (November 1911): 417–30; Alfred H. Brooks, "Geography in the Development of the Alaska Coal Deposits," *Annals of the Association of American Geographers* 1 (1911): 85–94; Alfred H. Brooks, "The Future of Alaska Coal," in *Report of the Proceedings of the American Mining Congress: Fourteenth Annual Session* (Denver, CO: American Mining Congress, 1911), 291.

64. James L. Penick, *Progressive Politics and Conservation: The Ballinger-Pinchot Affair* (Chicago: University of Chicago Press, 1968); Char Miller, *Gifford Pinchot and the Making of Modern Environmentalism* (Washington, DC: Island Press, 2001), 206–26; John E. Lathrop and George Kibbe Turner, "Billions of Treasure: Shall the Mineral Wealth of Alaska Enrich the Guggenheim Trust or the United States Treasury?" *McClure's* 34, no. 3 (1910): 341; see also, "Alaska's Contribution to Our Coal Supply," *Review of Reviews* 41, no. 4 (1910): 483–84.

65. Anderson, *Benton MacKaye,* 76–77; "Are You with Us on This Great Project? It's to Help Seattle" *Seattle Star,* August 5, 1913, 1; Gilson Gardner, "Senator's Plan Would Compel Production and Sale of Coal at the Very Lowest Price Possible," *Minneapolis Daily News,* July 28, 1913; Gilson Gardner, "Coal Miner, Consumer and Uncle Sam to Be Partners Who'll Share Equally in the Development of Alaska," *Columbus (OH) Citizen,* July 26, 1913 (this same story ran in the *Cleveland Press* on July 26, 1913 with the unreassuring headline "Peo-

ple, Not Trusts, to Exploit Alaska"); Gilson Gardner, "When U.S. Opens Alaska the Coal Will Fly," *Chicago Day Book*, August 1, 1913, 29; Glison Gardner, "The Coal Will Fly When Uncle Sam Opens Alaska," *Minneapolis Daily News*, August 4, 1913; see also "Alaska Bill Sweeping: Poindexter Proposed Development of Coal," *Oregonian*, July 11, 1913, 9, "Poindexter Bill Novel Features: Government Operation of Mines, Steamships and Railroads in Alaska," *Springfield (IL) Union*, July 12, 1913, 2, and "Would Split Gains with Coal Miners," *Trenton (NJ) Evening Times*, July 11, 1913, 4.

66. Coal Board to commanding officer of the *Maryland*, September 10, 1913, and anonymous miner to Navy Department, August 8, 1913, file 25320 (13)-(105), box 1019, GCSN; H.R. Ex. Doc. No. 876, 63rd Cong, 2nd sess. (1914), 60–61, 121. Miners suspected that recently resigned secretary of the interior Walter Fisher preferred the "Matanuski" coal field to facilitate development of the government railroad from Fairbanks to the coast. This Russian inflection on "Matanuska" suggests either a perception of foreignness among the miners with respect to interference in Alaskan matters or else simply reflects the Russian derivation of the river's name from "mednovtsy," meaning "copper people," the name Russian traders bestowed upon the Athabaskan natives of a region thought to contain many copper deposits (Andrei A. Znamenski, *Through Orthodox Eyes: Russian Missionary Narratives of Travels to the Dena'ina and Ahtna, 1850s–1930s* [Fairbanks: University of Alaska Press, 2003], 71).

67. S. Ex. Doc. No. 26, 64th Cong., 1st sess. (1915), 1–76; *Hearing of the House Committee on Naval Affairs*, 63rd Cong., 3rd sess., December 14, 1914, 755–77; *Hearing of the House Committee on Naval Affairs*, 63rd Cong., 3rd sess., December 16, 1914, 965–66; *Hearing of the House Committee on Naval Affairs*, 64th Cong., 1st sess., January 28, 1916, 776–78; Sumner Kittelle, "Navy Coal in Alaska," February 26, 1920, file 25320 (131)-(155:5), box 1019, GCSN; Sumner Kittelle to secretary of the navy, June 9, 1919, file 25320 (171:12), box 1020, GCSN; "Alaskan Coal Situation: Extracts from Reports of Captain Kittelle," 1919, file 25320 (156)-(156:29), box 1019, GCSN.

68. A typical collier carrying 11,500 tons of coal required about eighty-two days to "load, carry, and discharge" coal from Virginia's Hampton Roads, the navy's principle coal depot, through the Panama Canal to the naval station at Puget Sound, Washington. From Anchorage, the voyage to Puget Sound required less than half that time, a mere forty days. It took sixty-six days for a collier to get from Hampton Roads to San Francisco, but from Anchorage, it took just thirty-six. The trip from Hampton Roads to Pearl Harbor took 73.5 days, while it took 37.5 from Anchorage. The navy added an additional thirty-seven days to voyages from Hampton Roads should ships be forced to steam around Cape Horn in the event of a catastrophe at the Panama Canal (frequently noted as a threat during wartime). When peace prevailed, the cost of shipping coal from the East coast to the Pacific was relatively low and how long it took for a collier make the journey was not especially important. During war, the entire coal supply at the navy's Pacific bases was estimated to last a mere three months, and the speed and ease of replenishing the coal supply exceeded any question of cost. "The successful development of the Matanuska Coal Field by the Navy is of inestimable value," asserted one navy report. "The coal itself has passed Naval tests and fills every requirement. It is the only good steaming coal the Navy has access to on the Pacific in time of national emergency. Every effort should be exhausted here to obtain a future supply of coal for the Navy. It is surely wise policy to push this enterprise as rapidly as machinery and labor permit" (Philip J. Weiss, memorandum, July 1, 1921, box 2, NACC-Alaska).

69. Josephus Daniels, "Summary of the Opinion of the Navy Department in Regard to the Plan Which Should Be Pursued in So Far as Relates to the Navy's Interest in These Fields," March 12, 1917, file 25320 (131)-(155:5), box 1019, GCSN.

70. Hugh Rodman to secretary of the navy, June 18, 1920, policy file, box 4, NACC-Alaska; memorandum, August 4, 1920, plans for development file, box 2, NACC-Alaska.

71. Otto Dowling to William C. Cole, August 1, 1921, and April 11, 1922, file 25320 (156:68), box 1020, GCSN.

72. Otto Dowling to William C. Cole, August 1, 1921, file 25320 (156:68), box 1020, GCSN.

73. Otto Dowling to William C. Cole, November 25, 1921, file 25320 (156:68), box 1020, GCSN.

74. "The Coal Mining Operations of the Alaskan Engineering Commission," December 16, 1921, policy file, box 4, NACC-Alaska.

75. William C. Cole to Otto Dowling, February 6, 1922, file 25320 (156:68), box 1020, GCSN. As historian John G. Clark has observed, the 1919 "coal crisis" itself was a product of deep, structural transformations underway in America's industrial economy. The November–December miners' strike was preceded in September by a deadly strike in the steel industry, and in 1919 alone, roughly four million workers participated in thirty-six hundred strikes (*Energy and the Federal Government*, 112–17). James Johnson notes that these actions were frequently interpreted as "the opening volleys of revolution" whose participants were maligned as agents of Soviet intrigue (*The Politics of Soft Coal: The Bituminous Industry from World War I through the New Deal* [Urbana: University of Illinois Press, 1979], 81–84, 101–2).

76. Philip Weiss to William C. Cole, September 22, 1921, file 25320 (156:68), box 1020, GCSN; Sumner S. Smith to Otto Dowling, November 12, 1920, Otto Dowling to secretary of the navy, April 30, 1921, and Fredrick F. Mears to Albert B. Fall, October 7, 1921, wages file, box 4, NACC-Alaska. The first Alaskan wage increase raised wages from $6.50 to $7.10 per day; the second raised them still further to $8.60. Wages in Washington by the end of 1920 were $8.25 per day.

77. Fredrick F. Mears to Albert B. Fall, October 7, 1921, wages file, box 4, NACC-Alaska; Philip Weiss to William C. Cole, September 23, 1921, and George S. Rice to RDB, October 27, 1921, file 25320 (156:68), box 1020, GCSN.

78. Philip Weiss to William C. Cole, September 22 and September 23, 1921, Mines Committee to Dan Sutherland, September 23, 1921, and Sumner S. Smith to H. Foster Bain, September 28, 1921, file 25320 (156:68), box 1020, GCSN; Otto Dowling to Office of Naval Operations, September 25, 1921, file 25320 (156)-(156:29), box 1019, GCSN; Fredrick F. Mears to Albert B. Fall, October 7, 1921, wages file, box 4, NACC-Alaska.

79. E. C. Finney to Frederick Mears, September 27, 1921, Sumner S. Smith to H. Foster Bain, September 28, 1921, and E. C. Finney to William C. Cole, November 26, 1921, file 25320 (156:68), box 1020, GCSN; Edwin Denby to Otto Dowling, September 27, 1921, file 25320 (156)-(156:29), box 1019, GCSN.

80. Office of Naval Operations to Otto Dowling, February 25, 1922, telegram files, box 4, NACC-Alaska; chiefs of Bureau of Engineering and Bureau of Yards and Docks to secretary of the navy, March 10, 1922, file 25320 (171:12), box 1020, GCSN; W. P. T. Hill, "Inclosure 'B' of the Final Report of the Navy Alaskan Coal Commission to the Secretary of the Navy," 1922, Inclosure "B" file, 44, 47–48, 83, box 1, NACC-Alaska. A subsequent order in

March delayed the closure until May 1; this gave the navy time to shut down the mine and negotiate a deal to turn Chickaloon over to the Interior Department. See Edwin Denby to Otto Dowling, March 8, 1922, and Office of Naval Operations to Otto Dowling, March 18, 1922, telegram files, box 4, NACC-Alaska.

81. *Leases Upon Naval Oil Reserves: Hearings before the Committee on Public Lands and Surveys, United States Senate*, 68th Cong., 1st sess., November 30, 1923, 1:894–914, 916–62.

82. Political scandal is the lens, for example, in Burl Noggle, *Teapot Dome: Oil and Politics in the 1920's* (New York: Norton, 1962), and Laton McCartney, *The Teapot Dome Scandal: How Big Oil Bought the Harding White House and Tried to Steal the Country* (New York: Random House, 2008), not to mention Frederick Lewis Allen's enormously influential and still read *Only Yesterday: An Informal History of the Nineteen-Twenties* (New York: Harper and Brothers, 1931).

83. *United States v. Mammoth Oil Co.*, 5 F.2d 330 (D. Wyo. 1925), June 19, 1925.

84. Franklin D. Roosevelt, "The National Need of Naval Petroleum Reserves," box 40, FDRL.

85. "Urges Protection of Navy Oil Lands: Consulting Board Says Consumption Will Go Up More Than Tenfold by 1927," *New York Times*, December 10, 1916, 7; Lloyd N. Scott, *Naval Consulting Board of the United States* (Washington, DC: Government Printing Office, 1920), 62.

86. H.R. Ex. Doc. No. 1, 38th Cong., 2nd sess. (1864), 1096; H.R. Ex. Doc. No. 1, 40th Cong., 2nd sess. (1867), 175.

87. H.R. Ex. Doc. No. 3, 57th Cong., 2nd sess. (1902), 351–52; John Ise, *The United States Oil Policy* (New Haven, CT: Yale University Press, 1926), 62–64, 88; H.R. Ex. Doc. No. 3, 57th Cong., 1st sess. (1901), 2:1033; John R. Edwards, *Report of the U.S. Naval "Liquid Fuel" Board of Tests Conducted on the Hohenstein Water Tube Boiler* (Washington, DC: Government Printing Office, 1904), 434–35; John A. DeNovo, "Petroleum and the United States Navy before World War I," *Mississippi Valley Historical Review* 41, no. 4 (1955): 641–643. According to Bureau of Steam Engineering chief Ward Winchell, the mechanical problems stemmed largely from the fact that steam engineers burned oil in the same manner as coal (in bulk), even though the process of atomizing it first produced more satisfactory results (H.R. Ex. Doc. No. 3, 57th Cong., 2nd sess. [1902], 736).

88. Theodore Roosevelt, "Special Message of the President Transmitting the Report of the National Conservation Commission," in Henry Gannett, ed., *Report of the National Conservation Commission* (Washington, DC: Government Printing Office, 1909), 2; "The Prodigal Waste of Natural Resources: Undeveloped Wealth of the Nation, Supposed to be Inexhaustible, Found to be, According to Recent Survey, Nearing Its Limit," *New York Times*, November 10, 1907, SM10; Samuel P. Hays, *Conservation and the Gospel of Efficiency: The Progressive Conservation Movement, 1890–1920* (Cambridge, MA: Harvard University Press, 1959), 127–33.

89. David T. Day, "The Petroleum Resources of the United States," in *Papers on the Conservation of Mineral Resources*, United States Geological Survey Bulletin 394 (Washington, DC: Government Printing Office, 1909), 34–35, 44–46; David T. Day, A. C. Veatch, and Ralph Arnold to director of U.S. Geological Survey [George Otis Smith], November 11, 1908, in Max W. Ball, *Petroleum Withdrawals and Restorations Affecting the Public Domain* (Washington, DC: Government Printing Office, 1916), 117.

90. Peter A. Shulman, "'Science Can Never Demobilize': The United States Navy and Petroleum Geology, 1898–1924," *History and Technology* 19, no. 4 (2003): 365–85; J. Leonard Bates, *The Origins of Teapot Dome: Progressives, Parties, and Petroleum, 1909–1921* (Urbana: University of Illinois Press, 1963); George Otis Smith to secretary of the interior [James Garfield], February 24, 1908, in Ball, *Petroleum Withdrawals and Restorations*, 104.

91. H.R. Ex. Doc. No. 1480, 64th Cong., 2nd sess. (1916), 31–36.

92. H.R. Ex. Doc. No. 681, 63rd Cong., 2nd sess. (1913), 14–16; H.R. Ex. Doc. No. 1484, 63rd Cong., 3rd sess. (1914), 18–21; H.R. Ex. Doc. No. 20, 64th Cong., 1st sess. (1915), 62–65.

93. Franklin D. Roosevelt to Franklin K. Lane, September 16, 1916, box 1, NFOB.

94. Naval Fuel Oil Board, report, December 22, 1916, Naval Fuel Oil Board, report, appendix D, Naval Fuel Oil Board, ad interim report, September 21, 1916, and John Edwards to R. B. Owens, September 11, 1916, box 1, NFOB; Bates, *Origins of Teapot Dome*, 118–19. For a survey of the American and British naval transitions from coal to oil, see David Allan Snyder, "Petroleum and Power: Naval Fuel Technology and the Anglo-American Struggle for Core Hegemony, 1889–1922" (PhD diss., Texas A&M University, 2001).

95. H.R. Ex. Doc. No. 618, 65th Cong., 2nd sess. (1917), 57, 59, 264; H.R. Ex. Doc. No. 1450, 65th Cong., 3rd sess. (1918), 138; H.R. Ex. Doc. No. 994, 66th Cong., 3rd sess. (1920), 140.

96. Clark, *Energy and the Federal Government*, 92–106; Mark L. Requa, "The Necessity for Government Control of the Oil Industry," July 12, 1918, 4, 12, box 72, trays 139–40, USBM.

97. Bates, *The Origins of Teapot Dome*, 181–218; *Leases Upon Naval Oil Reserves*, 2:1634.

98. Robert S. Griffin to Thomas J. Walsh, October 23, 1923, box 211, TJWP. In Robison's telling, Stuart sought not only to preserve the status quo of naval control but also to further develop functions historically housed in Interior units like the Bureau of Mines or Geological Survey; that is, to have the navy employ geologists, petroleum engineers, and surveyors (*Leases Upon Naval Oil Reserves*, 1:895–900).

99. Frank H. Schofield to Thomas J. Walsh, October 31, 1923, box 213, TJWP; McCartney, *The Teapot Dome Scandal*, 85–68.

100. "Establishment of the Budget System," *Congressional Digest* 36, no. 5 (1957): 135; Donald T. Critchlow, *The Brookings Institution, 1916–1952: Expertise and the Public Interest in a Democratic Society* (DeKalb: Northern Illinois University Press, 1985), 28–40; Luther E. Gregory, "The Bureau of Yards and Docks and the Corps of Civil Engineers in Their Relations to War Problems," March 13, 1925, box 4, RG 15, USNWC.

101. *Leases Upon Naval Oil Reserves*, 1:894–914, 916–62.

102. For the Mammoth Oil contract, see *Cong. Record*, January 29, 1924, 1598–1603.

103. Phillips Payson O'Brien, *British and American Naval Power: Politics and Policy, 1900–1936* (Westport, CT: Praeger, 1998), 149–74; Richard W. Fanning, *Peace and Disarmament: Naval Rivalry and Arms Control, 1922–1933* (Lexington: University of Kentucky Press, 1995), 1–18; see also Roger Dingman, *Power in the Pacific: The Origins of Naval Arms Limitation, 1914–1922* (Chicago: University of Chicago Press, 1976).

104. Edwin Denby, memorandum, 1–2, 8, 11–12, box 2, EDP. The memorandum is undated but was likely written between February 1922 and August 1923.

105. "Reply of the Navy Department to Certain Questions Contained in House Resolution 286, Authorizing and Directing the Secretary of the Navy to Furnish the House of Representatives Certain Data on the Naval Strength of the American Navy, Etc.," *Cong. Record*, May 24, 1924, 9428.

106. Theodore Roosevelt Jr., "Comments on Article in the *New York Times* of April 28, 1924, Quoting Mr. W. B. Shearer," *Cong. Record*, May 24, 1924, 9433.

107. *Mammoth Oil Co. v. United States*, 275 U.S. 13 (1927).

108. Ibid.

109. *Leases Upon Naval Oil Reserves*, 1:379.

110. George Otis Smith, report, May 21, 1924, 4, box 39, TRJP. Smith's numbers reflect that the government received royalties of around 20 percent of Elk Hills production; of that amount, some two-thirds went to storage construction costs and only a third to the oil that would fill it.

111. See *Cong. Record*, January 29, 1924, 1594.

112. Theodore Roosevelt Jr. to Eliot Wadsworth, August 1, 1924, box 39, TRJP.

113. Theodore Roosevelt Jr. to Frederick Gillett, March 14, 1924, and Owen J. Roberts to Theodore Roosevelt Jr., March 3, 1924, box 39, TRJP. The special counsels, Atlee Pomerene and Owen J. Roberts, initially supported this plan under the assumption that Doheny's Pan American Petroleum and Transport Company would soon stop construction at Pearl Harbor and leave the work so far completed in jeopardy. Once they learned from Pan American that it would continue to construct the Pearl Harbor storage facilities despite legal proceedings, Pomerene and Roberts asked Roosevelt not to push Congress into voting on the emergency measure. The question underscored the uncertainty of what was, and what wasn't, in the national interest. Was the storage plan bad policy or was it reasonable policy conducted by illegal or corrupt means? By the summer, Ted Roosevelt was simultaneously asking that the government return to its previous conservation principles while continuing to operate under the legally suspect contracts with Pan American and Mammoth Oil (specifically that the navy continue to receive its oil royalties in oil certificates, presumably to be redeemed from the oil companies in the future). Roberts and Pomerene, however, had concluded that any pretense to the legality of the contracts undercut the government's case that they were, from the start, illegal and thus void. See A. E. Powell for Atlee Pomerene and Owen J. Roberts to Theodore Roosevelt Jr., March 15, 1924, Theodore Roosevelt Jr. to Atlee Pomerene and Owen J. Roberts, March 25, 1924, and Atlee Pomerene to Theodore Roosevelt Jr., June 13, 1924, box 39, TRJP. For more detailed accounting of the construction costs and oil royalties, see N. H. Wright, memorandum, May 23, 1924, box 39, TRJP.

114. Mark L. Requa, Van Manning, and George Otis Smith to Harry Garfield, February 28, 1919, roll 36, navy subject file, JDP.

115. Harry Garfield and representatives of the National Coal Association of the United Mine Workers of America, conference, February 11, 1919, roll 29; Harry Garfield to Josephus Daniels, March 14, 1919, roll 36, navy subject file, JDP.

116. Henry Doherty to George Otis Smith, June 8, 1924, box 1, USGS.

117. George Otis Smith to Hubert Work, August 19, 1924, box 1, USGS.

118. Unsigned [George O. Smith?] to Hubert Work, August 20, 1924, and George Otis Smith to Henry Doherty, August 27, 1924, box 1, USGS.

119. Calvin Coolidge to secretaries of war, navy, interior, and commerce, December 19, 1924, box 6, FOCB.

120. C. S. Baker, "Logistics—Its National Aspect," September 22, 1922, 1–2, folder 663, box 15, USNWC; "Course in Logistics: Outline of Logistics." November 1926, 2, folder 1206, box 29, RG 4, USNWC.

121. C. S. Baker, "The Modern Trend of Logistics," November 14, 1924, 1–9, 19, 26, folder 6, box 89, RG 8, USNWC.

122. Robert E. Coontz, "Logistic Element of the Grand Strategy," February 15, 1926, 1–3, 10–11, 16, XLOG 1926–27, folder 7, box 89, RG 8, USNWC.

123. Reuben E. Bakenhus, "Course in Logistics," December 1, 1926, 4–5, folder 1204, box 29, RG 4, USNWC.

124. Ibid., 7–9; "Course in Logistics: Outline of Logistics," November 1926, 3, 5–7, folder 1206, box 29, RG 4, USNWC.

125. "Course in Logistics," October 15, 1926, 1, folder 1207, box 29, RG 4, USNWC.

126. Chief of Naval Operations to fleet, force, squadrons, squadron and division commanders, and commanding officers of battleships, January 19, 1924, folder 4, box 45, RG 8, USNWC.

127. John R. D. Matheson, "Committee No. 5, Naval Supply: The Logistic Plan of the Orange War Plan," February 27, 1924, folder 7, box 49, RG 8, USNWC.

128. Ibid., 9–13.

Conclusion · Energy and Security in Perspective

1. *Petroleum Investigation: Hearings Before a Subcommittee of the Committee on Interstate and Foreign Commerce, United States House of Representatives*, 77th Cong., 1st sess., March 27, 1941, 3–29; John G. Clark, *Energy and the Federal Government: Fossil Fuel Policies, 1900–1946* (Urbana: University of Illinois Press, 1987), 333–34; "President Urges Bill for Pipeline," *New York Times*, May 21, 1941, 9; "Ickes Sees Oil Cut in East in Month," *New York Times*, June 6, 1941, 11; "Pipelines Backed by Oil Industry," *New York Times*, June 28, 1941, 8. On the politics of energy transportation infrastructures more generally, see Christopher F. Jones, *Routes of Power: Energy and Modern America* (Cambridge, MA: Harvard University Press, 2014).

2. *Petroleum Investigation*, 29–36, 37–41.

3. Paymaster general, monthly newsletter, May 1, 1941, 5, and June 5, 1941, 1, and "Annual Report of the Paymaster General of the Navy for the Fiscal Year 1941," 15, box 2, BSA; Bureau of Yards and Docks, *Building the Navy's Bases in World War II: History of the Bureau of Yards and Docks and the Civil Engineer Corps, 1940–1946*, 2 vols. (Washington, DC: Government Printing Office, 1947), 2:133–36; folders pertaining to the oil fuel situation in Japan, 1925–44, box 769, and the oil fuel situation in Germany, 1932–39, box 765, ROCNO, among intelligence files on coal and oil from other countries.

4. "Planning for Wartime Procurement of Supplies," 9–15, box 33, BSA; James E. Hewes Jr., *From Root to McNamara: Army Organization and Administration* (Washington, DC: Center of Military History, 1975), 12, 20, 51.

5. Though sufficient quantity of crude was never in much doubt, naval planners did worry about quality. The oil industry focused its refining on producing the most valuable oil products—gasoline and other special distillates. The navy was never certain that it would receive the full prioritization in an emergency that it believed necessary, nor was it sure that even with increased total production, the quantity of fuel oil (as opposed to gasoline) would be sufficient; see M. P. Refo Jr., "Navy Storage of Fuel Oil," January 19, 1939, box 29, BSA, and *Petroleum Investigation*, 29.

6. *Petroleum Facts and Figures*, 2nd ed. (Baltimore, MD: American Petroleum Institute, 1929), 47, 163, 188, 212; *Petroleum Facts and Figures*, 7th ed. (New York: American Petroleum

Institute, 1941), 13, 44, 53, 60; National Resources Planning Board, *A Report on National Planning and Public Works in Relation to Natural Resources and including Land Use and Water Resources with Finding and Recommendations*, vol. 4, *Report of the Planning Committee for Mineral Policy* (Washington, DC: Government Printing Office, 1934), 405.

7. On the subsequent diplomatic, political, and strategic history of oil during and immediately after World War II, see Aaron David Miller, *Search for Security: Saudi Arabian Oil and American Foreign Policy, 1939–1949* (Chapel Hill: University of North Carolina Press, 1980), Daniel Yergin, *The Prize: The Epic Quest for Oil, Money, and Power* (New York: Simon and Schuster, 1991), David S. Painter, *Oil and the American Century: The Political Economy of U.S. Foreign Oil Policy, 1941–1954* (Baltimore. MD: Johns Hopkins University Press, 1986), Stephen J. Randall, *United States Foreign Oil Policy Since World War I: For Profits and Security*, 2nd ed. (Montreal: McGill-Queen's University Press, 2005), and Robert Vitalis, *America's Kingdom: Mythmaking on the Saudi Oil Frontier* (Stanford, CA: Stanford University Press, 2007).

8. See, for example, Yergin, *The Prize*, Randall, *United States Foreign Oil Policy Since World War I*, Clark, *Energy and the Federal Government*, Martin V. Melosi, *Coping With Abundance: Energy and Environment in Industrial America* (New York: Knopf, 1985), Burl Noggle, *Teapot Dome: Oil and Politics in the 1920's* (New York: Norton, 1962), and J. Leonard Bates, *The Origins of Teapot Dome: Progressives, Parties, and Petroleum, 1909–1921* (Urbana: University of Illinois Press, 1963).

9. On the debate over international affairs in light of domestic political models, see David C. Hendrickson, *Union, Nation, or Empire: The American Debate over International Affairs, 1789–1941* (Lawrence: University Press of Kansas, 2009).

10. On the broader domestic transformations wrought by empire, see, for introductions to a now-vast literature, Alfred W. McCoy and Francisco A. Scarano, eds., *Colonial Crucible: Empire and the Making of the Modern American State* (Madison: University of Wisconsin Press, 2009), Amy Kaplan and Donald A. Pease, eds., *Cultures of United States Imperialism* (Durham, NC: Duke University Press, 1993), Paul A. Kramer, *The Blood of Government: Race, Empire, the United States, and the Philippines* (Chapel Hill: University of North Carolina Press, 2006), and Mary A. Renda, *Taking Haiti: Military Occupation and the Culture of U.S. Imperialism, 1915–1940* (Chapel Hill: University of North Carolina Press, 2000).

11. Clark, *Energy and the Federal Government*, 381–90.

12. "Transcript of the President's Address on the Energy Situation," *New York Times*, November 8, 1973, 32.

13. As of 2013, petroleum imports have actually declined by nearly half—to 33.0 percent of consumption—since their peak of 60.3 percent in 2005 (U.S. Energy Information Administration, *Monthly Energy Review*, February 2014, 41 [table 3.3a]).

14. "Transcript of President's News Conference on Foreign and Domestic Matters," *New York Times*, March 7, 1981.

15. See George Bush: "Remarks at the University of Nebraska in Lincoln," June 13, 1989, William J. Clinton, "Remarks Prior to Discussions with Prime Minister Atal Behari Vajpayee of India and an Exchange With Reporters," September 15, 2000, George W. Bush: "The President's News Conference," February 14, 2007, and Barack Obama: "Remarks at the Department of Energy," February 5, 2009, www.presidency.ucsb.edu, accessed January 20, 2014.

16. Thomas L. Friedman, *The World is Flat: A Brief History of the Twenty-First Century*, exp. ed. (New York: Picador, 2007), 382.

17. For critiques of simplistic calls for energy independence, see, for example, Yergin, *The Quest*, 267–68, Philip E. Auerswald, "Energy Conundrums: The Myth of Energy Insecurity," *Issues in Science and Technology* 22, no. 4 (2006): 65–70, Ian W. H. Parry and J. W. Anderson, "Petroleum: Energy Independence is Unrealistic," *Resources*, no. 156 (Winter 2005): 11–15, and Philip J. Deutch, "Energy Independence," *Foreign Policy*, November–December 2005, 20–25.

Archival Sources

Archival sources constitute the core of this book. Woven throughout are documents from the U.S. National Archives and Records Administration (NARA), the most important repository for this work and the archive in which I first began to explore the emergence of ideas about the origins of American energy security. Since this book straddles the nineteenth and twentieth centuries, and since it draws from the records of a variety of government agencies, I have made considerable use of both NARA 1 (in downtown Washington, D.C.) and NARA 2 (in College Park, Maryland). The locations of any given record group can be found on NARA webpages.

For naval records, I began with the indispensible files of RG 80, the General Records of the Department of the Navy. These files contain correspondence to and from Navy secretaries, subject file records, reports and hearings of the General Board, as well as material on the Naval Fuel Oil Board of 1916. The Naval Records Collection of the Office of Naval Records and Library, RG 45, includes microfilm series for letters received by the secretary of the navy (sorted by officer rank or miscellaneous senders), the subject file on fuel and fueling during the antebellum period, and the subject file for the operations of the Confederate navy. Additionally, some lesser-used record groups provide essential material on the routine functions of the department. Because the navy's Bureau of Naval Personnel originated, in part, in the older Bureau of Equipment and Recruiting, RG 24 contains records detailing Union coaling operations during the Civil War. The Records of the Office of the Chief of Naval Operations, RG 38, holds intelligence reports on rival powers' energy capabilities as reported by naval attachés overseas. The Records of the Bureau of Supplies and Accounts, RG 143, contains considerable material on fuel and logistics planning between the world wars.

Since energy and security are subjects that span bureaucratic boundaries, I cast my research net widely and made great use of several NARA record groups outside of naval records. Diplomatic collections included RG 59, the General Records of the Department of State, with its microfilmed series of instructions to special missions and the return correspondence of special agents (both of which I drew on for the voyage of Joseph Balestier to Brunei seeking coal concessions), as well as messages from American ministers in German states (on German funding for transatlantic mail steamers) and Colombia (on Ambrose Thompson's Chiriqui Improvement Company). For negotiations toward an Anglo-American postal treaty in the late 1840s, I used RG 84, the Records of the Foreign Service Posts of the Department of State. The Records of the U.S. Bureau of Mines, RG 70, provided material on World War I energy mobilization. The Records of the U.S. Geological Survey, RG 57, contain material on the formation of the naval petroleum reserves and debates over oil and security that led to the Federal Oil Conservation Board in 1924. The records of this board are in RG 232, the Records of the Petroleum Administrative Board (a Depression-era body that absorbed records from the Federal Oil Conservation Board).

Lastly from NARA, two records groups have received far less scholarly attention than they deserve: RG 46 holds the Records of the U.S. Senate and RG 233 the Records of the U.S. House of Representatives. Divided by congress and committee, they contain draft reports, correspondence, and great numbers of unpublished petitions and memorials. These files reveal one way nineteenth-century Americans engaged with their government.

I have woven NARA sources throughout this book, but many subjects depended on additional archives as well. Material on the history of mail steamers, besides that from NARA's House and Senate committee files, may be found in a variety of eclectically preserved personal and business records. Letters to Alfred Robinson in the Pacific Mail Steamship Company papers at UC Berkeley's Bancroft Library illuminate the early commercial development of San Francisco, as well as Robinson's role (as the Pacific Mail's Californian agent) in turning federal subsidies into dominance of Pacific steam communication. The microfilmed T. Butler King Papers of UNC Chapel Hill's Southern Historical Collection show the Georgia congressman's successful efforts to initiate federal subsidies to mail steam lines for communications and auxiliary national defense. The role of powerful Senate Post Office Committee chair Thomas Jefferson Rusk may be found in his papers at the Dolph Briscoe Center for American History at the University of Texas at Austin. The Papers of George Bancroft at the New York Public Library reveal the historian-diplomat's work negotiating a postal convention with Britain—a process sparked by steam subsidy competition between the two countries. Finally, the struggles of coaling early mail steamers—one of the main reasons that coal began attracting attention from Congress—may be found in the 1849–51 logbook of the Steamer *Oregon* held by the Huntington Library in San Marino, California, as well as the Letter Book of Marshall O. Roberts (a New York agent) in the U.S. Mail Steam Ship Company papers of the New-York Historical Society's Naval History Society Collection.

Sources on mid-nineteenth century steam engineering and coal expeditions may be found in several collections. On Ericsson's caloric engine, see the microfilm edition of the John Ericsson Papers from the American Swedish Historical Museum in Philadelphia, Pennsylvania, and the microfilm of the John Pendleton Kennedy Papers at the Enoch Pratt Free Library in Baltimore, Maryland. The nearby Maryland Historical Society houses the photostats of Benjamin Henry Latrobe's letters on his early experiments employing different fuels for railroad engines. For background material on Singapore consul and erstwhile diplomat Joseph Balestier, I relied on a circular letter of 1833 held by the Massachusetts Historical Society in Boston. On Balestier's failed negotiation with Brunei, I consulted the journal of George P. Welsh at the Library of Congress, which contains a travelogue of his voyage.

My account of coal and colonization draws most extensively from the extraordinarily rich R. W. Thompson collection at the Rutherford B. Hayes Presidential Center in Fremont, Ohio. Thompson was the attorney for colonization promoter Ambrose W. Thompson (no relation), and despite the efforts of Ambrose Thompson's heirs, the attorney managed to keep most Chiriqui Improvement Company records in his possession. Containing dispatches from agents in Costa Rica, Panama, and Colombia, these papers provide a compelling portrait of mid-nineteenth century racial thought, transnational infrastructure and colonization projects, and the porous national boundaries routinely crossed by Americans (and many others) in pursuit of profits. Other records on Chiriquí may be found in the collection of Chiriqui Improvement Company Papers at the Abraham Lincoln Presidential Library in Springfield, Illinois; the R. W. Thompson Correspondence at the Indiana State Library in Indianapolis; microfilm of the William Henry Seward Papers from the University of Rochester; and, most usefully, the papers of both Ambrose W. Thompson and Abraham Lincoln at the Library of Congress.

On southern attempts to fuel the Confederacy, I relied on the few but illuminating documents preserved in the Colin J. McRae Papers and the William Phineas Browne Family Papers held by the Alabama Department of Archives and History in Montgomery, Alabama, as well as a single (but fascinating) letter from William Quinn to Confederate navy secretary S. R. Mallory at the Virginia Historical Society in Richmond.

Massive archival collections cover Gilded Age foreign relations, touching on both public and private efforts toward annexation of distant territories large and small. Guided by my focus on territories sought as coaling stations, I made use of the William McKendree Gwin Papers at the Bancroft Library, which reveal the former senator's attempts to revive Ambrose Thompson's Chiriquí claims after the Civil War. The Samuel Latham Mitchill Barlow Papers at the Huntington Library show Barlow's efforts to annex Santo Domingo. The David D. Porter Family Papers at the Library of Congress expose Porter's views on annexing Hawaii in order to prevent other colonial powers from seizing it as a coaling station. The best source for John McAllister Schofield's 1872 mission to Hawaii are Schofield's papers at the Library of Congress.

Documents on the development of the science of logistics come foremost from the detailed records of the U.S. Naval War College Archives in Newport, Rhode Island. The school's Naval Historical Collection records groups covering publications (RG 4), intelligence and technical matters (RG 8), faculty and staff presentations (RG 14), and lectures (RG 15) collectively illuminate the pedagogical, intellectual, and bureaucratic process of integrating logistics into the center of modern warfare. I supplemented these papers with those of one-time War College president Caspar F. Goodrich in the New-York Historical Society's Naval History Society Collection. The Franklin D. Roosevelt Presidential Library and Museum in Hyde Park, New York, holds Roosevelt's Papers as assistant secretary of the navy, an understudied yet formative time in his career; he wielded considerable influence in this role, and his papers reveal much about the bureaucratization of naval logistics. The Papers of George Dewey at the Library of Congress reveal the disjuncture between popular reporting about the American seizure of Manila Bay and how it was experienced at the time, particularly as the seizure related to coaling. Also at the Library of Congress, the papers of Josephus Daniels and Samuel McGowan both show the influence of simultaneously moralistic and technocratic progressive reform on naval administration (including the influence of a shared tacit southern racism), while the papers of Theodore Roosevelt Jr. and Thomas J. Walsh provide essential details on Teapot Dome. The Edwin Denby Papers at the Bentley Historical Library of the University of Michigan shed light on the thinking of this navy secretary during and immediately after the Teapot Dome scandal.

On Alaska, the records of the Navy Alaskan Coal Commission are in NARA, RG 80, but half are stored in Washington, D.C., and half in the NARA facility in Anchorage, Alaska. A visit there allowed me to explore the remnants of the navy's coal camp and the surrounding Matanuska Valley; I was also able to have conversations with members of the indigenous Chickaloon Village. Regional planning visionary Benton MacKaye's views of the development of Alaska may be found in the MacKaye Family Papers at the Dartmouth College Library.

Lastly, for William D. Leahy's manuscript diary (lacking some details about the FDR-ibn Saud meeting of 1945 he later offered in his published memoir), see the William D. Leahy Papers at the Library of Congress.

Primary Source Databases

For the more than one hundred newspapers and periodicals appearing in this book, I have mostly relied on an ever-growing collection of digital databases. For newspapers, Readex's America's Historical Newspapers is now the largest and most comprehensive collection (and the one from which I cite most frequently throughout this book), but its many component series—ten, by September 2014—are available only by institutional subscription and at prohibitive rates for all but the most lavishly funded academic libraries. Fortunately, Readex's parent company, Newsbank, quietly makes this same content available for individual subscribers at affordable prices in the amateur genealogist-focused product, GenealogyBank. com. For my discussions of politics and political economy in mid-nineteenth century Panama and Columbia, I have also relied on Readex's database of Latin American newspapers. I have supplemented these proprietary databases with the publicly accessible Chronicling America database from the Library of Congress. This collection, which depends upon partnerships with libraries across the United States to digitize contributions, contains many important titles unavailable elsewhere, like the *New York Sun* and issues of the *New York Tribune* after the turn of the twentieth century. Articles from a number of major papers still published today only appear in their own databases, including the *New York Times* and the *Washington Post* (both available through ProQuest) and the *Times* of London (available through Cengage). Since these databases collectively overlap very little, it is important to query them all on every subject under study. For nineteenth- and early twentieth-century journals and magazines, ProQuest's American Periodical Series remains the best digitized repository, though much of their content now also appears in Google Book Search or HathiTrust databases.

Even with these vast databases, many newspapers and magazines remain available only on microfilm or in print. These titles often contain evidence that would be lost in the homogenizing digitization process. The issues of the Singapore *Free Press* held by the Library of Congress, for example, are stamped with the name of Joseph Balestier, the American consul, who likely sent the papers home to Washington as part of his official duties. This paper provided frequent reports about coal discoveries in Borneo, and Balestier likely shared these reports with a visiting Captain John Percival.

For public documents, the most valuable source was the ProQuest Congressional database (formerly LexisNexis Congressional), which contains the U.S. Serial Set (the official publications of the U.S. House and Senate) and an incomparably rich collection of published congressional committee hearings. These sources are familiar to historians and have long been available either in print or on microfiche (how I used them when this project began), but many run into thousands of pages, and until the advent of full-text searchability, great quantities of details and documents remained effectively hidden. I cite from some 120 published American public documents, mostly available from this collection. These range from annual reports of the War, State, Post Office, and especially Navy departments, along with accompanying documents, committee reports and compilations of requested executive documents, diplomatic papers, and even historical and scientific studies. I have supplemented this database with the Library of Congress's American Memory: A Century of Lawmaking for a New Nation, which contains easily accessible collections of nineteenth-century statutes, bills, and official journals from the Senate and House of Representatives. I have also made frequent reference to the *Congressional Globe*, the official (and sometimes even correct) transcript of the proceedings of the House and Senate. The *Globe* is fully available as part of Hein Online's massive collection of digitized legal and legislative documents. Other sources for essential public documents include the U.S. Naval War Records Office's *Official Records of the Union and Confederate Navies in the War of the Rebellion* (Washington, DC: Government Printing Office, 1897) and the corresponding War Department's *The War of the Rebellion: A Compilation of Records of the Union and Confederate Armies* (Washington, DC: Government Printing Office, 1890), both available online through Cornell University's Making of American project, the indispensible *Foreign Relations of the United States* series, available through 1960 in the University of Wisconsin Digital Collections and edging into 1980 through the U.S. State Department's Office of the Historian, and finally, Gerhard Peters's and John T. Woolley's American Presidency Project at the University of California, Santa Barbara, which provides a frequently updated collection of the public papers of modern presidents (a very useful tool in tracing public statements about energy independence).

Though brick-and-mortar libraries (and physical books) remain essential to historical research (mine included), the digital repositories of Google Book Search and HathiTrust have also become indispensable sources. I have found it particularly fruitful to access these two databases via the University of Michigan's library catalogue, Mirlyn, which indexes digital holdings alongside its collection of physical books. This portal is especially valuable when searching for individual volumes of multivolume serials, which are difficult to locate through simple Google Books or HathiTrust searches alone.

For other sources, I used Early English Books Online, which covers printed works in English from the fifteenth and sixteenth centuries and which proved helpful in tracing the early theorizing of great circle navigation. Cengage's The Making of the Modern World was useful in tracking down published ephemera, especially related to business ventures. I found the investment publications of would-be steamship proprietor William Wheelwright here, for example. Yet despite the ever-growing volume of digitized printed sources, much remains inaccessible except in print or microfilm and will likely remain so for a long time. The ease of searching vast corpuses must be balanced by remembering their incompleteness, as they often are missing the most important sources.

Secondary Sources

The history of energy does not fit neatly into any single subdisciplinary framework, and my thinking about the subject has been shaped by a number of key works straddling the history of technology, economic history, social history, diplomatic history, environmental history, and world history. On the rise of fossil fuels and modern energy systems in the United States, see Thomas P. Hughes, *Networks of Power: Electrification in Western Society, 1880–1930* (Baltimore, MD: Johns Hopkins University Press, 1983), Louis C. Hunter, *Steam Power*, vol. 2 of *A History of Industrial Power in the United States, 1780–1930* (Charlottesville: University Press of Virginia, 1985), Martin Melosi, *Coping with Abundance: Energy and Environment in Industrial America* (New York: Knopf, 1985), Daniel Yergin, *The Prize: The Epic Quest for Oil, Money, and Power* (New York: Simon and Schuster, 1991), Richard White, *The Organic Machine: The Remaking of the Columbia River* (New York: Hill and Wang, 1995), Brian Black, *Petrolia: The Landscape of America's First Oil Boom* (Baltimore, MD: Johns Hopkins University Press, 2000), David Nye, *Consuming Power: A Social History of American Energies* (Cambridge, MA: MIT Press, 2001), Sean Patrick Adams, *Old Dominion, Industrial Commonwealth: Coal, Politics, and Economy in Antebellum America* (Baltimore, MD: Johns Hopkins University Press, 2004), Paul Lucier, *Scientists and Swindlers: Consulting on Coal and Oil in America, 1820–1890* (Baltimore, MD: Johns Hopkins University Press, 2008), and Christopher F. Jones, *Routes of Power: Energy and Modern America* (Cambridge, MA: Harvard University Press, 2014).

On domestic energy policy, see John Ise, *The United States Oil Policy* (New Haven, CT: Yale University Press, 1926), Gerald D. Nash, *United States Oil Policy, 1890–1964: Business and Government in Twentieth Century America* (Pittsburgh, PA: University of Pittsburgh Press, 1968), John G. Clark, *Energy and the Federal Government: Fossil Fuel Policies, 1900–1946* (Urbana: University of Illinois Press, 1987), and Paul Sabin, *Crude Politics: The California Oil Market, 1900–1940* (Berkeley: University of California Press, 2005). On foreign energy policy, see David S. Painter, *Oil and the American Century: The Political Economy of U.S. Foreign Oil Policy, 1941–1954* (Baltimore, MD: Johns Hopkins University Press, 1986), Stephen J. Randall, *United States Foreign Oil Policy Since World War I: For Profits and Security*, 2nd ed. (Montreal: McGill-Queen's University Press, 2005), and Robert Vitalis, *America's Kingdom: Mythmaking on the Saudi Oil Frontier* (Stanford, CA: Stanford University Press, 2007).

Energy has also served as an organizing theme in large-scale, global histories like Vaclav Smil's *Energy in Nature and Society: General Energetics of Complex Systems* (Cambridge, MA: MIT Press, 2008) as well as Alfred Crosby's *Children of the Sun: A History of Humanity's Unappeasable Appetite for Energy* (New York: Norton, 2006). J. R. McNeill places twentieth-century developments in energy and industry into a larger environmental narrative in *Something New Under the Sun: An Environmental History of the Twentieth-Century World* (New York: Norton, 2000). Timothy Mitchell connects the organization of energy production with structures of political power in *Carbon Democracy: Political Power in the Age of Oil* (London: Verso, 2011).

For technology and American foreign relations more broadly, I have been influenced by David Paull Nickles, *Under the Wire: How the Telegraph Changed Diplomacy* (Cambridge, MA: Harvard University Press, 2003), Michael Adas, *Dominance by Design: Technological*

Imperatives and America's Civilizing Mission (Cambridge, MA: Belknap Press of Harvard University Press, 2006), and Jonathan Reed Winkler, *Nexus: Strategic Communications and American Security in World War I* (Cambridge, MA: Harvard University Press, 2008). Works on the European context may be found in the notes.

On the development of modern naval thought and logistics in particular, see Dirk Bönker, *Militarism in a Global Age: Naval Ambitions in Germany and the United States before World War I* (Ithaca, NY: Cornell University Press, 2012), Martin van Creveld, *Supplying War: Logistics from Wallenstein to Patton*, 2nd ed. (New York: Cambridge University Press, 2004), Edward Miller, *War Plan Orange: The U.S. Strategy to Defeat Japan, 1897–1945* (Annapolis, MD: Naval Institute Press, 1991), and Steven T. Ross, *American War Plans, 1890–1939* (London: Frank Cass, 2002). For the traditional view of World War I mobilization, see Robert D. Cuff, *The War Industries Board: Business-Government Relations during World War I* (Baltimore, MD: Johns Hopkins University Press, 1973). For the broader view, Paul A. C. Koistinen's *The Political Economy of American Warfare*, 5 vols. (Lawrence: University Press of Kansas, 1996–2012), spanning 1606 to 2011, is essential.

Hundreds of other works shaped this book. Specific references may be found in the notes.